Developments in Environmental Modelling, 16

Modelling in Ecotoxicology

Developments in Environmental Modelling

1. ENERGY AND ECOLOGICAL MODELLING edited by W.J. Mitsch, R.W. Bossermann and J.M. Klopatek 1981 839 pp.

2. WATER MANAGEMENT MODELS IN PRACTICE: A CASE STUDY OF THE ASWAN HIGH DAM by D. Whittington and G. Guariso 1983 xxii+246 pp.

3. NUMERICAL ECOLOGY by L. Legendre and P. Legendre 1983 xvi+419 pp.

4A. APPLICATION OF ECOLOGICAL MODELLING IN ENVIRONMENTAL MANAGEMENT PART A edited by S.E. Jørgensen 1983 viii+735 pp.

4.B APPLICATION OF ECOLOGICAL MODELLING IN ENVIRONMENTAL MANAGEMENT PART B edited by S.E. Jørgensen and W.J. Mitsch 1983 viii+438 pp.

5. ANALYSIS OF ECOLOGICAL SYSTEMS: STATE-OF-THE-ART IN ECOLOGICAL MODELLING edited by W.K. Lauenroth, G.V. Skogerboe and M. Flug 1983 992 pp.

6. MODELLING THE FATE AND EFFECT OF TOXIC SUBSTANCES IN THE ENVIRONMENT edited by S.E. Jørgensen 1984 viii+342 pp.

7. MATHEMATICAL MODELS IN BIOLOGICAL WASTE WATER TREATMENT edited by S.E. Jørgensen and M.J. Gromiec 1985 vi+802 pp.

8. FRESHWATER ECOSYSTEMS: MODELLING AND SIMULATION by M. Straškaba and A.H. Gnauck 1985 309 pp.

9. FUNDAMENTALS OF ECOLOGICAL MODELLING by S.E. Jørgensen 1986 389 pp.

10. AGRICULTURAL NONPOINT SOURCE POLLUTION: MODEL SELECTION AND APPLICATION edited by A. Giorgini and F. Zingales 1986 409 pp.

11. MATHEMATICAL MODELLING OF ENVIRONMENTAL AND ECOLOGICAL SYSTEMS edited by J.B. Shukla, T.G. Hallam and V. Capasso 1987 xii+254pp.

12. WETLAND MODELLING edited by W.J. Mitsch, M. Straskraba and S.E. Jørgensen, 1988 x+228 p.

13. ADVANCES IN ENVIRONMENTAL MODELLING edited by A. Marani 1988 691 pp.

14. MATHEMATICAL SUBMODELS IN WATER QUALITY SYSTEMS edited by S.E. Jørgensen and M.J. Gromiec 1989 408 pp.

15. ENVIRONMENTAL MODELS: EMISSIONS AND CONSEQUENCES edited by J. Fenhann, H. Larsen, G.A. Mackenzie and B. Rasmussen, 1990 xii+490 pp.

Developments in Environmental Modelling, 16

Modelling in Ecotoxicology

Edited by

S.E. Jørgensen

Langkaer Vaenge 9, 3500 Vaerløse, Copenhagen, Denmark

ELSEVIER
Amsterdam — Oxford — New York — Tokyo 1990

ELSEVIER SCIENCE PUBLISHERS B.V.
Sara Burgerhartstraat 25
P.O. Box 211, 1000 AE Amsterdam, The Netherlands

Distributors for the United States and Canada:

ELSEVIER SCIENCE PUBLISHING COMPANY INC.
655, Avenue of the Americas
New York, NY 10010, U.S.A.

Library of Congress Cataloging-in-Publication Data

Modelling in ecotoxicology / edited by S.E. Jørgensen.
p. cm. -- (Developments in environmental modelling ; 16)
Includes bibliographical references (p.).
ISBN 0-444-88699-0 (Elsevier Science Pub.)
1. Pollution--Environmental aspects. 2. Pollution--Environmental aspects--Simulation methods. I. Jørgensen, Sven Erik, 1934- II. Series.
QH545.A1M62 1990
574.5'222--dc20 90-35192
CIP

ISBN 0-444-88699-0

Printed in The Netherlands

TABLE OF CONTENTS

PART A

ECOTOXICOLOGICAL MODELS:

GENERAL CONSIDERATIONS

CHAPTER 1

INTRODUCTION

by
S.E. Jørgensen

1.1 NEEDS OF MODELS IN ECOTOXICOLOGY

Ecotoxicology is a compound of *ecology* and *toxicology. Ecology* is the science of living organisms and their interactions with their environment, while *toxicology* is the science of toxic substances and their effects on living organisms. This implies that *ecotoxicology* is the science of toxic substances in the environment and their impact on the living organisms.

During the last ten to fifteen years ecotoxicology has developed from ecology and toxicology. Concern about toxic substances in our environment probably started ten years earlier with the publication of Rachel Carson`s famous book "Silent Spring", although at that time, the concept of bio-accumulation was still not fully understood scientifically. Our knowledge of chemical reactions in the laboratory or in the chemical plants, where the intensive thermodynamic functions of temperature, pressure and humidity could be kept almost constant, was extensively developed during the two decades following World War Two, but our knowledge of reactions of abiotic (and also biotic) chemical compounds in the environment was still very limited. The problem was too complex, because in Nature nothing is constant and the study of a single process is impossible, as many interrelated processes take place simultanously.

Since the early seventies much research has been devoted to the fate and effect of chemical compounds in Nature, when the first environmental wave started in the developed countries. During the last fifteen years thousands of books and papers have been written about the processes involving toxic substances in Nature. Our knowledge has been greatly expanded in this area, although we are very far from a full understanding of this complex problem.

The chemical industry has developed very rapidly during the last 3-4 decades and today we use many more chemicals in everyday life than we did 30-40 years ago. Our knowledge of the fate and effect of such chemicals in the environment has not yet followed the rate of chemical innovation in spite of our expanding knowledge of ecotoxicology. Today about 50.000 chemicals are produced on an industrial scale, but we have only sufficient data to evaluate the environmental consequences of a few percent of these chemicals. Our need for ecotoxicological knowledge has never been more pronounced than today. Even more resources must be allocated to this field in the near future, if we are to be able to cope with the threat of more or less toxic chemical compounds in our environment.

The need for ecotoxicological research has increased enormously due to the following reasons:

1. We are using more chemicals than ever before.
2. We now realize the dangers, these chemicals display for all life on earth.
3. We now realize that problems related to chemicals in the environment are much more complex than we first thought.
4. In many countries the chemical industries have been forced to reveal the environmental consequences of all new products before production is initiated. Consequently, the further development of the chemical industry is dependent on more knowledge of the ecotoxicology.

Ecotoxicology deals with very complex problems. All compounds in the environment are exposed to many physical and chemical reactions: chemical reactions with oxygen, water and the many other chemical compounds in the environment; photolysis; biological decomposition; uptake and excretion by living organisms; adsorption; ion exchange; and exchange between two spheres - atmosphere, hydrosphere or lithosphere.

The forcing functions (or external variables) of the environment, such as precipitation, temperature, wind, input and output of chemical compounds, emigration and immigration of living organisms, and so on, are continuously variable. All these factors influence chemical reactions and their rates, and they may change randomly.

This is in contrast to the chemical reactions in the laboratory or in a chemical plant, where the reactions are pure, or almost pure, with only minor side reactions, and where the conditions are constant or only a few factors are changed according to a predetermined scheme.

This difference between environmental chemistry and more classical chemistry is illustrated in Fig. 1.1.

These differences between the application of chemistry to environmental processes and those of the laboratory or chemical plant may be summarized:

1. Many chemical compounds are present in the environment, while in the laboratory or chemical plant processes involve only a few compounds.
2. Many processes take place simultaneously in the environment, while in the laboratory or chemical plant attempts are made to keep processes

as pure as possible.

3. The conditions (forcing functions) are kept constant in the laboratory or chemical plant, or they follow a predetermined scheme, whereas forcing functions in the environment vary and these variations are to a certain extent random.
4. The chemistry of the environmental processes is further complicated by the presence of many species of living organisms which interact with the chemical compounds. In the laboratory or in a chemical plant one, or maybe a few, species (for instance by fermentation processes) may be involved in the chemical reactions.

Consequently, the chemistry and biology of toxic substances in the environment, their fate and effects, is extremely complex and a lot of data is needed to cope with the problem. Although much data has been provided during the last 10-15 years by intensive research in the field, we still have a great need of data. We are forced to estimate much of the data needed. Furthermore, we cannot gain an overview of the many components and their interactions by simple methods, but need to develop models able to synthesize our knowledge and provide the overview of the problems.

The application of models has therefore been a natural development in ecotoxicology and this book attempts to outline the state of the art of modelling the fate and effects of toxic substances in the environment. Modelling in ecotoxicology differs from modelling in other fields by the great lack of data. The quality of the models is very dependent on the parameters used, and as we do not have a wide knowledge of parameters in ecotoxicological processes, good parameter estimation methods are crucial for ecotoxicological models. A comprehensive review of available parameter estimation methods is therefore included in this volume. Model examples and case studies have also been included to illustrate the difficulties and short comings in practical modelling.

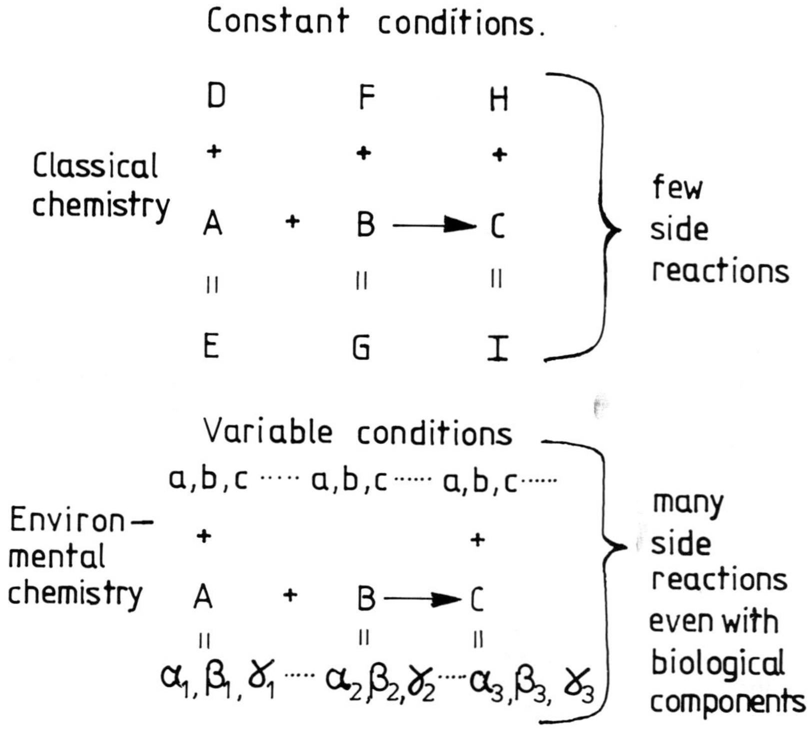

Fig. 1.1: The difference between classical and environmental chemistry.

CHAPTER 2

MODELLING CONCEPTS

by
S.E. Jørgensen

2.1 MODELLING AND ECOTOXICOLOGY

A model can be considered a synthesis of 'knowledge elements' about a system. The quality of models is therefore very dependent on the quality of 'knowledge elements' and the available data. If knowledge and data on a given problem related to a system are poor, the model cannot be expected to fill the holes in our knowledge or repair a poor set of data. On the other hand, models can provide new knowledge about the reactions and properties of the entire system. Since the model represents a synthesis of knowledge and data it can, as such, provide results particularly about the properties of the system.

Analysis of problems is a powerful tool in Science but up to now synthesis of knowledge has not been so widely used as a tool. However, the emergence of very complex environmental problems has provoked development of ecological and environmental modelling as a powerful synthezising tool, where the system reactions and properties are the focus.

Models are, on the one hand, a synthezising tool, but it should not be forgotten that models also may be used to analyse the properties of the entire system at the system level. We are therefore not using models in ecotoxicology only to take an overview of the problems, but also to reveal the reactions of the entire system to the impact from toxic substances, including the changes in the system due to the toxic substances.

The use of models as such is not new, since models have been used by mankind as a simplified picture of reality for solving problems. Newton's laws may be considered models of the impact of gravity on bodies. Models, even when used as a synthezising tool, will of course not contain *all* features of the real system, because then it would be the real system itself. But it is of importance that we extract the knowledge that is *essential* to the context of the problem to be solved or described.

Correspondingly, an ecotoxicological model must contain the features of interest for the management of the scientific problem that we wish to solve by use of a model. This statement is of particular importance for ecotoxicological models, as they can very easily become too complex to develop for practical use.

Models may either be physical or mathematical. Physical models contain the main components of the real system in an attempt to observe processes and reactions of the complex system - an ecosystem - by use of observations in the simpler system - the physical model. If, for instance, we want to study the interactions between a toxic substance and a system of plants, insects and soil in Nature, we may construct a simplified system, which contains these components and make our observations on the simpler system and thereby facilitate our interpretation of the data. Physical models are often named microcosmos, as they contain all major components of the larger system, but on a smaller scale.

This book focuses almost entirely on mathematical models, which are based on a mathematical formulation of the processes that are most important for the problem being considered.

The field of environmental modelling has developed very rapidly during the last decade due essentially to two factors:

1. The development of computer technology, which has enabled us to handle very complex mathematical systems.
2. A general understanding of pollution problem, including the problems related to the application of toxic substances

The idea behind the applications of ecotoxicological models in environmental management is illustrated in Figure 2.l. Toxic substances are released from man-made or man-controlled systems to the environment (ecosystems), where they are harmful to living organisms and may change the reactions, the function or even the structure of the entire system. Complete elimination of all emissions is impossible for post-industrial man and with a global human population of 5 billions to feed. But if we can relate an emission with its ecological implications for the environment, we shall be able to provide recommendations on which emissions we have to eliminate or reduce, and how much they have to be reduced to guarantee no, or almost no, adverse effect. The idea behind the use of model is to come up with the best possible estimation for the relation between emission and ecological consequences, by synthezising all the knowledge - or the most important parts of this knowledge - to be able to make an overview of the problem in focus.

The resulting recommendations may be either in form of emission limitations, or in form of a ban on the use of the toxic substance in question. This is a political decision. Although the model may give a rather clear answer, there is always economic cost involved in such decisions. It is possible in some instances to construct *ecological-economic* models, which consider also the economy of the problem, but those model are not yet sufficiently developed to give reliable guidelines in more than a few cases.

But *ecological-economic* models will most probably become more developed in the near future, and during the next decade they will be used most widely in environmental management.

The difficult part of modelling is not the mathematical formulation or the translation of the mathematics into a computer language. The introduction of personal computers and easily applicable software has made it much more easy to handle these steps of modelling. The more difficult part is to provide the necessary knowledge and be able to estimate which components and processes to include in the model. An ecologist or ecotoxicologist with some knowledge of mathematics and computer science is better fitted to construct ecotoxicological models than a mathematician with some knowledge of ecotoxicology and ecology .

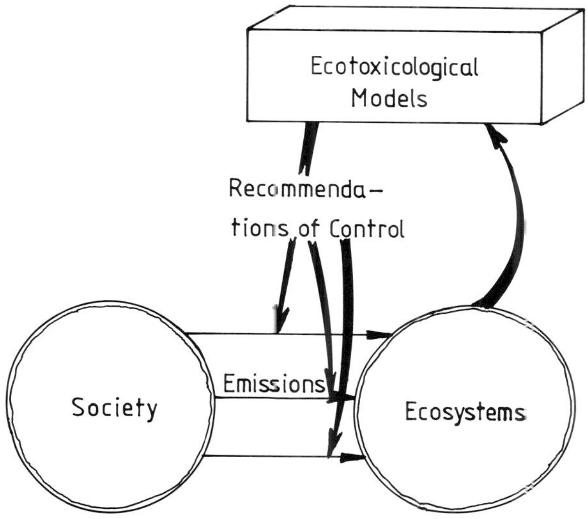

Fig. 2.1: The idea behind the use of ecotoxicologial models in environmental management.

2.2 DEFINITIONS OF MODELLING ELEMENTS

In its mathematical formulation an ecological or ecotoxicological model consists of five components:

1. Forcing functions or external variables, which are functions or variables of an external nature that influence the state of the ecosystem. In a management context the problem to be solved can often be reformulated as follows: if certain forcing functions are varied, what will be their influence on the state of the ecosystem? The model is used to predict what will change in the ecosystem when forcing functions are varied with time. The forcing functions that we can control are often called **control functions.** Such control functions in ecotoxicological models are the input of toxic substances to the ecosystems, while other forcing functions of interest could be climatic variables, which influence the biotic and abiotic components and the process rates.

2. State variables describe, as the name indicates, the state of the ecosystem. The selection of the state variable is crucial for the model structure, but in many cases the choice is obvious. If, for instance, we want to model the bioaccumulation of a toxic substance, the state variables should be the organisms in the most important foodchains and their concentrations of the toxic substance. When the model is used in a management context the values of the state variables predicted by changing the forcing functions can be considered as the results of the model, because the model will contain relations between the forcing functions and the state variables.

3. Mathematical equations are used to represent the biological, chemical and physical processes. They define the relationship between the forcing functions and state variables. The same type of process may be found in many different ecotoxicological contexts, which implies that the same equations can be used in different models. This does not imply, however, that the same process is formulated always by the same equation. Firstly the influence of other factors may mean another equation provides better fit for the process under consideration. Furthermore, the number of details needed for the model may differ because of a different degree of complexity of the system and/or the problem. Some modellers refer to the description and mathematical formulation of processes as submodels. Chapter 6 will present some of the most used submodels in ecotoxicological modelling.

4. Parameters are coefficients in the mathematical representation of processes. They may be considered constant for a specific ecosystem or part of ecosystem. In causal models the parameter will have a scientific definition, e.g. the excretion rate of cadmium from a fish. Many parameters are not shown in the literature as constants but as ranges, but even that is of great value in parameter estimation, as will be discussed further in the chapters 3, 4 and 5. A comprehensive collection of ecological and ecotoxicological parameters can be found in Jørgensen et al. (1979) and (1988).

5. Universal constants such as the gas constant and atomic weights are also used in most models.

Models can be defined as formal expressions of the essential elements of a problem in mathematical terms. The first recognition of the problem is often verbally. This may be recognized as an essential preliminary step in the modelling procedure, which will be treated in more detail in the next section. The verbal model is, however, difficult to visualize and it is therefore conveniently translated into a **conceptual diagram,** which contains the state variables, the forcing function and how these components are interrelated by mathematical formulations of processes.

Figure 2.2. illustrates a conceptual diagram of the nitrogen cycle in a lake. The state variables are: nitrate; ammonium (which is toxic to fish in the unionized form ammonia); nitrogen in phytoplankton; nitrogen in zooplankton; nitrogen in fish; nitrogen in the sediment; and nitrogen in detritus.

The forcing functions are: in- and out-flows; concentration of nitrogen components in the in- and out-flows; solar radiation, here indicated by the arrow PHOTO; and the temperature, which is not shown on the diagram, but which influences all process rates. The arrows in the diagram illustrate the processes and they are formulated by use of mathematical expressions in the mathematical part of the model.

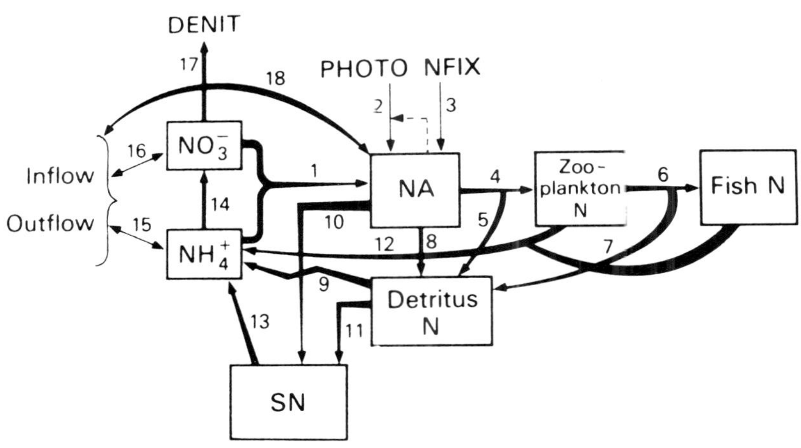

Fig. 2.2: The conceptual diagram of a nitrogen cycle in an aquatic ecosystem. The processes are: 1) uptake of nitrate and ammonium by algae; 2) photosynthesis; 3) nitrogen fixation; 4) grazing with loss of undigested matter; 5), 6) and 7) are predation and losses of undigested matter; 8) mortality; 9) mineralization; 10) settling of algae; 11) settling of detritus; 12) excretion of ammonium from zooplankton; 13) release of nitrogen from the sediment; 14) nitrification; 15), 16) and 18) input/output; and, 17) denitrification.

The terms of three steps in the modelling procedure should be defined in this section. They are calibration, verification and validation.

Calibration is an attempt to find the best accordance between computed and observed data by variation of some selected parameters. It may be carried out by trial and error, or by use of software developed to find the parameters that give the best fit between observed and computed values. In some static models, alsi in some simple models, which contain only a few and well defined or directly measured parameters, calibration may not be required.

Verification is a test of the *internal logic* of the model. Typical questions in the verification phase are: Does the model reacts as expected? Is the model stable in the long-term? Does the model follow the law of mass conservation? Verification is mostly a subjective assessment of the behaviour of the model. To a large extent verification will go on during the use of the model before the calibration phase mentioned above.

Validation must be distinguished from verification. Validation consists of an objective test on how well the model outputs fit the data. The selection of possible objective tests will be dependent on the scope of the model, but the standard deviation between model predictions and observations is often used, and a comparison of observed and predicted minimum or maximum values of a particularly important state variable. If several state variables are included in the validation, they may be given different weights

Further details on these important steps in modelling will be given in the next section where the entire modelling procedure will be presented.

2.3 THE MODELLING PROCEDURE

In this section a tentative modelling procedure is presented. The author of this volume has used this procedure successfully several times and strongly recommends the very careful use of all steps of the procedure. Other scientists in the field have published slightly different procedures, but a detailed examination of the differences reveals that these are only minor. The most important steps of modelling are included in all the recommended modelling procedures.

The primary focus of all research at all times is to define the problem. Only in this way can limited research resources be correctly allocated and not dispersed into irrelevant activities.

The first modelling step is therefore a **definition of the problem** and the definition will need to be bound by the constituents of **space, time and subsystems.** The bounding of the problem in space and time is usually easy and consequently more explicit than the identification of the ecotoxicological subsystems to be incorporated in the model.

Figure 2.3 shows the procedure used by the author, but it is important to emphasize that this procedure is unlikely to be entirely correct at the first attempt, so there is no need to aim at perfection in one step. The procedure should be considered as an interactive process and the main requirement is to get started (Jeffers, 1978).

It is diffcult at least in the first instance to determine the optimum number of subsystems to be included in the model for an acceptable level of accuracy as defined by the scope of the model. In many cases it become necessary at a later stage to accept a lower level than originally intended owing to lack of data or the need to provide additional data for improvement of the model.

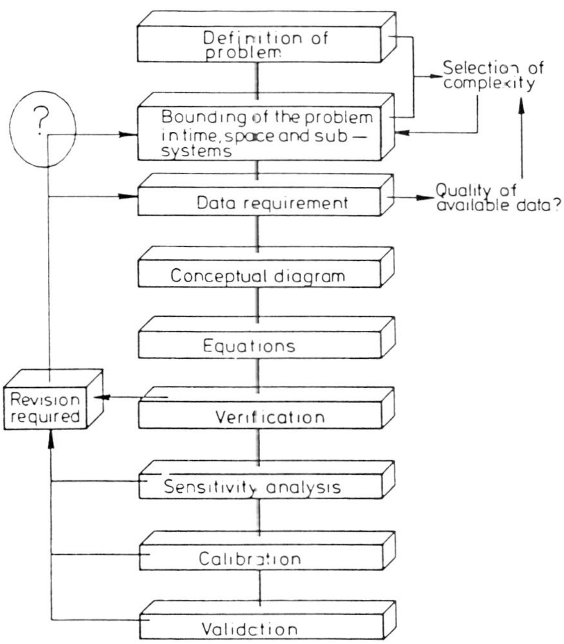

Fig. 2.3: A tentative modelling procedure, reproduced from Jørgensen (1988).

It has been argued many times that a complex model should account more accurately for the reactions of the real system, but this is not necessary true. Additional factors may be incorporated. A more complex model will also contain more parameters and with an increasing number of parameters in the model there will also be increasing uncertainty, because the parameters have to be estimated either by more observations in the field, by laboratory experiments, or by calibrations, which again are based on field measurements. Parameter estimations are never error free, and errors are carried through into the model, thereby contributing to the uncertainty of the model.

The problem of selecting **the right model complexity,** will be further discussed in section 3.1. It is a problem of particular interest for modelling in ecotoxicology.

A first approach to the data requirement can be given at this stage, although this is likely be changed at a later stage, with experience gained from the verification, calibration , sensitivity analysis and validation.

In principle data for all the selected state variables should be available and only in few cases would it be acceptable to omit measurements of selected state variables, as the success of the calibration and validation is closely linked to **the quality and quantity of the data.**

Once the **model complexity**, at least at the first attempt, has been selected, even on a first provisional basis, it is possible **to conceptualize the model** - for instance, in the form of a diagram as shown in Figure 2.2. It will give information on which state variables, forcing functions and processes are required in the model.

The next step is the formulation of the processes as **mathematical equations.** The survey of submodels given in chapter 6 will provide help for this step, but, as pointed out in section 2.2., many processes have more than one equation and it may be of great importance for the results of the final model to select the right one for the case under consideration.

Once the system of mathematical equations is available, **verification** can be carried out. As pointed out in section 2.2. it is an important step, which unfortunately is omitted by some modellers. At this stage it is recommended that attempts are made to answer the following questions:

1. Is the model stable in the long-term? The model is run for a long period with the same annual variations in the forcing functions to observe whether values of the state variables are maintained at approximately the same levels. During the first period state variables dependent on the initial values, and it is recommended the model be run also with initial values corresponding to the long-term values of the state variables. This procedure can also be recommended for finding the initial values if they are not measured or known by other means.

2. Does the model react as expected? If the input of toxic substances is increased, we should expected a higher concentration of the toxic substance in the top-carnivorous organisms. If that is not the case, it indicates that some formulations may be wrong and they should be corrected.

In general at this stage some playing with the model is recommended. Through such exercises the modeller gets acquaintained with the model and its reactions to pertubations.

After verification there follows **the sensitivity analysis.** Through this the modeller gets a good overview of the most *sensitive components of the model.* The sensitivity analysis attempts to provide a measure of the sensitivity of either parameters, forcing functions or submodels to the state variables of greatest interest in the model. If a modeller wants to simulate a toxic substance concentration in, for instance, carnivorous insects as a result of the use of insecticides, he will obviously choose this state variable as the most important one, maybe in addition to the concentration of the toxic substance concentration in plants and herbivorous insects.

In practical modelling the sensitivity analysis is carried out by changing the parameter, the forcing function or the submodel and observing the corresponding response on the selected state variables. Thus, the sensitivity of a parameter, S, is defined as follows:

$$S = [\ \partial x/x\ ]/[\ \partial P/P\ ], \tag{2.1}$$

where x is the considered state variable and P is the examined parameter.

The relative change in the parameter value is chosen on the basis of our knowledge of the certainty of the parameters. If, for instance, the modeller estimates the uncertainty to be about 50% relatively, he will probably choose a change in the parameters of ± 10% and ± 50% and record the corresponding change in the state variable(s). It is often necessary to find the sensitivity at two or more levels of parameter change as the relation between a parameter and a state variable is rarely linear.

A sensitivity analysis of the submodels (process equations) can also be carried out. In this case the change in a state variable is recorded when the equation of the submodel is deleted from the model or changed to an alternative expression, for instance, with more details built into the submodel. Such results may be used to make structural changes in the model. If the sensitivity, for instance, shows that a given more detailed submodel is crucial for the model results this result should be used to change the model correspondingly. The selection of the complexity and structure of the model should therefore work hand in hand with the sensitivity analysis. In

Figure 2.3 this is shown as a feedback from the sensitivity analysis to the data requirements. This is in accordance with the order of model, presented by Halfon et al. (1979).

A sensitivity analysis of forcing functions gives an impression of the importance of the various forcing functions and tells us the accuracy required of the forcing function data.

The scope of the **calibration** is the improvement of the parameter estimation. Some parameters in causal ecological models can be found in the literature, not necessary as constants but as approximate values or intervals. If approximate values cannot be found in the literature, one of the estimation methods presented in Chapters 4 and 5 must be used. Jørgensen et al (1979) and (1988) contain more than 50,000 parameters of interest for ecological and ecotoxicological modellers. Tables in the Appendix illustrate such a collection of parameters values culled from the literature.

To cover all possible parameters for all possible ecotoxicological models, however, we need to know more than one billion parameters. Thus it is obvious that in ecotoxicological modelling there is a particular need for the use of **parameter estimation methods.** This will be discussed further in chapter 3, section 3.1. Under all circumstances it is a great advantage to be able to give even approximate values for the parameters, before calibration is started. It is, of course, much easier to search for a value between 1 and 10 than to search between 0 and $+\infty$.

Even where all parameters are known within intervals, either from the literature or from estimation methods, the model usually has to be calibrated. Several sets of parameters are tested by the calibration and the various model outputs of state variables are compared with measured values of the same state variables. The set of parameter giving the best accord between model output and measured values is chosen.

The need for calibration can be explained using the following characteristics of ecological models and their parameters:

1. Most ecotoxicological parameters are not known as exact values. Therefore all literature values for ecotoxicological parameters have a degree of uncertainty. The parameter estimation methods, that will be presented in chapters four and five, have in most cases an even greater uncertainty.
2. All ecotoxicological models are simplifications of nature. The most important components and processes may be included, but the model structure does not account for all detail. The influence of some components and processes of minor importance can to a certain extent be taken into account by the calibration. This will give the parameter values slightly different from the real, but unknown values of Nature, but this difference may partly account for the influence of the details

that have been omitted.

3. Most ecotoxicological models are 'lumped models', which implies that one parameter represents the average values for several species. As each species has its own characteristic parameter value, the variation in the species composition with time will inevitably give a corresponding variation in the average parameter used in the model.

A calibration cannot be carried out randomly, if more than a couple of parameters have been selected for calibration. If, for instance, ten parameters have to be calibrated and the uncertainty justifies the testing of ten values of each parameter, the model has to be run 10^{10} times which, of course, is an impossible task. Therefore the modeller will have to learn the behaviour of the model by varying one or two parameters at a time and observing the response of the most crucial state variables. In some cases it is possible to separate the model into a number of submodels, which can be calibrated more or less independently. Although the calibration described is based to some extent on a systematic approach, it is still a trial and error procedure. However, procedures for automatic calibration are available. Table 2.1. gives the characteristics of a software named PSI intended for automatic calibration. This does not mean that the trial and error calibration described above is redundant. If automatic calibration should give a satisfactory result within a limited time frame, only 4-8 parameters need to be calibrated at the same time. Under all circumstances it is easier to find the optimum set of parameters, the smaller the uncertainties in the parameters before the calibration is started.

Table 2.1
Automatic calibration by use of PSI

The user gives:

1. Initial guesses for parameter
2. Ranges of parameter variation
3. A set of measured state variable
4. An acceptable maximum value for the standard deviation between modelled and measured values

The software tests various combinations of parameters and calculates the standard deviation. The combinations are selected by use of the Marquardt method which combine the Gauss-Newton method and steepest descend method. It continues until the acceptable maximum value for the standard deviation has been achieved.

In the trial and error calibration the modeller has more or less intuitively set up some calibration criteria. For example, he may want to simulate rather accurately the minimum oxygen concentration for a stream model and/or the time at which the minimum occurs. When he is satisfied with these model results, he may want to simulate the shape of the oxygen concentration versus time curve properly, and so on. He calibrates the model step by step to achieve these objectives step by step.

If an automatic calibration procedure is applied, it is necessary to formulate objective criteria for the calibration. A possible function would be based on the term for the calculation of the standard deviation:

$$Y = [(\Sigma (x_c - x_m)/ x_{m,a})/n]^{1/2} \qquad (2.2)$$

where x_c is the computed value of a state variable, x_m is the corresponding measured value, $x_{m,a}$ is the average, measured value of a state variable, and n is the number of measured or computed values.

Y is followed and computed during the automatic calibration and the goal of the calibration is to obtain as low a Y-value as possible.

In many cases, however, the modeller is more interested in good accord between model output and observations for one or two state variables, and less interested in good accord for the other state variables. In that case he may choose weights for the various state variables to account for the emphasis he puts on each state variable in the model. For a model of the fate and effect of an insecticide he may put emphasis on the toxic substance concentration of the carnivorous insects while considering the toxic substance concentrations in plants, herbivorous insects and soil as less important. He may therefore choose a weight of ten for the first state variable and only one for the last three.

If it is impossible to calibrate a model properly, it is not necessarily due to an incorrect model. It may be due to the low quality of the data set for the quality of the data is crucial for the calibration. It is furthermore of great importance that **the observations reflect the dynamics of the system.** If the objectives of the model is to give a good description of one or a few state variables, then it is essential that the data is able to show these dynamics. The frequency of the data collection should therefore reflect the dynamics of the state variables in focus. Unfortunately this rule has often been violated in modelling. It is strongly recommended that the dynamics of all state variables are considered before the data collection program is determined in detail. Some state variables often have particularly pronounced changes in specific periods - often the spring - and it may be a great advantage to have a particular dense data collection in this period; see Jørgensen et al (1981), where it is shown how a dense data collection program in a certain period can be applied to give an additional certainty in the determination of some important parameters.

From these considerations it is possible to offer recommendations for a calibration of an ecotoxicological model:

1. Find **as many parameters as possible from the literature.** Even *wide* ranges for the parameters should be considered very valuable, as even approximate initial guesses for all parameters are needed.
2. If some parameters cannot be found in the literature, which is often the case for ecotoxicological parameters, the **estimation methods** mentioned in chapter 4 and 5 should be used, or, where some crucial parameters are involved, these may be determined by experiment in situ or in the laboratory.
3. A **sensitivity analysis** should be carried out to determine which parameters are most important to know with high certainty.
4. The use of a **dense data collection program** for the most important state variables should be considered, to provide better estimations for the most crucial parameters.
5. At this stage a **calibration** should be carried out using the not yet applied data. The most important parameters are selected and the calibration is limited to these, or, at the most, to eight to ten parameters. At first the calibration involves using the trial and error method to get acquainted with the model reaction to changes in the parameters. An automatic calibration procedure is used afterwards to polish the parameter estimation.
6. These results are used in **a second sensitivity analysis**, which may give results different from the first sensitivity analysis.
7. **A second calibration** is now used on the parameters shown to be most important by the second sensitivity analysis. As before both calibration methods may be used. After this final calibration the model can be considered calibrated and we can proceed to the next step: validation.

Calibration should always be followed by **validation.** By this step the modeller tests the model against an independent data set and observes how well model simulations fit this data set. It must, however, be emphasized that the validation only confirms the model behaviour under the range of conditions represented by the available data. Consequently, it is preferable to validate the model using data obtained from a period in which conditions prevailed other than those from the period of data collection for calibration. For instance when a model of the fate and effect of toxic substances is tested, it would be preferable to have a data set for the calibration and the validation, which differs by the level of toxic substance impact on the ecosystem. If an ideal validation cannot be obtained, it is, however, still important to validate the model. The model can always be used as a management tool, provided that the modeller presents the open questions of the model to the manager. As we gain more experience in use of the model, the number of open questions may be reduced.

The method of validation is dependent on the objectives of the model. A comparison between measured and computed data by use of the objective function (2.2) is an obvious test. However, often this is not sufficient, as it may not focus on the main objective of the model, but only on the general ability of the model to describe correctly the state variables of the ecosystem. It is therefore necessary to translate the main objectives of the model into a few validation criteria. They cannot be formulated generally, but are individual for the model and the modeller. For instanc, if we are concerned with the maximum concentration of a toxic substance in carnivorous insects, it would be useful to compare the measured and computed maximum concentrations of this level in the food chain.

The discussion on the validation, can be summarized into the following issues:

1. **Validation is always required,** to get a picture of the reliability of the model.
2. Attempts should be made to get data for the validation, which are **entirely different from those used in the calibration.** It is important to have data from a **wide range of forcing functions,** that are defined by the objectives of the model.
3. The validation criteria are formulated on basis of **the objectives of the model** and **the quality of the available data.**

2.4 TYPES OF MODELS

It is useful to distinguish between various types of model and discuss the selection of model types briefly. A more comprehensive treatment of this topic can be found in Jørgensen (1988).

Pairs of models are shown in Table 2.2. The first division of models is based on the application: **scientific and management models.** The next pair is: **stochastic and deterministic models.** A stochastic model contains stochastic input disturbances and random measurement errors, as shown in Fig. 2.4. If they are both assumed to be zero the stochastic model reduces to a deterministic model, provided that the parameters are not estimated in terms of statistical distributions. A deterministic model assumes, that the future response of the system is completely determined by a knowledge of the present state and future measured inputs. Stochastic models are rarely applied in ecotoxicology.

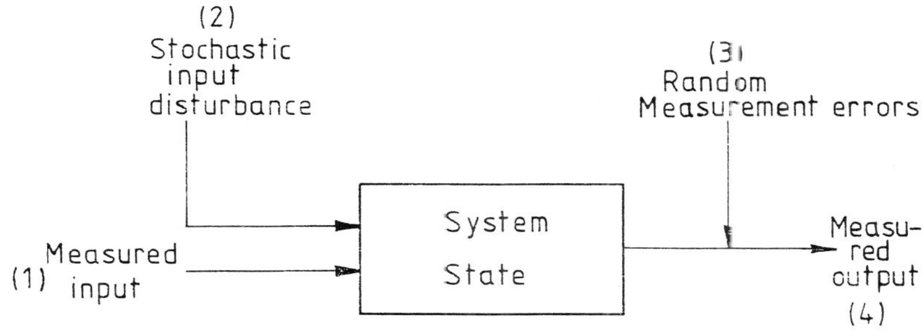

Fig. 2.4: A stochastic model considers (1) (2) and (3), while a deterministic model assumes that (2) and (3) are zero.

The third pair in Table 2.2 is **compartment and matrix models.** Compartment models are understood by some modellers to be based on the use of compartments in the conceptual diagram, while other modellers distinguish between the two classes of models entirely by the mathematical formulation, as indicated in the table. Both types of models are applied in ecotoxicology - the matrix model mainly in population dynamics.

The classification **reductionistic and holistic models** is based upon a difference in the scientific ideas behind the model. The reductionistic modeller will attempt to incorporate as many details of the system as possible to be able to capture its behaviour. He believes that the properties of the system are the sum of the details. The holistic modeller, on the other hand, attempts to include in the model properties of the ecosystem working as a system by use of general principles. In this case the properties of the system, and not the sum of all the details, are considered, the holistic modeller presuming that the system posseses some additional properties because the subsystems are working as a unit. Both types of models may be found in ecotoxicology, but in general the ecotoxicologist must adopt a holistic approach to problems to get an overview, because ecotoxicological problems are very complex. The idea behind many of the parameter estimation methods presented in chapter four and five are holistic in their basic ideas.

Table 2.2
Classification of Models (pairs of model types)

Type of models	Characterization
Research models	Used as a research tool
Management models	Used as a management tool
Deterministic models	The predicted values are computed exactly
Stochastic models	The predicted values depend on probability distribution
Compartment models	The variables defining the system are quantified by means of time-dependent differential equations
Matrix models	Use matrices in the mathematical formulation
Reductionistic models	Include as many relevant details as possible
Holistic models	Use general principles
Static models	The variables defining the system are not dependent on time
Dynamic models	The variables defining the system are a function of time (or perhaps of space)
Distributed models	The parameters are considered functions of time and space
Lumped models	The parameters are within certain prescribed spatial locations and time, considered as constants
Linear models	First degree equations are used consecutively
Nonlinear models	One or more of the equations are not first degree
Causal models	The inputs, the states and the outputs are interrelated by use of causal relations
Black box models	The input disturbances affect only the output responses. No causality is required
Autonomous models	The derivatives are not explicitly dependent on the independent variable (time)
Nonautonomous models	The derivatives are explicitly dependent on the independent variable (time)

Dynamic systems may have four classes of states. *The initial state* changes through *transient states* to a state, where the system oscillates round a *steady state,* as shown in Fig. 2.5. The transient phase can only be described by use of a **dynamic model**, which uses differential or difference equations to describe the system response to external factors. Differential equations are used to represent continous changes of state with time, while difference equations use discrete time steps. The steady state corresponds to the situation where all derivatives are equal to zero. The oscillations round the steady state are described by use of a dynamic model, while steady state ifself can be described by use of **a static model.** As all derivatives are equal to zero in steady states the static model is reduced to *algebraic* *equations.* Some dynamic systems have no steady state; for instance, systems which show *limit cycles.* This fourth state possibility obviously requires a dynamic model to describe the system behaviour. In this case the system is always non-linear, although there are non-linear systems, which have steady states.

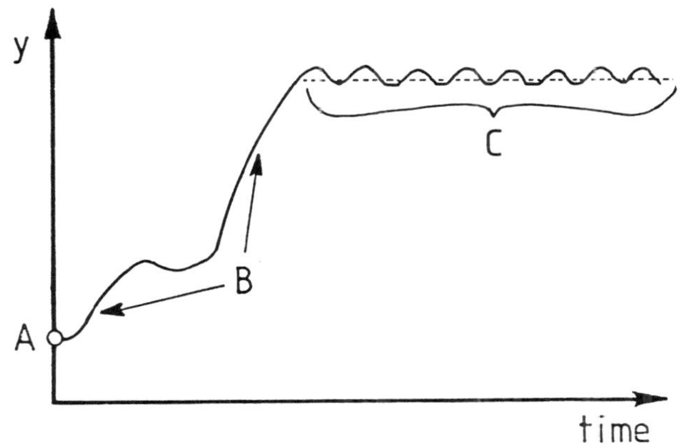

Fig. 2.5: Y is a state variable expressed as a function of time. A is the initial state, B transient states and C the oscillation around the steady sta:e. The dotted line corresponds the steady state, which can be descibed by use of a static model.

A static model assumes, conseqJently, that all variables and parameters are independent of time. The advantage of the static model is its potential for simplifying subsequent computational effort through the elimination of one of the independent variables in the model relationship.

A **distributed model** accounts for variations of variables in time and space. A typical example would be an advection-diffusion model for transport of dissolved toxic substance along a stream. It might include variations in the three orthogonal directions. The analyst might, however, decide on the basis of prior observations, that gradients of dissolved material along one or two directicns are not sufficiently large to merit inclusion in the model. The model would then be reduced by that assumption to a lumped parameter model. Whereas the **lumped model** is frequently based upon ordinary differential eqLations, the distributed model is usually defined by *partial differential equations.*

The causal or internally descriptive model characterizes how the inputs are connected to states and how the states are connected to each other and to the outputs of the system, wheras **the black box model** reflects only what changes the input will effect in the output response. In other words the causal model provides a description of the internal mechanisms of process behaviour. The black box model deals only with what is measurable: the input and the output. The relationship may be found by a statistical analysis . If on the other hand the processes are described in the model by use of equations, which give the relationship, the model will be causal.

The modeller may prefer to use black box descriptions in the cases, where his knowledge about the processes is rather limited. The disadvantage of the black box model is, however, that it is limited in application to the considered ecosystem or at least to a similar ecosystem. If general applicability is needed it is essential to set up a causal model. The latter type is much more widely used in ecotoxicology than the black box model, mainly due to the understanding that the causal model gives to the user of the functioning of the system, including the many chemical, physical and biological reactions.

Autonomous models are not explicitly dependent on the time (the independent variable):

$$dy/dt = a^*y^b + c^*y^d + e \tag{2.3}$$

Nonautonomous models contain terms, g(t), that make the derivatives dependent on time. For instance:

$$dy/dt = a^*y^b + c^*y^d + e + g(t) \tag{2.4}$$

Table 2.3 shows another classification of models. The differences between the three types of models are the choice of components used as state variables. If the model aims at the description of a number of individuals, species or classes of species, the model will be called **biodemographic.** A model, which describes the energy flows is named **bioenergetic** and the state variables will typically be expressed in kW or kW per unit of volume or area. The **biogeochemical** models consider the flow of material and the state variables are indicated as kg of kg per unit of volume or area.

Table 2.3
Identification of Models

Type of models	Organization	Pattern	Measurements	Examples
Biodemographic	Conservation of species or genetic information	Life cycles	Number of individuals or species	Chapter 6
Bioenergetic	Conservation energy	Energy flow	Energy	Chapter 4
Biogeochemical	Conservation of mass	Element cycles	Mass or concentrations	Chapter 7

2.5 CHARACTERISTIC FEATURES OF ECOTOXICOLOGICAL MODELS

An increasing interest in management of toxic substance pollution has emerged during the last decade and this has caused an equally large interest in toxic substance modelling.

Toxic substance models attempt to model the fate and effect of toxic substances in ecosystems. They are characterized (see the classification in Table 2.2) by being management models, deterministic models, more frequently compartment models than matrix models, holistic models, lumped models (because the trouble to find parameters is greater than for ecological modelling in general), most frequently nonlinear, causal autonomous models.

If we turn to the classification in Table 2.3, toxic substance models are most often biogeochemical models, because they attempt to describe the mass flows of the considered toxic substances, although there are models of the population dynamics, which include the influence of toxic substances on the birth rate and/or the mortality, and therefore should be considered a toxic substance model.

Toxic substance models differ from other ecological models by.

1. The need for *parameters* to cover all possible toxic substance models is great, and general estimation methods are therefore used quite widely. See chapter 4 and 5.
2. The *safety margin should be high* when, for instance, expressed as the ratio between the actual concentration and the concentration that gives undesired effects.
3. The possible inclusion of *an effect component,* which relates the output concentration to its effect. It is quite easy to include an effect component in the model; it is, however, often a problem to find a well

examined relationship to base it on.

4. The possibility and need of *simple models,* because of point 1 and 2, and of our limited knowledge of process details, parameters, sublethal effects, antagonistic and synergistic effects.

It may be an advantage to clarify several questions before developing a toxic substance model in accordance with the procedure presented in section 2.3:

1. Obtain the best possible knowledge about the possible **processes** of the considered toxic substances in the ecosystem. As far as possible knowledge about the quantitative role of the processes should be obtained.
2. Attempt to get **parameters from the literature and/ or from own experiments** (in situ or in the laboratory)
3. **Estimate** also all parameters by the methods presented in chapters four and five.
4. Compare the results from 2) and 3) and attempt to explain **discrepancies.**
5. Estimate which processes and state variables it would be feasible and relevant to include into the model. If there is the slightest doubt then include at this stage **too many processes and state variables rather than too few.**
6. Use a **sensitivity analysis** to evaluate the significance of the individual processes and state variables. In many cases this will lead to further **simplification.**

The description of the chemical, physical and biological processes will, in general, be in accordance with the equations presented in chapter six. The processes involved in the interaction between an organism and a toxic substance are shown Figure 2.6. The organism takes up toxic substances, either from the feed, or directly from the environment (air or water). The latter uptake is described by the concentration factor, which gives the ratio between the concentration of the toxic substance in the the air or water, while the first uptake is dependent on the amount of the toxic substance, which is digested and not digested.

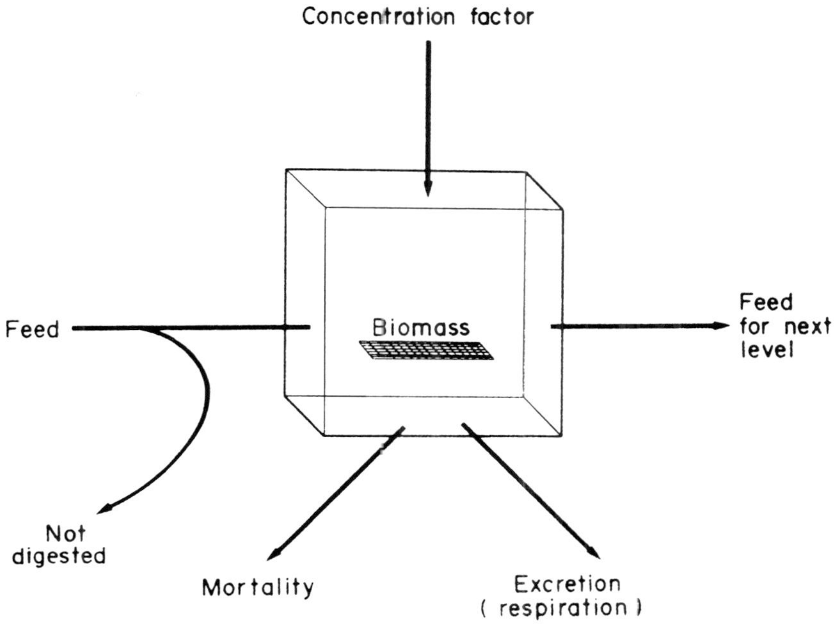

Fig. 2.6: Principle for modelling the concentration of a toxic substance at a trophic level. (Jørgensen, 1979).

CHAPTER 3

USE OF ECOTOXICOLOGICAL MODELS IN MANAGEMENT

by
S.E. Jørgensen

3.1 COMPLEXITY OF ECOTOXICOLOGICAL MODELS

The literature of ecological modelling contains several methods, which are applicable to the selection of the model complexity. A rather comprehensive discussion is presented in Jørgensen (1988), where references are given to the following papers devoted to this question: Halfon (1983) and (1984), Halfon et al (1979), Costanza and Sklar (1985), Bosserman (1980) and (1982) and Jørgensen and Mejer (1977).

It is clear from the discussion in chapter two, that selection of the model complexity is *a matter of balance.* On the one hand, it is of course necessary to include the state variables and the processes that are essential for the problem in focus. On the other hand, it is - as already pointed out - of importance not to make the model more complex than the data set can bear. Our knowledge about processes, state variables and our data set determine the model. If our knowledge is poor the model will not be able to give many details and it will have a relative high uncertainty. If we have a profound knowledge about the problem that we want to model, we can construct a more detailed model with a low uncertainty. Many researchers claim that a model cannot be developed before one has a certain level of knowledge and that it is wrong to attempt to construct a model in a data poor situation. In my opinion this is wrong, because the model can always assist the researcher by a synthesis of his knowledge and by a vizualisation of the system. But the researcher must, of course, always present the shortcomings and the uncertainties of the model and not try to pretend that the model is a real picture of reality in all its details. In the hand of the researcher a model will often be a fruitful instrument to test hypotheses.

It should not be forgotten that models have always been applied in science. The difference between the present models and previous ones is only that today we are able to work with very complex models due to the modern computer technology. Sometimes it has also been a temptation to construct models that are too complex - it has been so easy to add more equations and state variables, but much harder to obtain the data needed for calibration and validation of the model.

Even if we have very detailed knowledge about a problem, we are unlikely able to develop a model which is capable of accounting for the complete input-output behaviour of the real ecosystem, and still be valid in all frames (Zeigler 1976). This model is named the base model by Zeigler, and it would be very complex and require such a great number of computational resources, that it would be almost impossible to simulate. The base model of an ecotoxicological problem will never be fully known, because of the complexity of the system and the impossibility of observing all states. However, given the experimental frame of current interest, a modeller is likely to find it possible to construct a relatively simple model that is workable in that frame.

Up to a point a model may be made more realistic by adding more and more connections After that point addition of new parameters after that point does not contribute to improved simulations, but on the contrary more parameters imply more uncertainty, because of the possible lack of information about the flows which the parameters quantify. Given a certain amount of data, the addition of new state variables or parameters beyond a certain model complexity does not add to our ability to model the ecosystem, but only adds unaccounted uncertainty. These ideas are visualized in Figure 3.1.

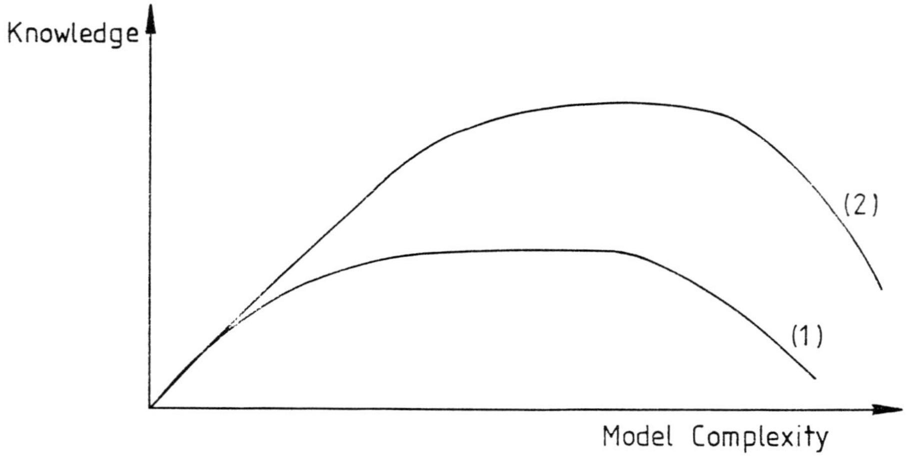

Fig. 3.1: Knowledge plotted versus model complexity measured, for instance, by the number of state variables. The knowledge increases up to a certain level. Increased complexity beyond this level will not add to ones knowledge about the modelled system. At a certain level ones knowledge might even be decreased. (2) corresponds to an available data set, which is more comprehensive or has a better quality than (1). (Costanza and Sklar (1985).

The relationship between knowledge gained through a model and its complexity is shown for two levels of data quality and quantity. The question under discussion can be formulated with relation to this figure: **How can we select the complexity and the structure of the model to assure that we are on the optimum for knowledge gained or the best answer to the question posed to the model?**

We shall not here discuss the methods available for selecting a good model structure, but refer again to the publications mentioned at the start of the chapter. If a rather complex mode is to be developed it is recommended that one of the methods presented in these publications is used; but for simpler models it is often sufficent to go for a model of balanced complexity, as discussed above.

Constanza and Sklar (1985) have examined 88 different models and they were able to show that the more theoretical discussion behind Figure 3.1, actually is valid in practice. Their results are summarized in Figure 3.2, where effectiveness is plotted versus articulation. Effectiveness is understood as a product of how much the model is able to tell us and with what certainty; while articulation is a measure of the complexity of the model in respect to number of components, time and space. It is clearly seen by comparison of Figures 3.1 and 3.2 that they show the same relationship.

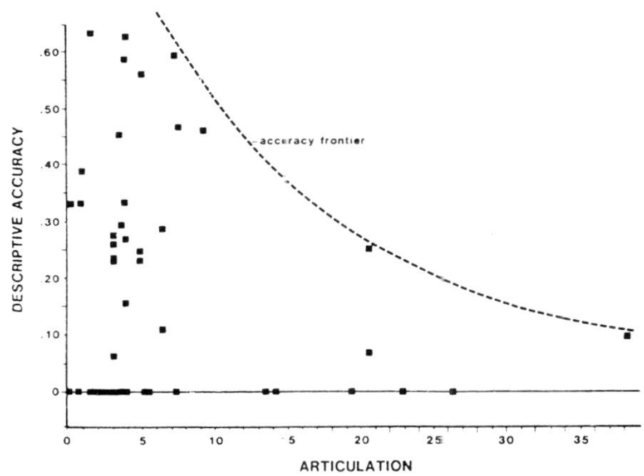

Fig. 3.2: Plot of articulation index versus descriptive accuracy index for the models reviewed by Costanza and Sklar (1985). The figure is reproduced from this paper.

Selection of the right complexity is of great importance in ecological modelling and it is even more important when ecotoxicological problems are involved. An ecotoxicological model may easily be too complex.

The characteristic features of ecotoxicological models in relation to complexity may be summarized as follows:

1. Our knowledge about ecotoxicological processes *is in general smaller* than our knowledge about ecological processes.
2. *The number of parameters we need to know is very large, if we want to cover all possible combinations of toxic substances, their processes and their interactions with living organisms.* If we consider that there are about 50,000 toxic substances produced in such quantities that they may threaten the environment, that the number of parameters for each toxic substance is at least 10, that each of these parameters are different for different species and that we to get a representative picture of all species we should at least know the interaction with, let us say 1000 species, **then we would require to know 50,000*10*1000 parameters = 500 millions parameters.**
3. Fortunately, *in most cases we do not need to model ecotoxicological problems very accurately,* because we aim for the application of **a large safety factor.**

3.2 CHARATERISTIC FEATURES OF ECOTOXICOLOGICAL MODELS

Ecotoxicological models differ from general ecological models by

1) being most often **more simple,**
2) require **more parameters,**
3) **use of estimation methods more widely,**
4) possibly inclusion of **an effect component.**

Ecotoxicological models can be divided into five classes. The classification presented here is based on differences in the modelling structure. The decision as to which model class to apply is based upon the ecotoxicological problem that the model is going to solve. This classification is different from the classifications presented in chapter two where the mathematics or the philosophy behind the models were the governing principles, whereas the present classification is based upon the actual problem that is modelled. The definitions of the model classes are given below and, also indicated where it is most appropriate to use each of the model types.

1. Food chain or food web dynamic models.

This class of models considers the flow of toxic substances through the food chain or food web. Such models will be relative complex and contain many state variables. The models will furthermore contain many parameters, which often have to be estimated by one of the methods presented in chapter four and five. This type of model will typically be used when a great number of organisms are affected by the toxic substance, or the entire structure of the ecosystem is threaten by the presence of a toxic substance. Because of the complexity of these models, they have not been used widely. They are similar to the more complex eutrophication models which consider the flow of nutrients through the food chain or even through the food web. In some instances they are even constructed as submodels of a eutrophication model, see for instance Thomann et al 1974. Figure 3.3. shows a conceptual diagram of a ecotoxicological food chain model for lead. The flow of lead from atmospheric fall out and waste water to an aquatic ecosystem, where it is concentrated through the food chain - the so called 'bioaccumulation'. A simplification is hardly possible for this model, because it is the aim of the model to describe and quantify the bioaccumulation through the food chain.

2. Static models of the mass flows of toxic substances.

If the seasonal changes are minor, or of minor importance, a static model of the mass flows will often be sufficient to describe the situation and even indicate the expected changes if the input of toxic substances is reduced or enlarged. This type of model is based upon a mass balance as clearly seen from the example in Figure 3.4. It will often, but not necessarily, contain more trophic levels, and the modeller often is concerned with the flow of the toxic substance through the food chain. The example in Figure 3.4 considers only one trophic level. If there are some seasonal changes, this type, which in most cases is simpler than type one, can still be an advantage to use, for instance, if the modeller is concerned with the worst case and not the changes.

3. A dynamic model of a toxic substance in a trophic level.

It is often only the toxic substance concentration in one trophic level which is of concern. This includes the zero trophic level, which is understood as the medium - either soil, water or air.

Figure 3.5 gives an example. It is a model of copper contamination in an aquatic ecosystem. The main concern is the copper concentration in the water, as it may reach a toxic level for the phytoplankton. Zooplankton and fish are much less sensitive to copper contamination, so the alarm clock rings first at the concentration level that is harmful to phytoplankton. However, only the ionic form is toxic and it is therefore necessary to model the partition of copper in ionic form, complex bound form and adsorped form. The exchange between copper in the water phase and in the sediment is also included, because the sediment is able to accumulate relatively large amounts of heavy metals. The amount released from the sediment may be significant under certain circumstances - for instance under low pH.

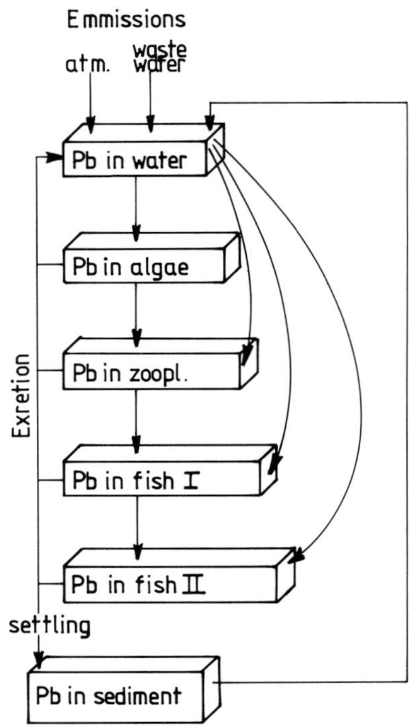

Fig. 3.3: Conceptual diagram of the bioaccumulation of lead through a food chain in an aquatic ecosystem.

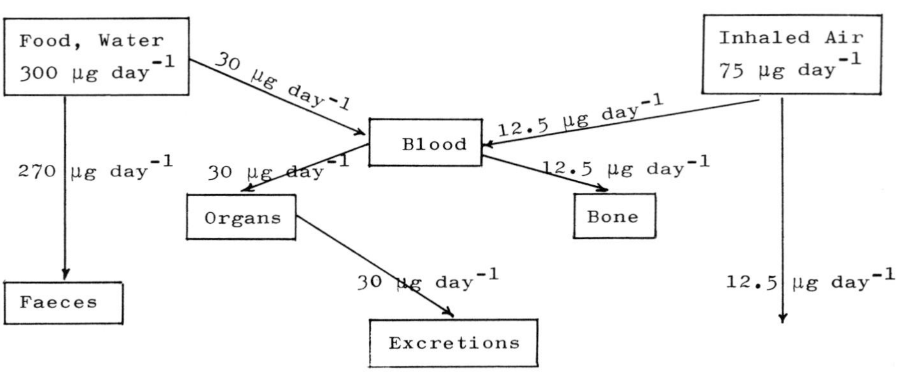

Fig. 3.4: A static model of the lead uptake by an average Dane.

Figure 3.6 gives another example. Here the main concern is the DDT concentration in fish, where these may be such high concentration of DDT that, in accordance with WHO's standards, they are unfit for human consumption. The model can therefore be simplified by not including the entire food chain but only the fish. Some physical-chemical reactions in the water phase is however of importance and they are included as shown on the conceptual diagram Figure 3.6.

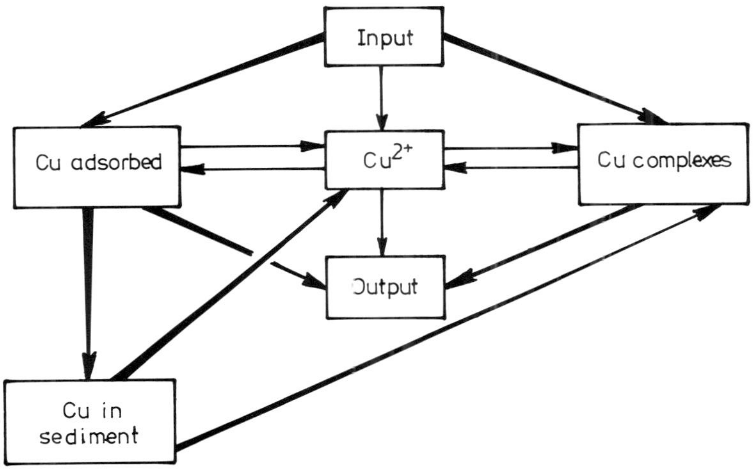

Fig. 3.5: Conceptual diagram of a simple copper-model.

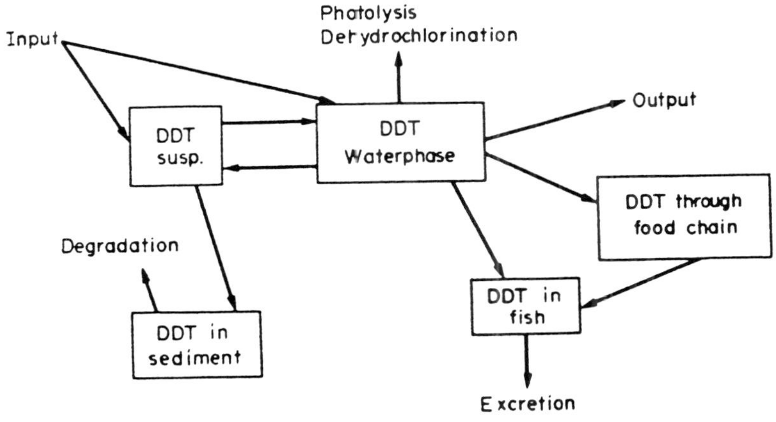

Fig.3.6: Conceptual diagram of a simple DDT-model.

As seen from these examples simplifications are often feasible when the problem is well defined, including which component is most sensitive to toxic matter, and which processes most important for concentration changes.

4. Ecotoxicological models in population dynamics.

Population models are biodemographic models and have therefore numbers of individuals or species as state variables. The simple population models consider only one population. The growth of the population is a result of the difference between natality and mortality:

$$dN/dt = B^*N - M^*N = r^*N, \qquad (3.1)$$

where N is the number of individuals, B is the natality i.e. the number of new individuals per unit of time and per unit of population, M is the mortality, i.e. the number of organisms that died per unit of time and per unit of population; and, r is the increase in the number of organisms per unit of time and per unit of population, and = B - M. The concentration of toxic substance in the environment or in the organisms influences the natality and the mortality, and if the relation between a toxic substance concentration and these population dynamic parameters are included in the model, it becomes a ecotoxicological model of population dynamics.

Population dynamic models may include two or more trophic levels and ecotoxicological models will include the influence of the toxic substance concentration on natality, mortality and interactions between these populations. In other words, an ecotoxicological model of population dynamics is a general model of population dynamics with the inclusion of the relation between toxic substance concentrations and some of the model parameters.

5. Ecotoxicological Models with Effect Components.

Though class 4 models already include relations between concentrations of toxic substances and their effects, these are limited to population dynamic parameters. In comparison class 5 models include more comprehensive relations between toxic substance concentrations and effects. These models may include not only lethal and/or sublethal effects but also effects on biochemical reactions or on the enzyme system.

Figure 3.7 gives a conceptualization of this type of model. As shown on this figure, the effects may be considered on various levels of the biological hierarchy from the cells to the ecosystems.

In many problems it may be necessary to go into more detail on the effect to answer the following relevant questions:

1. Does the toxic substance accumulate in the organism?

2. What will be the long term concentration in the organism when uptake rate, excretion rate and biochemical decomposition rate are considered?
3. What is the chronic effect of this concentration?
4. Does the toxic substance accumulate in one or more organs?
5. What is the transfer between various parts of the organism?
6. Eventually will decomposition products cause additional effects?

A detailed answer to all these question may require a model of the processes that take place in the organism, and a translation of concentrations in various parts of the organism into effects. This implies, of course, that the intake = (uptake by the organism)*(efficiency of uptake) is known. Intake may either be from water or air, which also may be expressed by a concentration factor. In appendix 3 a list of concentration factors is given - which are the ratios between the concentration in the organism and in the air or water.

However, if all the above mentioned processes should be taken into consideration for just a few organisms, the model will easily become too complex, contain too many parameters to calibrate and require more detailed knowledge than it is possibleto provide. Therefore, most models in this class will not consider too many details of the partition of the toxic substances in organisms and their corresponding effects, but rather be limited to the simple accumulation in the organisms and their effect.

In most cases accumulation is rather easy to model and the following simple equation is often sufficiently accurate:

$$d\,C/d\,t = (e_f{*}C_f{*}F + e_m{*}C_m{*}V)/W - Ex{*}C = (INT)/W - Ex{*}C \qquad (3.2)$$

where C is the concentration of the toxic substance in the organism; e_f and e_m are the efficiencies for the uptake from the food and medium respectively (water or air); C_f and C_m are the concentration of the toxic substance in the food and medium respectively; F is the amount of food uptake per day; V is the volume of water or air taken up per day; W is the body weight either as dry or wet matter; and Ex is the excretion coefficient (1/day). As seen from the equation INT covers the total intake of toxic substance per day.

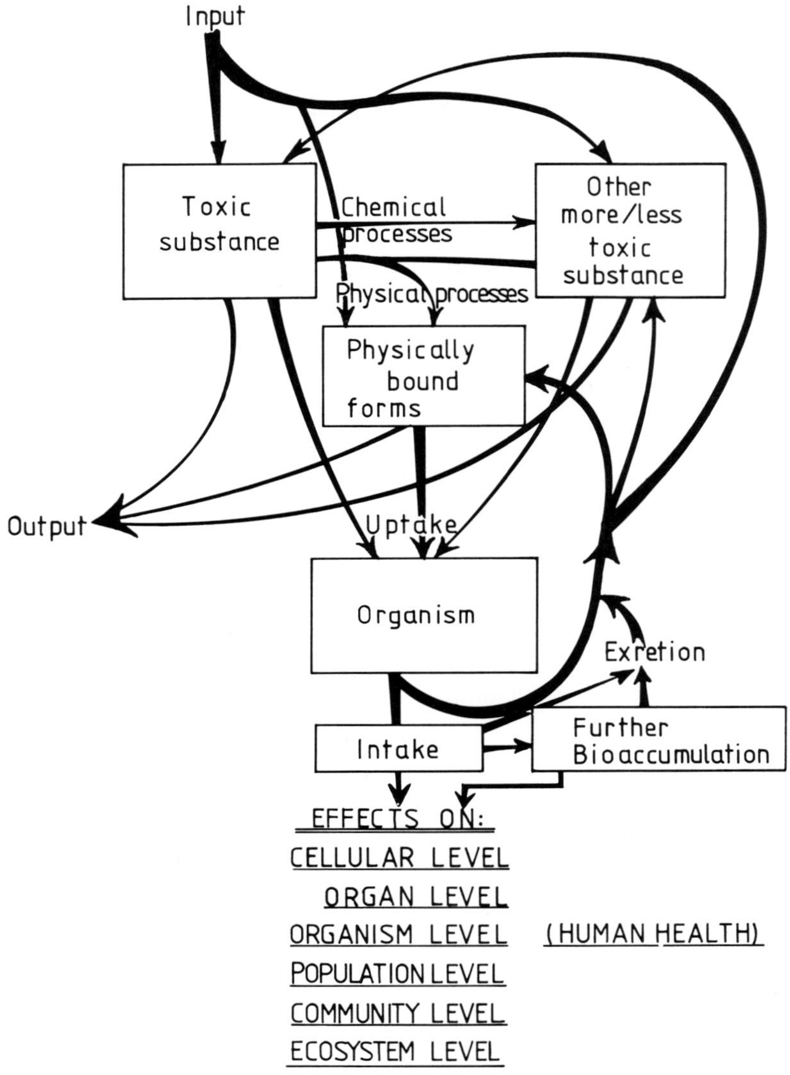

Fig. 3.7: Conceptual diagram of the principle in ecotoxicological models with effect components.

This equation has a numerical solution, and the corresponding plot is shown in Figure 3.8:

$$C/C(max) = (INT*(1 - exp(Ex*t)))/(W*Ex) \qquad (3.3)$$

where C(max) is the steady state value of C:

$$C(max) = INT/(W*Ex) \qquad (3.4)$$

The presentation of the five classes of models above, clearly show the advantages and limitations of ecotoxicological models. The simplifications used in classes two and three often offer great advantages. They are sufficiently accurate to give a good and applicable picture (overview) of the concentrations of toxic substances in the environment, due to the application of great safety factors. The application of the estimation methods, presented in chapter four and five, renders it feasible to construct such models, even our knowledge of the parameters is quite limited. The estimation methods have obviously a high uncertainty, but the great safety factor helps in accepting this uncertainty. On the other hand our knowledge about the effects of toxic substances is very limited - particularly at the organism and organ level. It must not be expected, therefore, that models with effect components give more than a first rough picture of what is known today in this area.

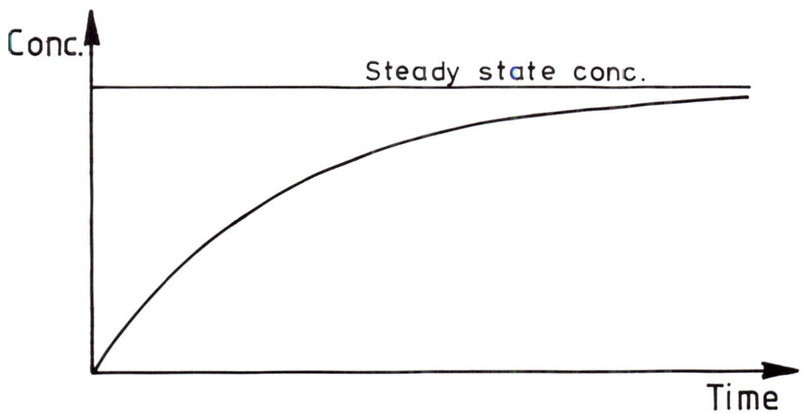

Fig. 3.8: Concentration of a toxic substance in an organism versus time.

Because of the character of ecotoxicological mode s it is recommended that a few questions be clarified before entering the modelling procedure, as presented in Chapter 2: (see also section 2.5)

1. Obtain the best possible knowledge about *the processes* of the toxic substance under consideration.
2. Attempt to get *parameters* of the toxic substance processes in the environment from the literature.
3. *Estimate all parameters* using methods presented in Chapters 4 and 5.
4. *Compare* the results from 2 and 3 and attempt to explain *discrepancies*, if present.
5. Use widely *sensitivity analysis to estimate* which processes and state variables it would be feasible and relevant to include into the model.

3.3 AN OVERVIEW: THE APPLICATION OF MODELS IN ECOTOXICOLOGY

A number of toxic substance models is reviewed in Table 3.1. to give an impression of the types of model available today. Most models reflect the proposition that good knowledge of the problem and ecosystem can be used to make reasonable simplifications. Model characteristics indicated in the table are state variables and/or processes considered in the model. Note in the table the number of modelled toxic substances and the processes taken into account.

As well be seen models of classes one, two, three and five are represented in the table.

There is only one class four model included in the table. Ecological modelling has been approached from two sides: population dynamics and biogeochemical flow analysis. As the second approach has been most involved in environmental isssues, it has been most naturally also to approach the toxic substance problem from this angle. The few class four models are, together with the example included in Table 3.1, population dynamic models, with a few additional equations to account for the influence of toxic substances on natality and mortality. Provided that these relations are available, it should be relatively easy to construct this type of model.

A study of these examples, and of the case studies presented in Chapter 8, illustrates clearly that the most difficult part of modelling the effect and distribution of toxic substances is to obtain the relevant knowledge about the behaviour of the toxic substances in the environment, and to use this knowledge to make the feasible simplifications.

Table 3.1
Examples of Toxic Substance Models

Toxic Substance	Model Characteristics	Reference
Cadmium	Food chain similar to a eutrophication model	Thomann et al., 1974
Mercury	6 state variables: water, sediment, suspended matter, invertebrates, plant and fish	Miller, 1979
Vinyl chloride	Chemical processes in water	Gillett et al., 1974
Methyl parathion	Chemical processes in water and benzothiophenemicrobial degradation, adsorption, 2-4 trophic levels	Lassiter, 1978
Methyl mercury	A single trophic level: food intake, excretion metabolism growth	Fagerstrøm & Aasell, 1973
Heavy metals	Concentration factor, excretion, bioaccumulation	Aoyama et al., 1978
Pesticides in fish DDT & methoxychlor	Ingestion, concentration factor, adsorption on body, defecation, excretion, chemical decomposition, natural mortality	Leung, 1978
Zinc in algae	Concentration factor, secretion hydrodynamical distribution	Seip, 1978
Copper in sea	Complex formation, adsorption sublethal effect of ionic copper	Orlobn et al., 1980
Lead	Hydrodynamics, precipitation, toxic effects of free ionic lead on algae, invertebrates and fish	Lam and Simons, 1976
Radionuclides	Hydrodynamics, decay, uptake and release by various aquatic surfaces	Gromiec & Gloyna, 1973
Radionuclides	Radionuclides in grass, grains, vegetables, milks eggs, beef and poultry are state variables	Kirschner & Whicker, 1984
SO_2, NO_x and heavy metals on sprucefir forests	Threshold model for accumulation effect of pollutants. Air and soil	Kohlmaier et al., 1984
Toxic environmental chemicals in general	Hazard ranking and assessment from physiocochemical data and a limited number of laboratory tests.	Bro-Rasmussen & Christiansen, 1984
Heavy metals	Adsorption, chemical reactions, ion exchange	Several authors
Polycyclic aromatic hydrocarbons	Transport, degradation, bioaccumulation	Bartell, Gardner & O'Neill, 1984
Persistent toxic organic substances	Groundwater movement, transport and accumulation of pollutants in groundwater	Uchrin, 1984

Table 3.1 (continued)

Toxic Substance	Model Characteristics	Reference
Cadmium, PCB	Hydraulic overflow rate (settling), sediment interactions, steady state food chain submodel	Thomann, 1984
Hydrophobic organic compounds	Gas exchange, sorption/desorption, hydrolysis, photolysis, hydrodynamics	Schwarzenbach & Imboden 1984
Mirex	Water-sediment exchange processes, adsorption, volatilization, bioaccumulation	Halfon, 1984
Toxins (aromatic hydrocarbons, Cd)	Hydrodynamics, deposition, resuspension, volatilization, photooxidation, decomposition, adsorption, complex formation, (humic acid)	Harris et al., 1984
Heavy metals	Hydraulic submodel, adsorption	Nyholm, Nielsen & Pedersen, 1984
Oil Slicks	Transport and spreading, influence of surface tension, gravity and weathering processes	Nihoul, 1984
Acidic rain (soil)	Aerodynamic, deposition	Kauppi, 1984
Acidic rain	C, N and S cycles and their influence on acidity	Arp, 1983

CHAPTER 4

ESTIMATION OF PHYSICAL-CHEMICAL PARAMETERS IN ECOTOXICOLOGY

by
S.E. Jørgensen

4.1 USE OF BENCHMARK CHEMICALS

EPA (Environmental Protection Agency, U.S.) has selected 12 chemicals to represent all the organic chemicals produced on an industrial scale. The idea is that these 12 chemicals indicate a wide range of properties, from the very volatile to the very persistent. The properties of these 12 chemicals are well known, and other chemicals of similar structure, molecular weight and so on should have properties which could be derived from the similarities and differences with one of the benchmark chemicals. Because these chemicals represent a wide spectrum of organic chemicals, they have also been conveniently applied to test the accuracy of the various estimation methods.

About 50,000 chemicals are produced on a large scale (1000 kg or greater per year), but only about 4,000 chemicals are produced at a rate of 1 million kg per year, or greater. Approximately one third of these chemicals are well-defined organics and the 12 Benchmark Chemicals were selected in particular to represent these c. 1300 organic structures.

The important properties that we would like to know for all circumstances are listed in Table 4.1, while in Appendix 1 some of these properties are given for Benchmark chemicals. The properties given in Appendix 1 should also be considered as indicators for the range of values to be expected for other chemicals. The conclusions from these considerations can therefore be formulated in the following two points:

1. Check estimation methods on the benchmark chemicals, particularly for new and still unevaluated estimation methods.
2. Compare the result of any estimation methods with the spectrum of values listed in Appendix 1 to evaluate the probability of the result.

Table 4.1.
Important Specific Properties of Chemicals.

A. Physical Properties: vapour pressure, water solubility, Henry`s constant, adsorption coefficient, partition coefficient octanol/ water, mass transfer coefficient, density, viscosity, diffusion coefficient in air and water, boiling and melting point.

B. Chemical Properties: rate of hydrolysis, rate of oxidation and/or reduction, rate of photolysis, acid dissociation coefficient(s).

C. Biological Properties: bioconcentration factor, rate of biodegradation in soil and water under aerobic or anaerobic conditions.

4.2 ESTIMATION OF CRITICAL TEMPERATURE AND BOILING POINT

The critical temperature and the boiling point are important parameters in chemical engineering and as a result many estimation methods are available. We shall not present all the estimation methods here, only a few classical methods.

The relation between the critical temperature and the boiling point is formulated in **Guldberg's Rule**:

$$T_c = 1.5^* T_b \tag{4.1}$$

T is the temperature in Kelvin degrees and the indices c and b mean the critical temperature the boiling point, respectively.

The constant 1.5 may be estimated more accurately in the following way:

$$T_c = T_b / \Omega \tag{4.2}$$

where Ω can be estimated from the sum of contributions listed in Table 1, Appendix 2, or from the following equations:

$$\Omega = 0.574 + \Sigma \Delta T \tag{4.3}$$

$$\Omega = 0.567 + \Sigma \Delta T - (\Sigma \Delta T)^2 \tag{4.4}$$

The values for ΔT in equation (4.3) are found in accordance with Riedel (1952) and the values for equation (4.4) are found in accordance with Lydersen (1955), see Table 2 Appendix 2.

Thodos (1955) has given evaluation methods using the van der Waals constants a and b and proposed the use of the following equation for estimating the critical temperature:

$$T_c = 8a/ 27R^*b, \quad (4.5)$$

where R is the gas constant (= 82.05 cm^3 atm/K° g*mole), and a and b the above mentioned van der Waals constants obtained from the van der Waals equation. a and b are estimated for aliphatic hydrocarbons by means of the values in Table 3, Appendix 2. Notice that this table gives $a^{0.626}$ and $b^{0.76}$.

The boiling point may also be estimated by a very simple method which, however, not is very accurate unless the molecular weight, M, is close to 200. On the other hand, this method is recommended for use if no other methods are available. The method relates the boiling point to the moleular weight:

$$\log T_b = 3 - 4/\sqrt{M} \quad (4.6)$$

A slightly more accurate method based on the molecular weight, as well as on constants determined from the structure and the atomic composition, can also be recommended:

$$T_b = ¥ * \Omega [M/(ß +0.33)^2]^A, \quad (4.7)$$

where ¥ is found from Table 4 in Appendix 2, Ω from Table 2 in Appendix 2, ß from Table 5 in Appendix 2, and A from Table 6 in Appendix 2.

4.3 ESTIMATION OF HENRY`S CONSTANT AND THE VAPOUR PRESSURE

Henry`s law relates partial pressure,p, with the molecular fraction, x:

$$p = H * x, \quad (4.8)$$

where H is Henry`s constant.

For a dilute solution x is proportional to the molarity, and as we are mostly interested in cases with low x-values, we may use Henry's constant in relation to the following equation: p = H * m, where m is the molarity of the solution. In equation (4.8) H is in terms of unit atm, while in the latter equation it will be in unit atm m^3/mol.

A useful method of estimating H is based on a relation between the boiling and melting point: (Mackay et al 1982):

$$\ln H = 10.6^*(1 - T_b/T) + 6.8^*(1 - T_m/T) + 0.0318^*(T_b - 273) - 5.15 \text{ (atm m}^3\text{/mol)} \quad (4.9)$$

If it is not available the melting point, T_m, can be found from the following relation between melting and boiling point:

$$T_m = 0.5839^* \; T_b \quad (4.10)$$

The vapour pressure in atm, P, can be estimated quite accurately by the equation derived by Mackay el al 1982a:

$$\ln P = -(4.4 + \ln T_b) \; [1.803 \; (T_b/T - 1) - 0.803 \ln(T_b/T)] - 6.8 * (T_m/T - 1) \quad (4.11)$$

As can be seen this method estimates the vapour pressure as a function of the temperature, T. The last term in equation (4.11) is ignored for liquids i.e. when $T > T_m$

In accordance with the Clausius-Clapeyron correlation, a plot of ln P versus 1/T should produce a straight line. However, experimental data shows a deviation from a straight line and the relation in (4.11) gives a better accord with experimental data.

If the vapour pressure is known by two different temperatures, it may be better to use the following equation as a substitution for the Clausius-Clapeyron correlation:

$$\ln P = B - A/ (T - 43) \quad (4.12)$$

A and B are constants to be found from two related p.T-values.

4.4 ESTIMATION OF SOLUBILITY AND PARTITION COEFFICIENT

As it will be demonstrated in the next chapter estimation of solubility is extremely important in ecotoxicology, where many biologically important parameters are estimated from the water solubility or partition coefficient. However, in many cases the estimation of these two important physical-chemical parameters require four other parameters to be estimated: the critical pressure, the critical volume, a so called parachor, and molar refraction, as they are the basis for estimation of either water solubility or the oil or octanol/water partition coefficient.

The parachor, PCH, can be estimated by use of Table 7 in Appendix 2, where the atomic and structural contributions to the calculation of the parachor are listed.

The critical pressure can be found from the following equations:

$$P_c = M / (ß + 0.33)^2 \tag{4.13}$$

where M is the molecular weigth, and ß can be found from Table 5, Appendix 2. Another possibility would be to use the following relation between the critical temperature and the critical pressure:

$$P_c * T_c = ¥ \tag{4.14}$$

where ¥ can be found in Table 4, Appendix 2. A third possibility is to use a relation between the critical pressure and the parachor:

$$\log P_c = C + D \log (PCH), \tag{4.15}$$

where C and D are constants which are listed for groups of compounds in Table 8, Appendix 2, and PCH symbolizes the parachor.

The critical volume in cubic centimeters per gram mole, V_c, may be found from one or more of the following four equations:

$$T_c^{0.3} * V_c = k * (PCH)^{1.2} \tag{4.16}$$

$$V_c = 0.55 (1.5 * PCH + 9 - 4.34 * RD)^{1.155} \tag{4.17}$$

$$V_c = PCH/ 0.77 \tag{4.18}$$

$$V_c = 3 * s * b \tag{4.19}$$

where k = 3.34 for substances containing the functional groups -C=O, -C=N-, -COOH, -OH, and k = 2.92 for all other compounds; RD is the molar refraction, which can be found by using an additive method similar to that by which the parachor is calculated (see Table 9, Appendix 2); b is one of the two constants in the Van der Waals equation and can be found from Table 3 in Appendix 2; and, s may be calculated as follows:

1. For normal saturated aliphatic hydrocarbons: s = 0.7849 - 0.01337 * N, where N is the total number of carbon atoms.
2. For branched-chain saturated aliphatic hydrocarbons: s = 0.8100 - 0.0138 * N.
3. For unsaturated aliphatic hydrocarbons, s is found for the corresponding and the result multiplied by 0.9931.

The critical volume can furthermore be found directly from atomic and atomic-group contributions by use of Table 10 Appendix 2.

The molal volume, V_m, can be found from the critical data. Two equations, relating the molar volume with the critical data, should be mentioned:

$$V_m = 0.044 * V_c / R \tag{4.20}$$

$$1 / V_m = (0.422 \log P_c + 1.981) / V_c \tag{4.21}$$

where R is an empirical expansion factor which can be found from Fig. 4.1, when the reduced temperature T_r (i.e. the temperature relative to the critical temperature) and the reduced pressure P_r (i.e. the pressure relative to the critical pressure) are known. The molal volume is usually expressed in unit of cubic centimeter per mole.

The water solubility, S, and the octanol or oil/ water partition coefficient, K_{ow}, can be estimated from either the molecular weight, the molal volume or the molar refraction. The latter two are estimated before hand in accordance with methods given above. Table 11, Appendix 2, gives correlation equations of various calculation methods for log K_{ow} and log S. The correlation coefficients are sufficiently high to allow a good estimation by use of these equations.

The water solubility and the partition coefficent are also related. Figure 4.2. gives this relationship and the corresponding correlation equation: log K_{ow} = 7.10 - 1.05 log S for a range of organic compounds, including aliphatic and aromatic hydrocarbons, acids and organochlorine and organophophorous insecticides. S is indicated in μmol/l determined at 10-30°C, while the partition coefficent is determined at room temperature (20°C).

A number of regression equations - S versus K_{ow} - are shown in Appendix 1, Tables 16-17.

Cramer (1980 and 1980a) has proposed a relationship between K_{ow} and 5 constants based upon the molecular structure:

$$\log K_{ow} = 1.604 + 3.65*B - 7.66*C - 5.47*D - 0.31*E + 5.09*F \tag{4.22}$$

B, C, D, E and F can be found from Table 18, Appendix 2. r = 0.996 for this equation.

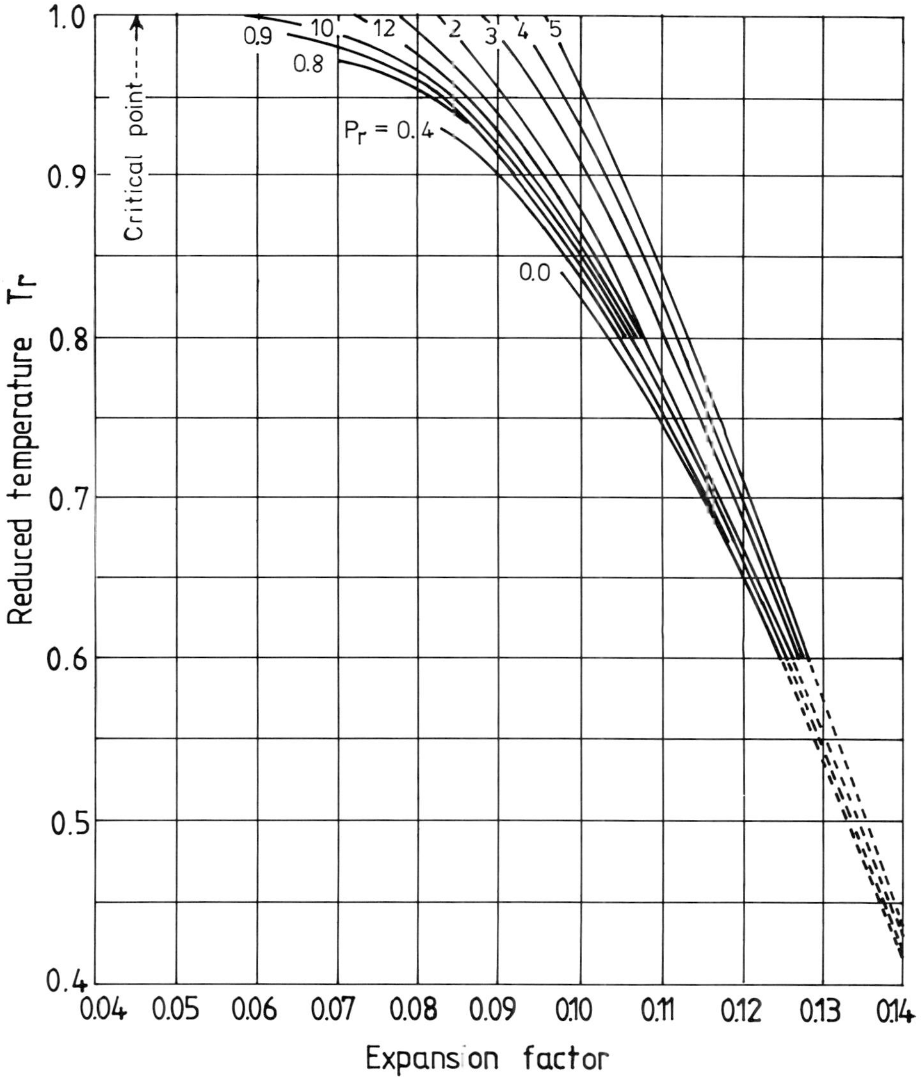

Fig. 4.1: Expansion factor - temperature plot.

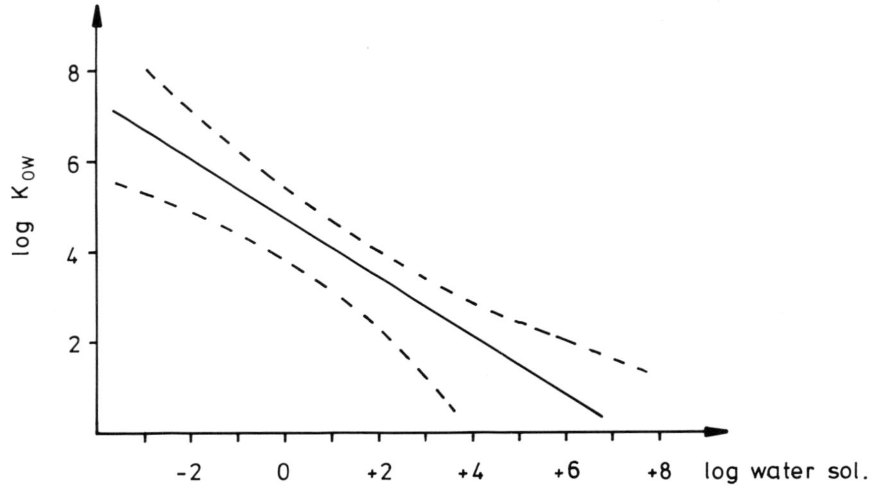

Fig. 4.2: Relationship between the logarithm of the water solubility and logarithm of the partition coefficient. K_{ow} is the n-octanol:water partition coefficient, determined at room temperature and S is the solubility in water (μmol/l) at 10-30°C (Chiou et al., 1977).

4.5 ESTIMATION OF ADSORPTION ISOTHERMS

Sorption equilibria are commonly described by use of adsorption isotherms, which are plots of pollutant concentrations in the sorbed phase, C_s,versus pollutant concentrations in the solution, usually water, C_w, at constant temperature.

In sufficiently dilute systems it is possible to replace activity with concentrations and the following equation can be set up with good approximation:

$$C_s = C_w{}^* K_a \tag{4.23}$$

As will be seen this equation assumes that there is a constant ratio between the concentrations in the two phases.

From numerous studies of the distribution between water and soil or sediment, it has been found that the amount and type of clay, pH, cation-exchange capacity (CEC), and particle size affect the sorption significantly. However, it has been shown that organic carbon dominates the sorption of uncharged organic compounds; see, for instance, Karickhoff et al (1979),

Chou et al (1982), Swarzenbach and Westall (1981) and Rao et al (1982). It is therefore advantageous to relate the concentration of organic compounds in the sorbed phase to the organic carbon fraction. The corresponding adsorption coefficient is named K_{ac} and it replaces K_a in equation 4.23.

K_{ac} can be estimated from K_{ow}, as shown by the equations for various classes of organic compounds in Table 12, Appendix 2. It may also be found from the mole fraction solubility, SOL, which is easily computed from S, when M is known, by use of the following equation:

$$\log K_{ac} = - 0.83 \log SOL - 0.01 (T_m - 25) - 0.93 \qquad (4.24)$$

In general the estimation methods above must be used cautiously, as they are extremely simplistic considering that soil and sediment are very complex mixtures of many organic and inorganic compounds.

4.6 ESTIMATION OF AIR/SOIL EXCHANGE COEFFICIENTS

The air side mass transfer coefficient above a soil or solid surface may be defined as:

$$J = ka * (Cs - Ca), \qquad (4.25)$$

where J is the flux rate of chemical perpendicular to the surface in g/m^2s; Cs is the chemical concentration in the air layer adjacent to the surface in g/m^3; Ca is the concentration of the chemical far from the surface. The difference between the two concentrations represents the driving force and ka is the exchange coefficent in m/s.

ka can be estimated by one of the following two equations:

$$ka = 2 (Da\ V/\pi\ L)^{0.5} \qquad (4.26a)$$

$$ka = 0.677 * Da * Re^{0.5} * Sc^{0.33} / L` \qquad (4.26b)$$

In these equations Da is the molecular diffusivity of the volatile chemical in air (m^2/s), V is the average velocity of air (m/s) parallel to the surface of length (m). Re is Reynolds number equal to V*L/ñ, where ñ is the kinematic viscosity of air at the considered temperature. Sc is the Schmidt number =ñ/Da.

As seen from these equations an estimation of Da is needed. Here the following equations may be used:

$$Da = 1.86 * 10^{-7} * T^{3/2} [(M + 29)/29 *M]^{0.5} / [P*q^2*\Omega] \quad (4.27)$$

$$Da = 4.3 *10^{-7} * T^{3/2} [(M + 29)/29*M]^{0.5} / P (V_b^{1/3} + V_{ba}^{1/3})^2 \quad (4.28)$$

In these equations the following symbols are used:
T is the absolute temperature
P is the pressure in atmospheres
M is the molecular weight
q is defined as the aritmetic mean of $0.833*V_c^{1/3}$ (V_c is the critical volume) for the considered component and air. q is 3.617 for air.

Ω is called the collision integral and may be computed as the arithmetic mean of the values found in Table 13, Appendix 2, on basis of 1.30* T_r. The one component is the considered chemical compound the other is air. Values of Ω are listed in Table 14, Appendix 2.

V_b is the molal volume at the normal boiling point for the considered component. Table 15, Appendix 2, contains the additive-volume increments to calculate this value.

V_{ba} is the molal volume at the normal boiling point for air.

4.7 ESTIMATION OF THE AIR/WATER EXCHANGE COEFFICIENT

The exchange between water and air may be described by the *two-film model of mass transfer*, which is generally applied in chemical engineering for numerous calculations of mass transfer. Fig. 4.3 illustrates the major features of this approach. The water phase is assumed to be well mixed, so that any volatile compound is of uniform concentration, cw, except in the vicinity of the interface, where the concentration is ci. A stagnant liquid film of thickness, l, separates the bulk of the water phase from the interface. Movement of a volatile component through this film is due to diffusion and the rate at which the component is transported across the film is:

$$J = kl (cw - ci), \quad (4.29)$$

where kl is a mass-transfer coefficient (m/s). There is also a stagnant film on the air side and the flux through this film may be described by use of equation (4.25). The concentrations at the interface are related at equilibrium by the following expression:

$$Cs/Ci = H/R * T, \tag{4.30}$$

where H is Henry's constant.
Algebraic manipulations to eliminate ci and cs yields:

$$J = KL\ (cw - ca*R*T/H) = KA(\ cw*H/R*T - ca) \tag{4.31}$$

where

$$1/KL = 1/kl + R*T/(ka*H) \tag{4.32}$$

and

$$1/KA = 1/ka + H/(R*T*kl) \tag{4.33}$$

KA and KL are the so-called overall mass transfer coefficients, which include both resistances. When H is small the contribution from the air resistance becomes dominating, and when H is large the air resistance becomes small and insignificant. ka usually has a value of 10^{-3} m/s, while kl has a value about 10^{-5} m/s

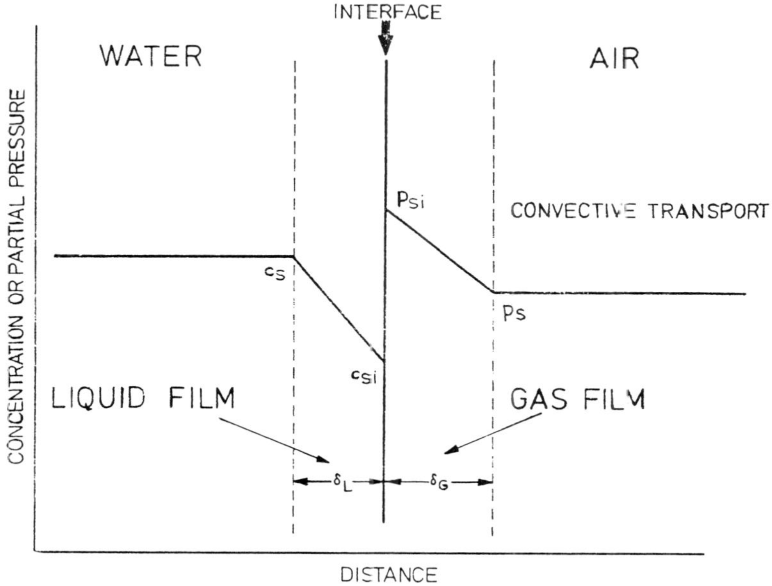

Fig. 4.3: The two-film model of mass transfer between water and air.

ka and kl may be estimated from the following equations:(from Wolff and Van der Heidje 1982)

$$ka = 1.0*10^{-3} + 46.2*\ 10^{-3}*\ U * Sca^{-0.67} \tag{4.34}$$

$$kl = 1.0*10^{-3} + 34.1*\ 10^{-4}*\ U * Scw^{-0.50} \tag{4.35}$$

$$kl = 1.0*10^{-3} + 144*\ 10^{-4}*\ U * Scw^{-0.50} \tag{4.36}$$

where U is the air-phase friction velocity, which is related to the actual wind velocity, measured at a height of 10 m in the unit m/s, U10:

$$U = 10^{-2}\ U10\ (6.1 + 0.63*U10)^{0.5} \tag{4.37}$$

(4.35) is valid for $U > 0.3$ m/s, while (4.36) is applicable for $U < 0.3$m/s.

Sca and Scw are the Schmidt numbers for air and water respectively. They are defined as $\mu/d*D$, where μ is the air or water viscosity in kg/m/s, d is the air or water density (kg/m^3), and D is the molecular diffusivity of the chemical in air or water (m^2/s). Sca is typically in the range 0.6 - 2.5, while Scw is the range 500-2000; see also MacKay et al. (1982).

How it is possible to estimate the molecular diffusivity in air was discussed in section 4.6.

For large unhydrated molecules with a molecular weight > 1000, Polson (1950) has suggested the use of the following simple expression at room temperature for estimating diffusion coefficients in dilute solutions, Dl:

$$Dl = 2.74*10^{-9}*M^{-1/3} \tag{4.38}$$

where M is the molecular weight.

Dl may also be found from the following equation:

$$Dl = 25.2*\ 10^{-5}*T/\mu*V_b^{1/3} \tag{4.39}$$

where T is the absolute temperature, μ is the viscosity of water at that temperature in centipoise, and V_b is the molal volume of solute at normal boiling point in cm^3/ mole. Notice that Dl is found in cm^2/s and the result must therefore be multiplied by 10^{-4} to obtain the molecular diffusivity in the right units for application in the equations (4.36) and (4.37).

4.8 ESTIMATION OF EVAPORATION RATE

The rate of evaporation is of importance when the response to a spill of a toxic chemical is considered.

Nabert and Schön (1970) have published an equation for times for complete evaporation relative to diethyl ether. E. Sonnich Thomsen has used this equation to calculate the absolute evaporation rates:

$$Q = 0.001315 * (p^{4/3} * M)^{0.60327} \tag{4.40}$$

Q is the rate of evaporation in g/s per m^2, p is the vapour pressure in mm Hg, and M is the mass of one mole of the compound (g/mol).

This equation is valid at standard conditions, i.e. at a wind velocity of 4 m/s, a spill size of 75 m^2 and a temperature of 20°C; see Farthing 1982, DOT 1978 and 1980).

The spill area, A, and the wind velocity u are taken into account by multiplying Q with a factor F, found from the following equation:

$$F = (A * 4/ 75 * u)^{0.643} \tag{4.41}$$

Temperature influences the evaporation rate mainly by the strong temperature-dependance of the vapor pressure. The estimation methods presented in section 4.2 may be used to find this relationship. As an alternative method the following relationship between temperature and vapor pressure may be used:

$$\log p = d - (e / f+t), \tag{4.42}$$

where p is the vapour pressure in mm Hg and t is the temperature in °C, while d, e and f are constants. For 592 compounds tabulated by Lange (1956), the following values were found:

$e = 1553 \pm 415.$
$f = 211 \pm 25.$

d is found from (4.42) by use of one set of corresponding p and t values.

4.9 ESTIMATION OF THE HYDROLYSIS RATE COEFFICIENTS

Hydrolysis is the reactions with water, hydroxide ions and/or hydrogen ions. It is considered an important environmental process that acts on many organic chemicals.

It is in generally observed that hydrolysis of organic compounds in water follows a first order reaction scheme. This is of course an oversimplification for most reactions . Thus the rate constant is a pseudo first-order constant and may include contributions from a number of different species.

It is a very difficult task to find good estimations for hydrolysis rate constants and the applicability of estimation methods is obviously particularly limited for complex media such as soil, sediment, etc., because the hydrolysis rate may be influenced by the presence of other compounds, acting as either inhibitors or catalysts for the process. However, to a limited extent the estimation methods are available with good approximations for water solutions or dispersions.

The **Hammett Equation** is suitable for estimating the rate constant for some aromatic compounds, see Exner (1972).It is usually expressed in the following form:

$$\log kh = \partial^* R + \log kho, \tag{4.43}$$

where kh is the hydrolysis rate constant for the considered chemical, ∂ is a so-called 'substituent constant', R a reaction constant, and kho the rate constant for the unsubstituted compound. Tables of ∂, R and kho (based on Jaffe, 1953) are shown in Appendix 2, Table 19. The application of equation (4.43) may be extended to other than aromatic compounds, but this approach should be used with prudence.

For a series of compounds with the same reactive group, the following expression is often valid:

$$\log kh = m^* pK_a + B, \tag{4.44}$$

which indicates that log kh is linearly related to pK_a for the reactive group. m and B are constants. It means that if we know kh and pK_a for two compounds of the series, we can find kh for other compounds, provided that pK_a is known.

4.10 ESTIMATION OF PHOTOCHEMICAL REACTION RATE COEFFICIENTS

Numerous organic chemicals introduced into the environment undergo transformation to other species by photochemical reactions. The rate is dependent on the concentration of the chemical compound undergoing photochemical reaction and the light intensity.

Lambert-Beers law for light absorption is useful in this relation:

$$\log (I_0/I) = e^*C^*pl, \tag{4.45}$$

where I_0 is the incident light, I is the light intensity emergent from the solution, C is the molar concentration, e is an absorption constant, and pl is the pathlength.

The absorption of light in solution due to photochemical reactions takes place in a narrow wavelength region. If this absorption is called a, and added to the absorption considered in equation (4.45), one gets:

$$\log (I_0/I) = e^*C^*l + a \tag{4.46}$$

or in intergrated form, separating the amount of absorped light:

$$I - I_0 = I_0 \left(1 - 10^{-(a + e^*C^*l)} \right) \tag{4.47}$$

For photochemical processes the rate at which photons are delivered to a reaction determines the reaction rate. Therefore it is useful for photochemical reactions to express the light flux in einsteins $/cm^{-2}/sec^{-1}/nm^{-1}$. Notice that the energy units generally applied in the literature is wavelength dependent, which *is not the case* for this unit.

The kinetics of photochemical processes may be expressed by the following equation:

$$- dC/dt = \Pi^*e^*L_w^*[C] = kp^*[C] \tag{4.48}$$

where Π is the reaction quantum yield, defined as (number of moles chemical reacted)/(number of einsteins adsorbed), L_w is the light intensity at wavelength w, and kp is a constant defined in equation (4.45). Table 2, Appendix 3, gives the values of L_w as a function of w for latitudes 20° and 40°.

Equation (4.45) refers to the rate of photochemical reactions at one wavelength, but may be used to estimate ks (kp summed up for all wavelengths) from experiments in sunlight covering all wavelength from 299 to 800 nm. If it is assumed that Π is constant over all wavelength, we can also find ks as:

$$ks = \Pi^* \sum e^*L_w \tag{4.49}$$

Table 3, Appendix 3, contains Π and sunlight half-lifes for several photochemical processes. It is hardly possible to set up methods of estimation, based on the molecular weight or the chemical structure, for the parameters that are needed for a quantative description of these processes.

In some cases it is possible to estimate roughly the reaction rate, when the structure including the reactive groups are considered and compared with similar compounds. For further and more detailed treatment of this important environmental process can be refered to Jørgensen and Gromiec (1988).

4.11 SUMMARY

The relationship between two or more physical-chemical parameters are shown in Fig. 4.4. It includes all the relations (the arrows) presented in this chapter.

If the same parameter for a given chemical compound can be found by more than one method, it is recommended that all the available methods are applied and a weighted average of the estimated values of the same parameter is used. The weights are estimated from the relative accuracy or reliability of the methods applied.

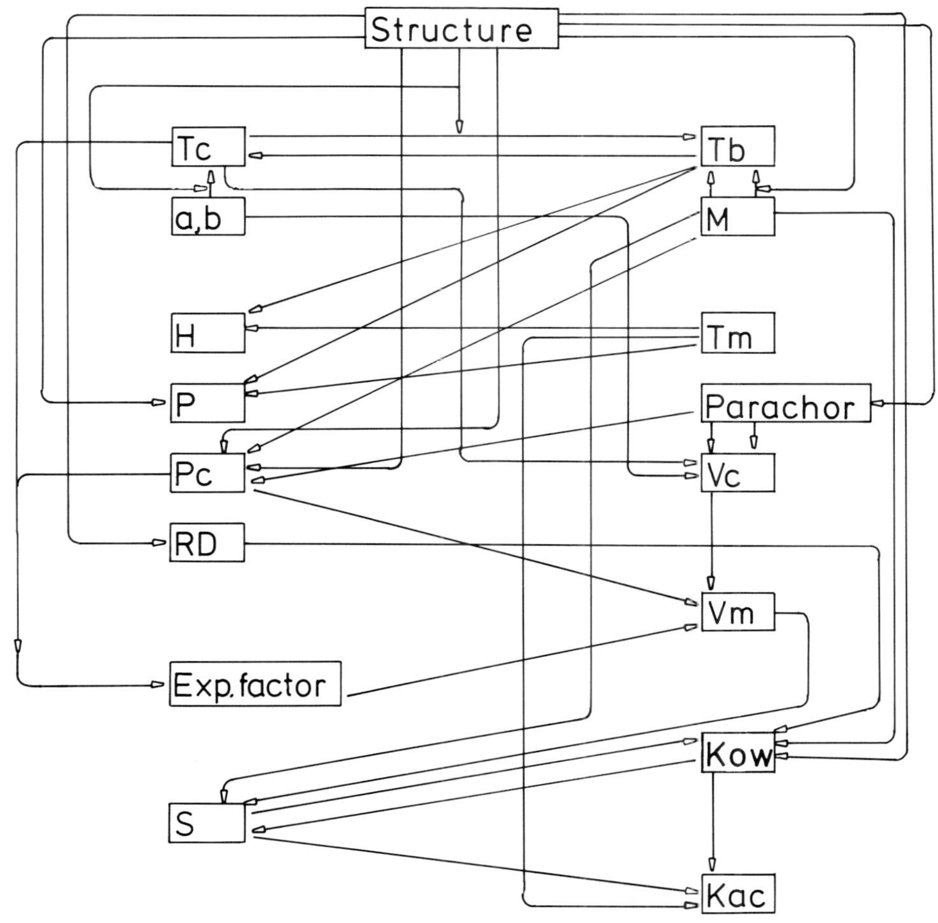

Fig. 4.4: Relationships between physical-chemical parameters as they are presented in this chapter.

CHAPTER 5

ESTIMATION OF BIOLOGICAL PARAMETERS IN ECOTOXICOLOGY

by
S.E. Jørgensen

5.1 ESTIMATION OF BIOCONCENTRATION AND BIOACCUMULATION

A wide variety of terms is used in an inconsistent and confusing manner to describe uptake and retention of pollutants by organisms using different paths and mechanisms. However three terms are now widely applied and accepted for these processes:

1. *Bioaccumulation* is the uptake and retention of pollutants by organisms via *any* mechanism or pathway. It implies that both the direct uptake from air or water and the indirect uptake from food are included. This process is often described by means of a so-called **biological concentration factor,** see below.

2. *Bioconcentration* is uptake and retention of pollutant by organisms directly from the water through gills and epithetial tissue. This process is often described by means of a **concentration factor;** see below.

3. *Biomagnification* is the process whereby pollutants are passed from one trophic level to another and it exhibits increasing concentrations in organisms related to their trophic status. This process may be described by use of the so-called **biological or ecological magnification factor, which gives the magnification from one level to the next.**

As indicated from these definitions, three different factors are used and they all indicate a ratio between two concentrations:

$$K = C_O / C \qquad (5.1)$$

where C_O is the concentration in the organism, while

1) C is the concentration in the medium (water or air), when K is **BCF = biological concentration factor** which has to be found from the direct pathway from the medium and indirect one thorough the food.

2) C is the concentration in the medium (water or air), when K is **CF = concentration factor** which has to be found only from the direct pathway from the medium.

3) C is the concentration in the food, when K is **EM = ecological magnification factor** which has to be found as the magnification from one trophic level to the next.

Equations relating BCF, CF and EM with the water solubility or the partition coefficient, are given in Table 5.I.

Table 5.1
Regression Equations for Estimation of Concentration and Bioconcentration Factors

Indicator	Relationship	Correlation coefficient	Range (Indicator)
K_{ow}	log CF = -0.973 + 0.767 log K_{ow}	0.76	$2.0{*}10^{-2}$ - $2.0{*}10^{6}$
K_{ow}	log CF = 0.7504 + 1.1587 log K_{ow}	0.98	$7.0{*}10^{0}$ - $1.6{*}10^{4}$
K_{ow}	log CF = 0.7285 + 0.6335 log K_{ow}	0.79	$1.6{*}10^{0}$ - $1.4{*}10^{4}$
K_{ow}	log CF = 0.124 + 0.542 log K_{ow}	0.95	$4.4{*}10^{2}$ - $4.2{*}10^{7}$
K_{ow}	log CF = -1.495 + 0.935 log K_{ow}	0.87	$1.6{*}10^{2}$ - $3.7{*}10^{6}$
K_{ow}	log CF = -0.70 + 0.85 log K_{ow}	0.95	$1.0{*}10^{0}$ - $1.0{*}10^{7}$
K_{ow}	log CF = 0.124 + 0.542 log K_{ow}	0.90	$1.0{*}10^{2}$ - $5.0{*}10^{7}$
K_{oc}	log BCF = 2.024 + 1.225 log K_{oc}	0.91	$0.4{*}10^{0}$ - $4.3{*}10^{4}$
K_{oc}	log BCF = 1.579 + 1.191 log K_{oc}	0.87	$3.2{*}10^{0}$ - $1.2{*}10^{6}$
S (mg l^{-1})	log BCF = 2.183 - 0.629 log S	- 0.66	$1.7{*}10^{-3}$ - $6.5{*}10^{5}$
S (µg l^{-1})	log BCF = 3.9950 - 0.3891 log S	- 0.92	$1.2{*}10^{0}$ - $3.7{*}10^{7}$
S (µg l^{-1})	log BCF = 4.4806 - 0.4732 log S	- 0.97	$1.3{*}10^{0}$ - $4.0{*}10^{7}$
S (mg l^{-1})	log BCF = 2.791 - 0.564 log S	- 0.72	$1.7{*}10^{-3}$ - $6.5{*}10^{5}$
S (µmol/l)	log BCF = 3.41 - 0.508 log S	- 0.96	$2.0{*}10^{-2}$ - $5.0{*}10^{3}$
	log EM = -0.4732 log S + 4.4806	-	-
	log EM = -0.3891 log S + 3.995	-	-

Animal	Number of chemicals	References
Fish species	36	Kenaga and Goring, 1978
Mosquito fish	9	Metcalf et al., 1975
Mosquito fish	11	Lu and Metcalf, 1975
Trout	8	Neely et al., 1974
Fish species	26	Kenaga and Goring, 1978
Fathead minnow	59	Veith et al., 1979
Fathead minnow, bluegill, mosquito fish, rainbow trout, green sunfish	59	Lassiter, 1975
Fish species	22	Kenaga and Goring, 1978
Fish species	13	Kenaga and Goring, 1978
Fish species	50	Kenaga and Goring, 1978
Mosquito fish	11	Lu and Metcalf, 1975
Mosquito fish	9	Metcalf et al., 1975
Fish species	36	Kenaga and Goring, 1978
Trout	7	Chiou et al., 1977
Mosquito fish, whole		Metcalf et al., 1975
Mosquito fish, whole		Lu and Metcalf, 1975

The use of all three concentration factors mentioned above are simplifications, which however often are sufficiently accurate, particularly at small concentrations. The concentration ratios are in many cases constants with good approximations, provided that a steady state situation has been achieved, i.e. the organisms have had sufficient time to respond to the concentration of the pollutant in the medium, with C, and C (time) constant. This is also reflected in the literature, where a wide range of factors have been published. Appendix 3, Table 1 gives various factors found in the literature.

However the use of constant factors cover the results of several processes, including direct uptake from the medium with a certain efficiency, uptake from the food with another efficiency, and excretion of the toxic substance - often as a first order reaction. An analysis of the interactions of these processes leads to the dynamic model, that is presented in equation (3.2.) and the solution of this differential equation is given in Fig. 3.8.

The factors which can be found in accordance with the equations in Table 5.1, or in the Table 1 of Appendix 3, are valid for specific species or classes of species. However, the earth has million of species and even were we to represent them all by, for instance, 10,000 different species, it will still be an enormous amount of parameters - see also the discussion in section 3.1.

Consequently, for biological parameters which include CF, BCF, EM, excretion rate coefficient, food consumption, need of air or water and uptake efficiencies through water, air and/or food, we need to be able to calculate them for at least, let us say, 10,000 selected species. This is possible, provided we have those parameters for other species.

Uptake efficiencies are to a certain extent independent of the size of the organism, but for other parameters mentioned above this is not so and it is possible to set up a relationship between the size of the organisms and the parameters.

Physics and chemistry have attempted to solve this problem by setting up some general relationships between the properties of the chemical compounds and their composition. If required data are not to be found in the literature such relationships are widely used as the second best approach to the problem; see also chapter 4.

If we draw a parallel with ecology, we need some general relationships which give us some good first estimations of the required parameters.

The application of such general relationships in chemistry provides quite acceptable estimations in many cases. In many ecological models used in the environmental *context the required accuracy is not very high.* In models for many toxic substances we need only to know whether we are far from, or close to, toxic levels. However, more experience with the application of such general relationships are needed before a more general use can be recommended.

In this context it should be emphasized that in chemistry such general relationships are used very carefully.

Modern molecular theory provides a sound basis for the predictions of reliable quantitative data on the chemical, physical and thermodynamic properties of pure substances and mixtures. The biological sciences are not based upon a similar comprehensive theory, but it is, to a certain extent, possible to apply laws of basic biochemical mechanisms to ecology. Furthermore, the basic biochemical mechanisms are the same for all plants and animals. The spectrum of biochemical compounds is of course wide, but considering the number of species and the number of possible chemical compounds it is, on the other hand, very limited. The number of different protein molecules is significant, but they are all constructed from only 24 different amino acids.

This explains why the elementary composition of all species is quite similar. All species need, for their fundamental biochemical functions, a certain amount of carbohydrates, proteins, fat and other compounds, and as these groups of biochemical substances are constructed from relatively few simple organic compounds, it is not surprising that the composition of living organisms varies only very little (see tables in Jørgensen et al., 1979).

The biochemical reaction pathways are also general, which is demonstrated in all textbooks on biochemistry. The utilization of the chemical energy in the food components is basically the same for microorganisms and mammals. It is, therefore, possible to calculate the energy, E_1, released by digestion of food, when the composition is known:

$$E_1 = 9 \frac{\text{fat\%}}{100} + 4 \frac{(\text{carbohydrates + proteins})\%}{100} \tag{5.2}$$

The law of energy conservation is also valid for a biological system, see Fig. 5.1. The chemical energy of the food components is used to cover the energy needs for growth, respiration, assimilation, reproduction and losses. As it is possible to set up relations between these needs on the one side, with some fundamental properties of the species on the other, it is possible to put a number on the items on Fig. 5.1 for different species. This is a rather general but valid approach to parameter estimation in ecological modelling.

The surface area of the species is a fundamental property. The surface area indicates quantitatively, the size of the boundary to the environment. Loss of heat to the environment must be proportional to this area and to the temperature difference, in accordance with the law of heat transfer. The rate of digestion, the lungs, hunting ground, etc. are, on the one hand, determinant for a number of parameters, and on the other hand, they are all dependent on the size of the animal.

It is therefore not surprising that a number of parameters for plants and animals are highly related to the size, which implies that it is possible to get very good first estimates for a number of parameters based only upon the size. Naturally, the parameters are also dependent on a number of characteristic features of the species, but their influence is minor compared with the size, and good estimates provided is valuable in many models, at least as a starting value in the calibration phase.

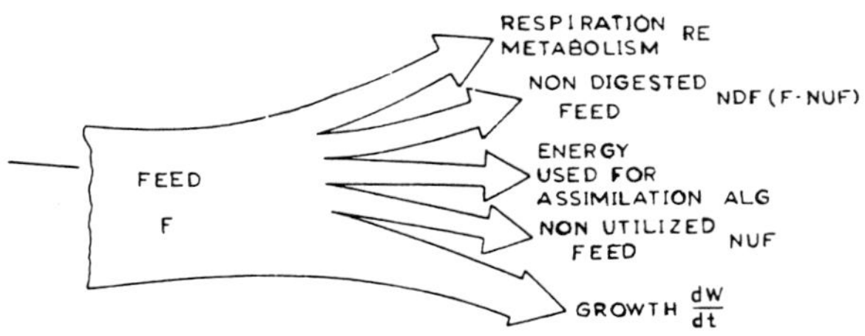

Fig. 5.1: The principle of the model of fish growth. The feed is used for respiration, is non-digested, is used for the assimilation processes, not utilized or is used for growth. Notice that ALG is equal to (F-NUF) (!-NDF) ALC and only (F-NUF) (!-NDF) (1-ALC) is available for respiration and growth. ALC is a parameter (see Jørgensen, 1979)

The conclusion of these considerations must therefore be, that there should be a number of parameters which might be related to simple properties, such as size of the organisms, and that such relations are based upon fundamental biochemistry and thermodynamics.

Above all there is a strong positive correlation between size and generation time, Tg, ranging from bacteria to the biggest mammals and trees (Bonner, 1965). The relationship is illustrated in Fig. 5.2. This relationship can be explained by use of the relationship between size (surface) and total metabolic action per unit of body weight mentioned above. It implies that the smaller the organism the greater the metabolic activity. The per capitum rate of increase, r, defined by the exponential or logistic growth equations:

$$dN/dt = rN \tag{5.3}$$

respectively:

$$dN/dT = rN(1-N/K) \tag{5.4}$$

is again inversely proportional to the generation time. This is shown in Fig. 5.3 where r and T_c are plotted on a log/log scale. The relationship falls, as can be seen, into a narrow straight band with the slope -1.

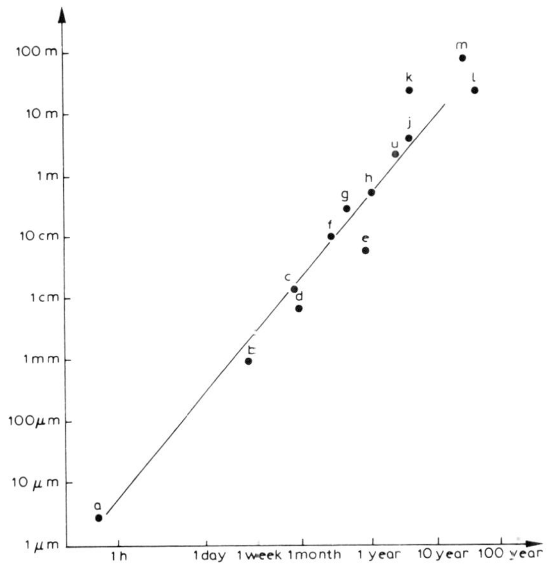

Fig. 5.2: Length and generation time plotted on log-log scale: (a) pseudomonas, (b) daphnia, (c) bee, (d) house fly, (e) snail, (f) mouse, (g) rat, (h) fox, (i) elk, (j) rhino, (k) whale, (l) birch, (m) fir.

This implies that r is related to the size of the organism, but, as shown by Fenchel (1970), actually falls into three groups: unicellular, poikilotherms and homeotherms, see Fig. 5.4. Thus the metabolic rate per unit of weight is related to the size as shown in Fig. 5.5. The same basis is expressed in the following equations, giving the respiration, feed consumption and ammonia excretion for fish when the weight, W, is known:

$$\text{Respiration} = \text{constant} * W^{0.80} \tag{5.5}$$
$$\text{Feed Consumption} = \text{constant} * W^{0.65} \tag{5.6}$$
$$\text{Ammonia Excretion} = \text{constant} * W^{0.72} \tag{5.7}$$

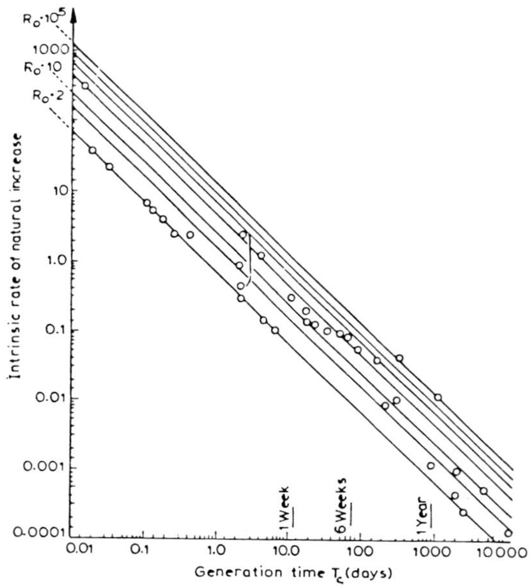

Fig. 5.3: Intrinsic rate of natural increase plotted against generation time, with diagonal lines representing net reproduction rate from 2 to 105 for a variety of organisms.

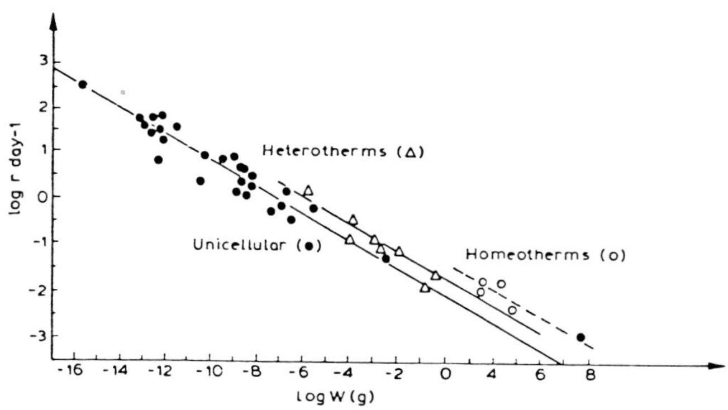

Fig. 5.4: Intrinsic rate of natural increase against weight for various animals. Metabolic rate versus weight for various animals.

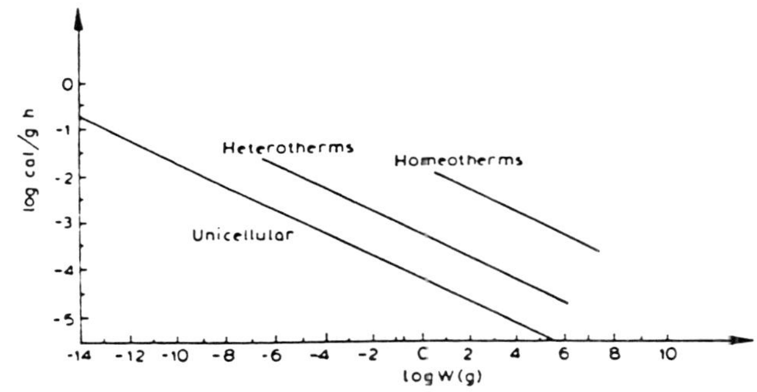

Fig. 5.5: Relationship of metabolic rate to we ght for various animals.

This is also expressed in Odum's equation (Odum, 1959, p. 56):

$$m = k W^{-1/3} \tag{5.8}$$

where k is roughly a constant for all species, equal tc about 1.4 kcal/g$^{2/3}$ day, and m is the metabolic rate per weight unit.

Similar relationships exist for other animals. The constants in these equations might be slightly different due to differences in shape, but the equations are otherwise the same.

All these examples illustrate the fundamental relationship in organisms between size (surface) and the biochemical activity. The surface determines the contact with the environment quantitatively, and thereby the possibility of taking up food and excreting waste substances.

The same relationships are shown in Figs. 5.6 - 5.8, where biochemical processes involving toxic substances are involved.

These figures are constructed from data in the literature, and, as can be seen, the excretion rate and uptake rate (for aquatic organisms) follow the same trends as the metabolic rate (Fig. 5.5). This is of course not surprising, as excretion is strongly dependent on metabolism and the direct uptake dependent on the surface.

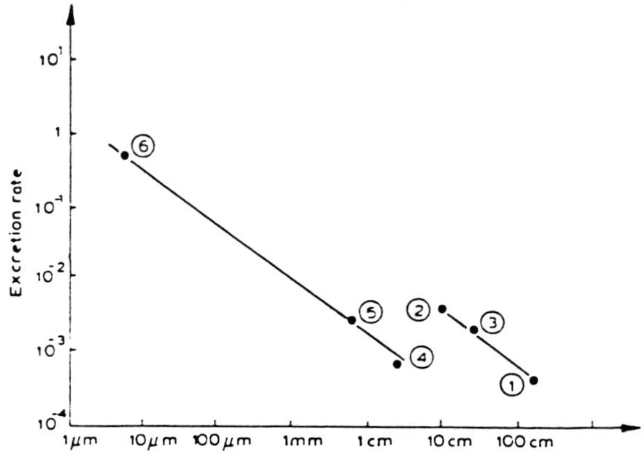

Fig. 5.6: Excretion of Cd $(24\ h)^{-1}$ plotted against the length of various animals: (1) Homo sapiens, (2) mice, (3) dogs, (4) oysters, (5) clams, (6) phytoplankton.

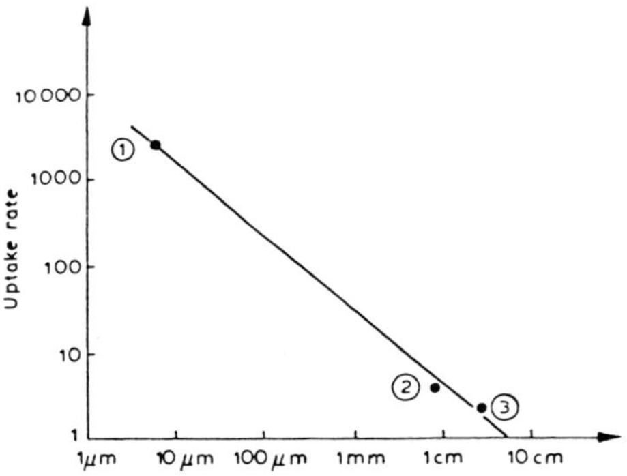

Fig. 5.7: Uptake rate (μg/g 24 h) plotted against the length of various animals (CD): (1) phytoplankton, (2) clams, (3) oysters. (Jørgensen, 1984)

The concentration factor indicating concentration in the organism vis-a-vis concentration in the medium, also follows the same lines; see fig. 5.9. By equilibrium the concentration factor can be expressed as the ratio between the uptake rate and the excretion rate, as shown in Jørgensen (1979). As most concentration factors are determined by the equilibrium, the relationship found in Fig. 5.9 seems reasonable. Intervals for concentration factors are here indicated for some species, in accordance with the literature (see Jørgensen et al., 1979).

The principles illustrated in Figs. 5.6 - 5.9 can be applied generally. In other words, it is possible to find the uptake and excretion rates, and concentration factor, provided these parameters are available for the element or compound under considerationfor one, but preferably several, species. When a plot similar to Figs. 5.6 - 5.9 is constructed it is possible to read the parameters when the size of the organism is known.

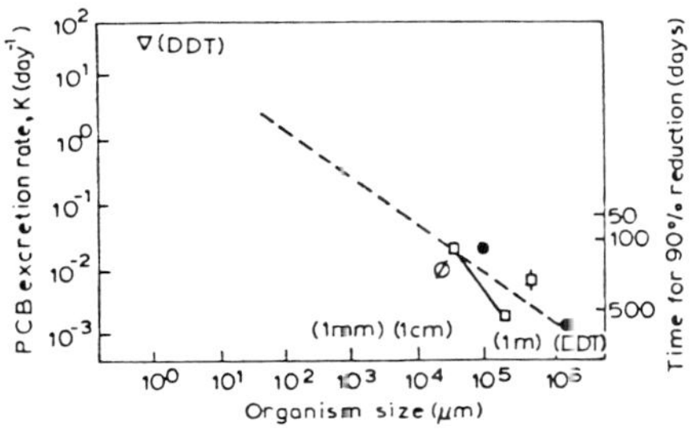

Fig. 5.8: Excretion of PBC and DDT versus organism size. (Thomann, 1984)

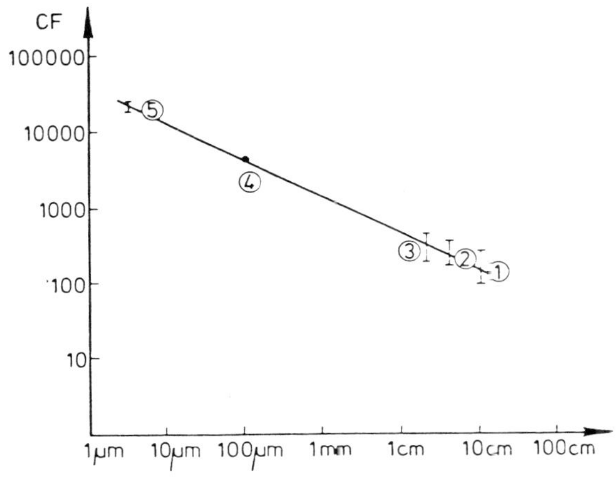

Fig. 5.9: CF for Cd versus size: (1) goldfish, (2) mussels, (3) shrimps, (4) zooplankton, (5) algae (brown - green).

5.2 ESTIMATION OF THE BIODEGRADATION RATE

The general mathematical expression used to relate the rate of microbiological growth was first suggested by Monod (1949):

$$dX/dt = \mu_{max} *X*S /(K_m + S), \quad (5.9)$$

where μ_{max} is maximum specific growth rate; K_m is the so-called half-saturation constant, which is the concentratin of S corresponding to the half of the maximum growth rate, S is the concentration of substrate; and X is the concentration of microorganisms.

The substrate concentration is low, in most cases of ecotoxicological interest, and (5.2) may be changed to second-order or a pseudo first-order reaction scheme. The latter is valid when the concentration of micro-organisms is constant, for instance under constant environmental conditions.

However, it is rather difficult to estimate the reaction rate of bio-degradation in ecotoxicology, because a large number of factors influence the reaction considerably and thereby also the concentration of micro-organisms. The most important of these factors are:

1) **Temperature.** This factors may be taken into consideration by use of the following equations:

$$K_t = K_{20} * K^{(t - 20)} \quad (5.10)$$

$$K_t = K_{t-20} * 2, \quad (5.11)$$

where K_t is the biodegradation coefficient at the temperature t. It might be expressed in g, or moles per unit of time or in g or moles per unit of time and per kg biomass. The latter units are used, when the influence of the microorganism concentration (X) is considered.

2) **Adaptation.** Previous exposure of a natural microbial population to a particular chemical compound has often a great influence on the degradation rate. Rather detailed examination is required to model the adaptation period and in most cases it is therefore necessary to presume that the adaptation is either complete or has not yet been initiated.

3) **Sorption.** This process was considered in chapter 4 and its influence on biodegradation is simply considered by inclusion of sorption isotherms in the model. The biodegradation takes place on the dissolved and/or suspended concentrates, while the sorption isotherms expresses the equilibria

between those concentrations and the adsorped amounts.

4) **Redox potential.** This environmental property determines which terminal electron acceptor is used and thereby the composition of microorganisms. Significantly increased (see f. inst. Larson et al 1981) as well as decreased (see f. inst. Pritchard et al., 1979) rates of biodegradation have been observed in anaerobic environments.

5) Ionic strength. The degradation rate will in general decrease as the ionic strength increases.

6) Nutrients. Microbial growth is affected by major nutrients as well as by essential micronutrients, including trace metals. The composition of microorganisms indicates that the biomass contains approximately 0.5 -1.0% phosphorus and 5-9% nitrogen. The effect of nitrogen and phosphorus limitation on the biodegradation of hydrocarbons in seawater is significant for the solution of this ecotoxicological problem; see Atlas 1981.

7) Moisture of Soil is one of the most important factors that influence the fate of chemicals in the terrestrial environment. Moisture affects the availability of a compound by controlling adsorption and diffusion, and has a significant influence on soil microorganisms. Walker has developed an empirical equation to account for the effects of moisture on biodegradation rates:

$$t_{0.5} = a * M^{-b}, \tag{5.12}$$

where M is the moisture contant, $t_{0.5}$ is the biological half-life, and a and b are constants.

8) Soil Composition. Rates of biodegradation vary considerably for various types of soil due to differences in availability of organic nutrients for microroganisms.

Some qualitative rules for the relation between molecular structure and rate of biodegradation may be set up:

1 *Aromatic compounds* are in general less biodegradable than aliphatic compounds.
2) The higher *the molecular weight,* the less biodegradable is the compound.
3) The *more substitutes* the molecule contains the less biodegradable is the compound.
4) *Double bonds* will in general increase the biodegradability.
5) *-O- and -N=* will in general decrease the biodegradability.

The possible relations between toxicity and structure will be presented in the next section. The rule that the higher the toxicity the less the biodegradability may be used in combination with the relation between toxicity and structure, if all other rules are unapplicable, but the rule must be used very careful.

A quantitative relation between biodegradation kinetics and structure exists, but it is weak, particuarly because the environmental factors mentioned above play a major role.

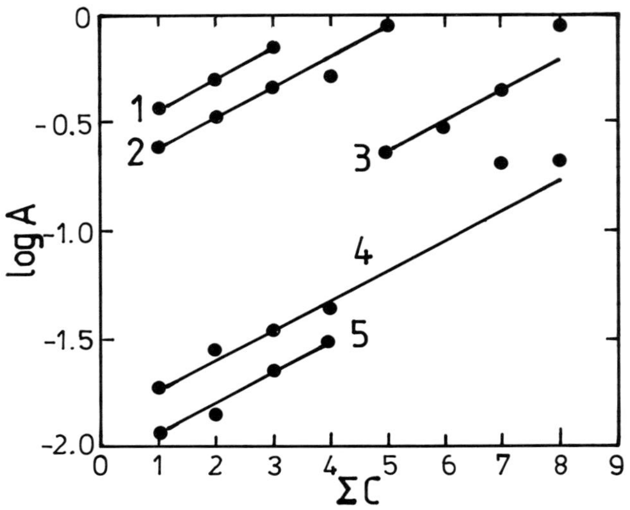

Fig. 5.10: Relationship between the logarithm of the thermodynamic activity required for equieffective biological action and the length of carbon chain in homologous series.
I. alkyl acetates: haemolysis of ox blood; II, alcohols: bactericidal action; III, aliphatic hydrocarbons: narcosis of mice; IV, alcohols: inhibition of development of sea-urchin eggs; V, alcohols: tadpole narcosis. (From Badger, 1946)

Fig. 5.10 shows the possibilities of relating the molecular weight, or rather the number of carbon atoms, to the first-order reaction constant of biodegradation for a homologous series. If the rate constant is known for biodegradation of one compound such a relationship may be used to find the rate constant for other compounds in the homologous series. Obviously, it is preferable to know the rate constant for the environment under examination or for a similar environment. The relation illustrated in Fig. 5.10. is in accordance with **Ferguson's principle**, that the activity A to produce equal effects is constant for a homologous series. The activity, A, is in this relation defined by:

$$A = P/P_s \text{ or } = C/C_s \tag{5.13}$$

where P is the partial pressure for the considered compound, P_s is the saturated vapour pressure at 20°C, C is the concentration of the substance, and C_s is the solubility. The solubility, the toxicity curve and the biodegradability curve are often all parallel for a homologous series.

In the next section it will be mentioned that the activities required to produce equal biological effects increase with increasing molecular size for a homologous series, although the increase in activity is not very pronounced. This implies that Ferguson's principle should not be used uncritically, it being better to plot as many as possible biodegradation rates versus the molecular weight or the number of carbon atoms, to get the relationship for the homologous series under consideration. However, in this context the influence of the 8 factors mentioned above must not be forgotten, and their effects should be included in the interpretation of the data.

The following procedure is recommended if a biodegradation rate for an organic compound is not available:

1) *Find as many biodegradation rates as possible* for other compounds with a similar structure or for compounds in the same homologous series.
2) *If the temperature, soil moisture or any other* of the 8 factors mentioned above are different for these biodegradation rates, attempt to correct the rate coefficients to the same environmental conditions as those valid for the compound with the unknown biodegradation rate coefficient.
3) In the first instance use the biodegradation rate coefficients which are valid *for the same environmental conditions as the compound* with the unknown biodegradation coefficient.
4) *Plot these values versus the molecular weight or the number of carbon atoms.*
5) *Furthermore, plot the biodegradation rate coefficients relative to the solubility or the saturated vapour pressure, see equation (5.6).*
6) In the second instance use the values *valid for other* environmental conditions and, if there are deviations from the two plots, attempt then to explain these deviations, at least qualitatively.
7) Use now the plot, that gives a) the best fit and/or b) the best explainations for the deviations.
8) If only one value is available, *use Ferguson's principle* .

The basis for another possible approach is the **Hammett equation** (see equation 4.43):

$$\log (K / K_x) = \partial * R \qquad (5.14)$$

where K is the equilibrium constant, the rate of any reaction in which the compound takes part, including biodegradations; K_x is the same constant but for a similar compound in which a hydrogen of the aromatic ring has been substituted for by X; ∂ is a constant (the substituent constant) whose value depends only on the nature of substituent X; and R is a constant (the reaction constant) which depends only on the type of reaction shown by the group of compounds concerned, and not on the substituent. ∂ may be found in Table 20, Appendix 2.

This equation may be used to find an unknown biodegradation rate coefficient for a compound in the following way:

1) **Find** ∂ by comparison with the biodegradation rate coefficient for **two compounds, which differ by the substituent X,** characteristic for the compound with unknown biodegradation rate coefficient - or find ∂ in Table 20, Appendix 2.
2) **Find R** by comparison with the biodegradation coefficient of two compounds, **which differ by another substituent Y.** ∂ **should be known for Y** or easy to find from other comparisons, see also Table 20, Appendix 2. The biodegradation should be as close as possible to the one for which the rate coefficient is unknown - also in relation *to the environmental factors,* see above. The two compounds selected must furthermore **have a structure as close as possible to the compound under consideration.**
3) **Now - knowing** ∂ **and R - it is possible to use equation (5.14) directly.**

For estimations of ∂ and R, see also Jaffe (1953).

If the two methods for estimation of biodegradation rates given above cannot be used because the biodegradation rate for a related compound is not known the equations in Appendix 6 may be used. They give an estimation of the biodegradation rate based on the chemical composition at room temperature under optimum conditions of moisture, pH, redox potential etc. However, these equations should be used very cautiously.

5.3 ESTIMATION OF TOXICITY

The literature about estimation methods for toxicity data is very comprehensive and it is not possible here to present more than a few estimation methods. This book is concerned with the development of ecotoxicological models and many models attempt to translate observed concentrations to effects. It is, however, going too far to present more than a short summary of the methods mainly based on physical-chemical data which are to a certain extent already available from other estimation methods given above.

Zahradnik (1962) proposed to use an equation identical in form to the Hammett equation for the estimation of toxicity parameters. As toxicity is a very complex concept often involving more than one process, it is recommended that this estimation method is applied very carefully. It has been shown, however, that in many cases it is possible to apply this estimation method with an acceptable accuracy.

Table 1 in Appendix 4 contains an overview of the most frequently applied toxicity data. Toxicity data for several inorganic and organic compounds are given in Table 2, Appendix 4.

Tables 3-8 in Appendix 4 gives an overview of some of the most important estimation methods based mainly on physical-chemical data. In calculating toxicity data, all the available constants should be used. The mean value of the logarithms is calculated and then the anti-logarithm taken.

In many cases it is possible to find one or more of the required toxicity data in the literature, but others may be needed in ecotoxicological modelling. In this case there are several methods available for estimating toxicity from other, known, toxicity data. Such estimation methods are shown in Table 9, Appendix 4.

Inorganic substances are usually broken down into ions and interact as such with the consitutents of cells and tissues. The toxic effect produced by an inorganic component is therefore determined by the toxicity of its ions. An estimation method for the toxicity of inorganic compounds should therefore be expected to correlate with the characteristic properties of the ions.

The following properties have been proposed as correlated to the metal toxicity:

1) **The standard potential** (SP) of the metal (Matthews 1904).
2) **The stability of metal complexes** (Shaw 1961).
3) **The solubility of sulfides** (log S); see for instance Liublina 1965.
4) **Pauling's electronegativity.**

For the relationships 1) and 3) the following equations are suggested:

$$\log LD_{50} = -0.63 * SP - 1.0 \tag{5.15}$$

$$\log LD_{50} = -0.21 * \log S + 0.75, \tag{5.16}$$

where LD_{50} is the lethal doses for white mice. The correlations are significant and rather high.

Table 10, Appendix 4 gives information on biological activity and threshold limits for inorganic compounds.

5.4 SUMMARY

Fig. 5.11 shows the relationships between physical-chemical parameters and molecular structure on the one hand, and biological parameters on the other, as they are presented here in Chapter 5.

Again, it is recommended that use is made of as many of the available methods as possible, and of the weighted average.

Particularly for toxicity data, but to a certain extent also for other biological parameters, it is necessary to use the logarithmic average for such calculations. It is emphasized that the presented estimation methods sometimes have a low accuracy and their use is therefore only a helpful alternative when the values can not be found in the literature. On the other hand, in most cases their accuracy is fully acceptable in the modelling context, provided that the accuracy of the parameter estimation is reflected in the use of model results.

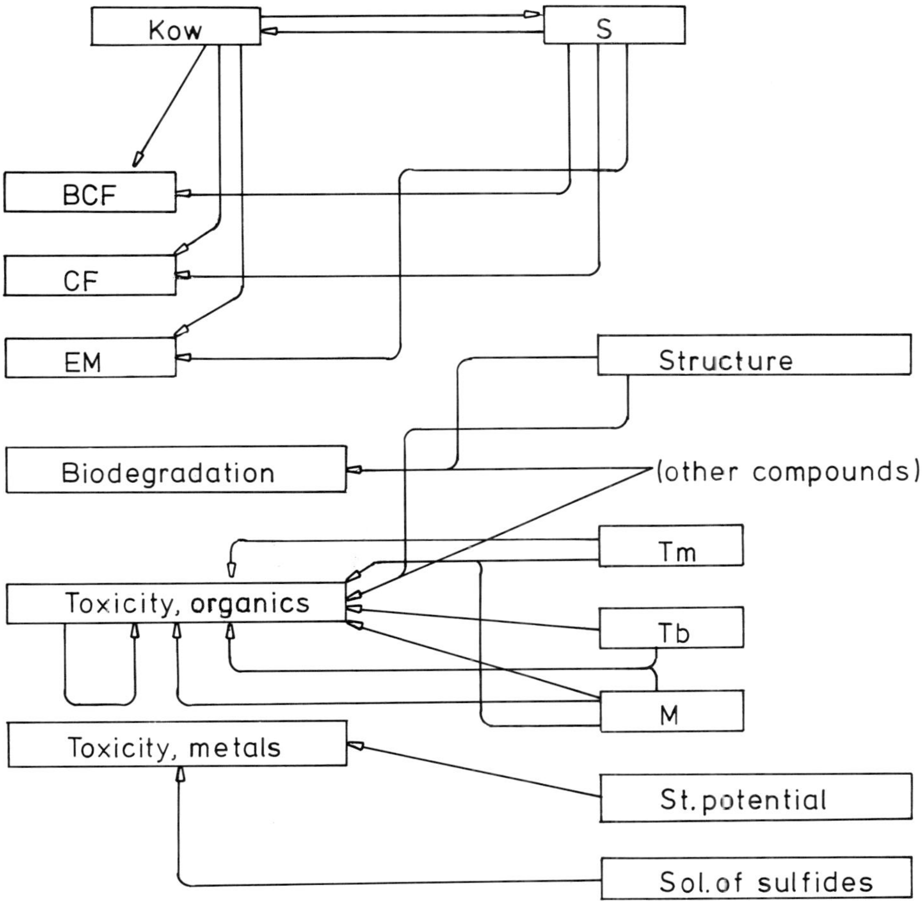

Fig. 5.11: Relationships between physical-chemical parameters and molecular structure on the one side and biological parameters on the other.

CHAPTER 6

TYPES OF MODELS OF PARTICULAR INTEREST IN EXOTOXICOLOGY

by

S.E. Jørgensen

6.1 APPLICATION OF MODELS IN ECOTOXICOLOGY

Models in ecotoxicology are in many respects similar to other environmental models. As other biogeochemical models, they are based on mass balances and the description of mass transfer processes. The difference lays more in the wide use of parameter estimation methods in ecotoxicology as presented in chapters 4 and 5, and less in the equations used in models for description of processes including mass transfers.

The next chapter will describe many good examples of ecotoxicological models based on the modelling techniques generally applied and based upon the submodels that are used widely in other environmental models.

The application of these techniques and submodels will not be repeated here - only illustrated in the next chapter. Those particularly interested in this topic can refer to Jørgensen (1983), Jørgensen and Mitsch (1983), and Jørgensen (1988).

However, two modelling considerations or modelling examples of particular interest for ecotoxicology and hardly used outside ecotoxicology are presented in this chapter.

The next section is devoted to the risk assessment, which often is an additional component in ecotoxicological models. The methods and considerations for the development of this modelling component will be presented.

Figure 6.1. shows the components (submodels) of a typical ecotoxicological modelling consideration. As seen on the figure there are 4 components (see Morgan 1984):

1) A fate or exposure model.

2) An effect model, translating the concentration into an effect.

3) A model for human perception processes.

4) A model for human evaluation processes.

The first two submodels are in principle "objective", predictive models, while the latter two are value orientated. The development of submodels (1) and (2) are based upon physical, chemical and biological processes and the parameter estimation methods presented in the chapters 4 and 5 are aimed at these submodels.

Submodel (1) is of the type of model, that is very similar to other environmental models and based upon mass transfer, mass balances, physical, chemical and biological processes.

The second submodel requires a good knowledge to the effects of the toxic components. This implies that the modeller must have sufficient toxicological knowledge of the considered component, or, alternatively uses the estimation methods presented in section 5.3.

The submodels (3) and (4) are different from the generally applied environmental management models and cannot be found in the literature quoted above, these are presented in some details in the next section of this chapter.

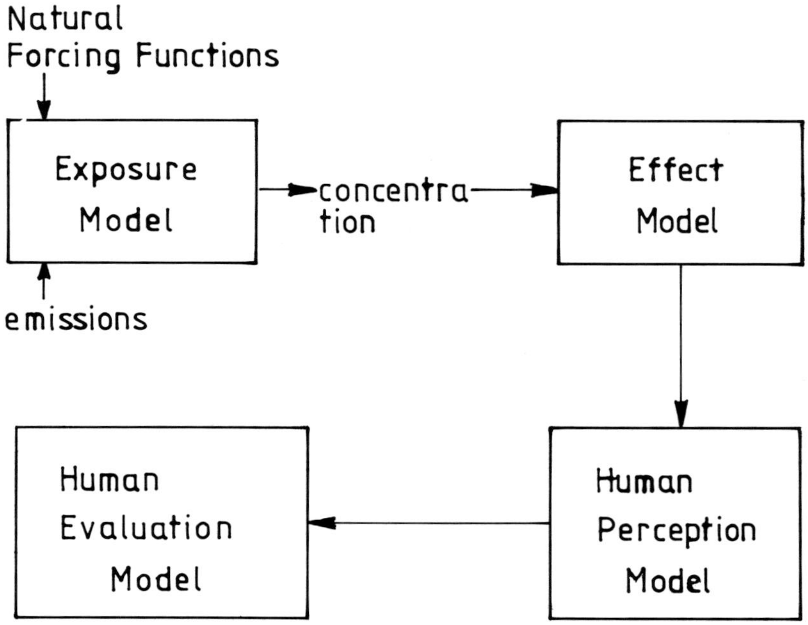

Fig. 6.1: The four submodels of a total ecotoxicological model are shown.

The third section of this chapter will present the so-called fugacity approach for modelling the distribution of chemicals in the six world compartments: air, water, soil, sediment, suspended sediment and biota. This modelling approach attempts to answer the following questions: In which of these six compartments can we expect the greatest problem for chemicals emitted to the environment? Which concentration will be expected in each compartment? What are the implications of this concentration? The two first questions are related to the first submodel, shown in Fig. 6.1, while the answer to the last question is the aim of submodel (2). The submodels (3) and (4), see Figure 6.1, may of course be added to the general fugacity model, but in this case is no different from other risk assessment submodels.

The fugacity model is to a large extent based on physical-chemical parameters and the use of these parameters to estimate the other required parameters. The applicability of the fugacity model is therefore very dependent on the use of estimation methods presented in Chapters 4 and 5.

6.2 MODELS WITH A RISK ASSESSMENT COMPONENT

The risk assessment component comprises human perception and evaluation processes. These submodels are explicitely value laden, but must of course build on objective information on concentrations and effects.

Factors that may be important to consider in this context are:

1. **Magnitude and time constant of exposure.**
2. **Spatial and temporal distribution of concentration.**
3. **Environmental conditions determining the process rates and effects.**
4. **Translation of concentrations into magnitude and duration of effects.**
5. **Spatial and temporal distribution of effects.**
6. **Reversibility of effects.**

The uncertainties relating to the information on which the model is based and the uncertainties related to the development of the model, are crucial in risk assessment. The uncertainty in risk problems may be classified into one or more of the following 5 categories:

1. *Good direct knowledge and statistical evidence* on the important components (state variables, processes and interrelations of the variables) of the model is available.
2. *Good knowledge and statistical evidence on the important submodels*

are available, but the aggregation of the submodels are less certain.

3. No good knowledge of the model components for the considered system is available, but *good data are available for the same processes from a similar system* and it is estimated that these data may be applied for use directly or with minor modifications, to the model development.
4. *Some, but insufficient, knowledge is available from other systems.* Attempts are made to use these data without the necessary transferability. Attempts are made to eliminate gaps in knowledge by use of additional experimental data as far as it is possible within the limited resources available for the project.
5. *The model is to a large extent based on the subjective judgment of experts.*

Unfortunately a very substantial proportion of risk assessment models fall into classes 3, 4 and 5.

The acknowledgement of the uncertainty is of great importance and may be taken into consideration, either qualitatively or quantatively. Another problem is of course: Where to take the uncertainty into account? Should the economy or the environment benefit from the uncertainty ? Unfortunately, up to now most decision makers have used the uncertainty to the benefit of the economy. This is of course completely unacceptable. The same decision makers would never, for a moment, consider for a civil engineering project whether uncertainty should be used for the benefit of the economy or the strength of a bridge.

Until 10-15 years ago the researchers had developed very little understanding of the processes by which people actually perceive the exposures and effects of toxic chemicals, but these processes are just as important for the risk assessment as the exposures and effects processes.

The characteristics of risks and effects are of importance for the perceptions of people. These characteristics may be summarized in the followings lists:

Characteristics of risk:

Voluntary or involuntary?

Are the levels known to the exposed people or to science?

Is it novel, or old and familiar?

Is it common or dreaded (for instance does it involve cancer)?

Does it involve death?

Are mishaps controllable?

Are future generations threathed?

Global, regional or local?

Function of time? How (whether for instance increasing or decreasing)?

Can it easily be reduced?

Characteristics of effects:
Immediate or delayed?
On many or a few people?
Global, regional or local?
Involve death?
Are effects of mishaps controllable?
Observable immediately?
Function of time?

A factor analysis has been performed by Slovic et al (1982) and it shows, among other results, a not surprising correlation between people's perception of dreadful and unknown risks.

Broadly speaking there are two methods of selecting the risks we will deal with.

The first may be described as the 'rational actor model', involving people that look systematically at all risks they face and make choices about which they will live with and at what levels. For decision making this approach would use some single, consistent, objective functions and a set of decision rules.

The second method may be named the 'political/cultural model'. It involves interactions between culture, social institutions and political processes for the identification of risks and determination of those which people will live with and at what level.

Both methods are unrealistic, as they are both completely impractical in their pure form. Therefore we must select a strategy for risk abatement founded on a workable alternative based on the philosophy behind both methods.

Several risk management systems are available, but no attempt will be made here to evaluate them. However, some recommendations should be given for the development of risk management systems:

1. *Consider as many of the characteristics listed above as possible* and include *the human perceptions of these characteristic* in the model.
2. Do not focus too *narrowly on certain types of risks.* This may lead to suboptimal solutions. Attempt to *approach the problem as broad-mindedly* as possible.
3. Choose strategies which are *pluralistic and adaptive.*
4. *Benefit-cost analysis is* an important element of the risk management model, but it is far from being the only important element and the *uncertainty in evaluation of benefit and cost should not be forgotten.* The variant of this analysis applicable to environmental risk management may be formulated as follows:

net social benefit =
social benefits of the project - "environmental" costs of the project (6.1)

5. *Use multi-attribute utility functions*, but remember that people in general have troubles in thinking about *more than 2-3, at the most 4, attributes* in each outcome.

6.3 FUGACITY MODELS

The application of fugacity in environmental modelling has been discussed by Makay and Paterson (1981 and 1982). Fugacity has dimensions of pressure and may be considered a measure of " escaping tendency" from the phase. When phases are at equilibrium their fugacities are equal.

Fugacity, f, is related to concentration by the following equation:

$$f = C / Z, \tag{6.2}$$

where C is the concentration (mol/m^3) and Z is the fugacity capacity (mol/m^3 Pa). As seen from this equation f has the unit of pressure or in this case, of Pa.

Equilibrium involves:

$$C_a / C_w = Z_a / Z_w = K_p , \tag{6.3}$$

where the indices a and w are used to indicate air and water, respectively, as examples for the phases. K_p is a partition coefficient.

Transfer rates between two phases by diffusion are expressed by the following equation (models per unit of area and time) :

$$N = D^* \Delta f, \tag{6.4}$$

where N is the rate of transfer, D is the diffusion coefficient and Δf is the difference in fugacity. D is the total resistence for the transfer consisting of the resistances of the two phases in series. Notice that D may be found as K*Z, where K is the transfer coefficient and Z is the fugacity capacity defined above.

The so-called 'unit world model' consists of six compartments: air, water, soil, sediment, suspended sediment and biota. This simplified model aims toward identification of the partition among these six compartment of toxic substances emitted to the environment. The conceptual diagram of the unit world model is shown in Figure 6.2.

The volumes and densities of the unit world and the definition of fugacity capacities are given in Table 1, Appendix 5.

It is also possible to use this approach regionally or for particular ecosystems. The volumes use in this latter case are listed in Table 2, Appendix 5.

The fugacity model may be used at four levels. The first level calculates the equilibrium distribution of a chemical between phases. It assumes that each compartment is well mixed and there is no reaction or advection into or out of the system.

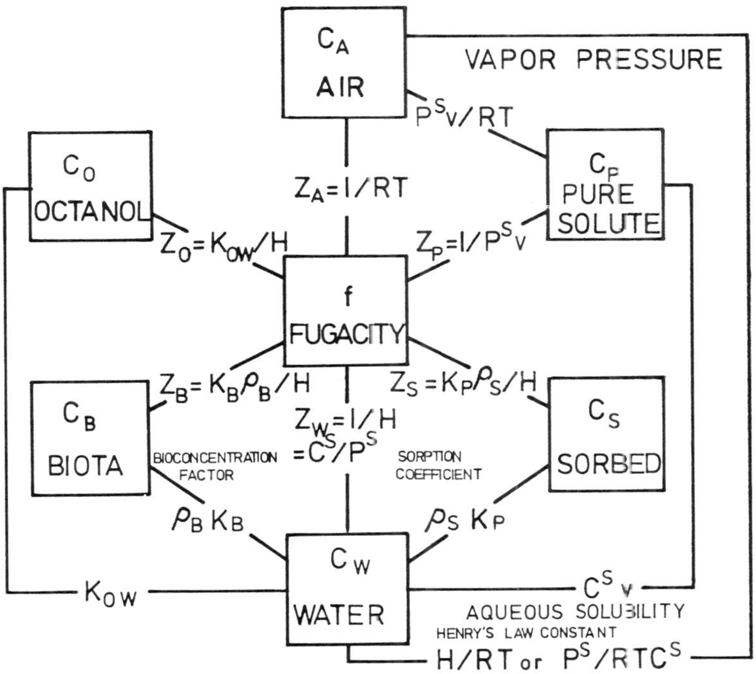

Fig. 6.2: Conceptual Diagram of the Fugacity Model.

If M is the total amount in moles, m_i and V_i are the amounts and volumes of each compartment, then the following equation is valid:

$$M = \Sigma m_i = \Sigma C_i{}^*V_i = \Sigma f_i{}^*Z_i{}^*V_i = f^*\Sigma Z_i{}^*V_i \qquad (6.5)$$

Since M is known, f may be calculated and the amount in each compartment is given by:

$$m_i = f^*V_i{}^*Z_i \qquad (6.6)$$

The percentage distribution is found as m/M and the concentrations are found by:

$$C_i = Z_i{}^* f \qquad (6.7)$$

Level two considers equilibrium but includes also reaction and advection. Reactions comprise photolysis, hydrolysis, biodegradation, oxidation and so on. All the processes are assumed to be first-order reactions. If the rate constants are not known or cannot be estimated, they are set to zero. Since process rates are expressed in first-order form they are additive and the total removal rate from a compartment can be found as:

$$\Sigma V_i{}^* C_i{}^* \Sigma k_j = \Sigma V_i{}^* C_i{}^* k_i \qquad (6.8)$$

As in level one a common fugacity,f, prevails and if we substitutes C by f*Z, we get for the total rate E is:

$$E = f^* \Sigma V_i{}^* Z_i{}^* k_i \qquad (6.9)$$

The average residence time, tr, due to reactions may be found by use of the following equation:

$$tr = M / E \qquad (6.10)$$

and the overall rate constant,K, is E / M or 1 / tr.

The third level is devoted to a steady state, nonequilibrium situation, which implies that the fugacities are different in each phase. Equation (6.4) is used to account for the transfer.

The D values may be calculated from quantities such as interphase areas, mass transfer coefficients (as indicated above D is the product of the transfer coefficient and the fugacity capacity: D = K * Z), release rate of chemicals into phases such as biota or sediment, and Z values, or by use of the estimation methods presented in chapters 4 and 5.

Level four involves a dynamic version of level three, where emissions and thus concentrations, are varying with time. This implies that differential equations must be applied for each compartment to calculate the change in concentrations with time, for instance:

$$V_i^* \; dC_i/ \; dt = E_i - V_i * C_i{}^*k_i - \Sigma D_{ij} * \Delta f_{ij} \quad (6.11).$$

This model level is similar in concept to the EXAMS model; see Mackay et al 1983.

Levels one or two are sufficient in most cases, but if the environmental management problem requires the prediction of 1) the time taken for a substance to accumulate to a certain concentration in a phase after emission has started, or, 2) the length of time for the system to recover after the emission has ceased, the fourth level must be applied.

This approach has been widely used and a typical example is given by Mackay (1984). It concerns the distribution of PCB between air and water at the Great Lakes. In this case H was 49.1 and the distribution coeffcient for air water (= H/R*T) therefore 0.02. The fugacity capacity for water = 1/H was found to be 0.0204 and the fugacity capacity for air = 1/ R*T = 0.000404. The distribution coeffecient between water and suspended matter in the water was estimated to be 100 000. As the concentration of suspended matter in the Great Lakes has been found to be $2*10^{-6}$ on a volume basis (approximately 4 mg/l ,the density beeing 2000 g/l), the fraction dissolved was 1 / (l + 0.2) = 0.833.

PCB concentration in water of the Great Lakes was found to be 2 ng/l, and in the air 2 ng/m^3. The fugacity can be calculated in water and air as C/Z and it was found that the fugacity in water is 17 times higher than in air, which implies that volatilization will occur.

If it is assumed that the transfer coefficient in water is 10^{-5} m/s and in air 10^{-3} m/s, the volatilization rate can be calculated from the traditional two-resistance model, using the relation D = K *Z to find the overall diffusion coefficient, D:

$$1 \,/\, D = 1/\, 10^{-5} * 0.0204 \;+ 1/10^{-3} * 0.000404 \quad (6.12)$$

D is found to be $1.36 * 10^{-7}$. N is calculated by use of (6.4):

$$N = D \,(\, f_w - f_a) = D \,(\, 2.8 * 10^{-7} - 1.53 * 10^{-8}\,) \quad (6.13)$$

N is found be to be $35.9 * 10^{-15}$ mol/m^2/s.

It can be shown that the transfer with precipitation is negligible compared with the volatilization rate, while the washout of particles and dry deposition are important processes. If these processes are considered, the net flux to the atmosphere becomes about 75% of the flux found above.

PART B

CASE STUDIES OF ECOTOXICOLOGICAL MODELS

OVERVIEW OF THE PRESENTED CASE STUDIES

Eigth case studies are presented in this part of the book. They illustrate the development and applicability of ecotoxicological models in environmental management, but presume all that the parameters are known either from the literature, from direct measurements in situ or in the laboratory, or by use of estimations methods as presented in Chapters 4-5.

The case studies could have been selected to illustrate environmental problems related either to the atmosphere, hydrosphere and lithosphere. Another possibility would be to select models concerned with pollution of inorganic and organic compounds. The combination of these two classifications could be illustrated by 3*2=6 case studies.

The case studies have, however, been selected to illustrate the spectrum of applicable models in various directions.

Chapter 7 presents the most simple model of the 7 case studies. It illustrates very clearly that it is possible to develop simple models useful in environmental management. The model relates chromium concentration in discharged waste water with chromium concentration in the sediment as function of the distance from the discharge point, and with the chromium concentrations in mussels. It is based on two equations only - a hydrodynamic equation and one describing the bioaccumulation from sediment to mussels. The model prognosis has been validated with surprisingly good results in spite of its simplicity.

Chapter 10 shows a mercury model based on a problem similar to the chromium model. Mercury is able to react with the organic matter in the sediment and bioaccumulates in pelagic as benthic food chains as well. It implies that a more complex model must be used to solve this problem. The model must account for the release of methyl- and dimethyl-mercury from the sediment and must furthermore consider the relationship between the weights of fish species and the mercury concentration.

A comparison between the chromium and the mercury model illustrates very clearly that a profound knowledge of the geobiochemistry of the considered heavy metal is needed to be able to select the right approximations. Mercury has a more complex goebiochemistry than chromium and it requires a more complex model than chromium under the same environmental conditions.

Chapter 9 shows a copper model and it differs from the chromium and mercury model by its more simple hydrodynamics. A more comprehensive description of the geochemistry of the copper species is however needed and the model is an illustrative example of a geochemical model. It is more complex than the chromium model in the description of the exchange processes of copper between sediment and water, but does not contain a hydrodynamic component as the chromium model does.

Chapter 8 is again a heavy metal model but this time for a terrestrial ecosystem. The aim of the model is to relate the input of cadmium and lead to a soil of known composition and pH to the cadmium and lead concentration in cultivated plants. The model illustrates the difference between modelling an aquatic and a terrestrial ecosystem. The flow processes in soil are very complex and a detailed description of these processes would require a very detailed knowledge of the permeability of soil at different sites and depths. Consequently these flow processes have been described very simply and the emphasis of the model has been on the factors that are regulating the uptake of heavy metals, including the solubility of the heavy metal ions. The case study illustrates very clearly that it is possible to simplify in very complex cases, provided the scope of the model is to describe an average situation.

Chapters 11 and 12 are concerned with models of an atmospheric pollutant: sulphur dioxide. Comparison between the models of aquatic ecosystems (chapters 7,9,10 and l3), of terrestrial ecosystems (chapter 8) and of pollutants in the atmosphere (chapters 11 and 12) shows that the differences are mostly related to the flow processes. Chapter 12 shows clearly that the flow processes in the atmosphere are significant for the distribution of the pollutants and, therefore, they require a very detailed and more accurate description, but also they are more difficult to describe properly. Chapter 12 shows very clearly how these obstacles may be overcome.

Chapter 11 is a very interesting case study, as it attempted to model the detailed processes of an air pollutant and its related effects on the plants. The case study illustrates how complex the effects on the plants are and how much more knowledge we need before we can model the effects of air pollutants properly. The scope of most ecotoxicological models is not to go into such details and the detailed description of this model is therefore not needed in most ecotoxicological models, however, if the aim of the model is to describe the ecotoxicological processes in plants, it can be done as this case study shows. The case study differs as seen form the other case study by the size of the system and the included details.

It would not be a proper modelling approach to link to the two models in chapters 11 and 12 to one very comprehensive model, but the two models may very well work together by use of the results from the distribution model as a forcing function in the plant model. Each level in the environmental hierarchy requires its own model, but it may cause difficulties to couple such model to one comprehensive model, as it will involve the use of too complex a model.

Chapter 13 presents a fate model of four chlorobenzenes in an aquatic ecosystem. It is an illustration of how to consider the many processes that characterize the environmental problems of organic pollutants. It differs from the heavy metal models by inclusion of other processes. The applied descriptions may also be used in other types of ecosystem and the model is therefore a general illustration of an ecotoxicological model concerned with organic pollutants.

Chapter 14 presents a quantitative evaluation of pesticide transformations and movement in soils. A comparison of several pesticides based upon their physical and chemical properties is carried out. This case study can hardly be called a model, but it has been included because it illustrates how important the knowlegde of parameters is, and how far it is possible to make quantitative evaluation on basis of a good knowledge of parameters. The chapter emphasizes, so to say, the significance of Chapters 4 and 5 on parameter estimation methods.

The eigth case studies illustrate - as seen from this brief overview - ecotoxicological problems related to heavy metals, gases and organic pollutants. They also show how a profound knowledge of the physical, chemical and biological processes are needed to make the right simplifications. They demonstrate the different significance of the flow processes in the three spheres, and the difference between a distribution and an effect model, and between models of different scales. The selected seven case studies should give the reader a good knowledge of the spectrum of models available today in ecotoxicology and related possibilities.

CHAPTER 7

MODELLING THE DISTRIBUTION OF CHROMIUM IN A DANISH FIRTH

by
S.E. Jørgensen

7.1 CASE STUDY DESCRIPTION

This case study has been presented in previous publications, see for instance Jørgensen (1988). It has been repeated in this context because:

1) The case study is very illustrative for what can be achieved **by use of simplified models.**
2) Recently it has been possible **to validate the prognosis** set up 8 years ago. Validations of models are not only important but absolutely necessary for development of reliable models. In this case it has been even possible to validate the model predictions. Unfortunately we have only very few cases of prognosis validations. Therefore it has been considered significant to include this case study now that the prognosis validations are available.
3) The model development shows clearly how important it is **to know the system and its processes, if the right model with the right simplifications shall be selected.**

A map of the system, Faaborg Firth, is shown in Figure 7.1. The numbers indicate sampling stations. The point close to the station 1 is the discharge point.

A tanning plant has for decades discharged waste water with a high concentration of chromium(III) into the firth. In 1958 production was expanded significantly and gave a pronounced increase in the chromium concentration in the sediment (see Mogensen and Jørgensen 1979). For further details see also Mogensen 1978.

It was the scope of this investigation to set up a model for the distribution of chromium in the firth on the basis of analysis of chromium in phytoplankton, zooplankton, fish, benthic fauna, water (dissolved as well as suspended) and sediment. Already during the first phase of the investigation it was clear, that the phytoplankton, zooplankton and fish were almost not contaminated by chromium, while the sediment and the benthic fauna showed a clearly raised concentration of chromium. This was easy to explain: the chromium(III) was precipitated as hydroxide by contact with the seawater which has a pH of 8.1 compared with 6.5 - 7.0 for the waste water.

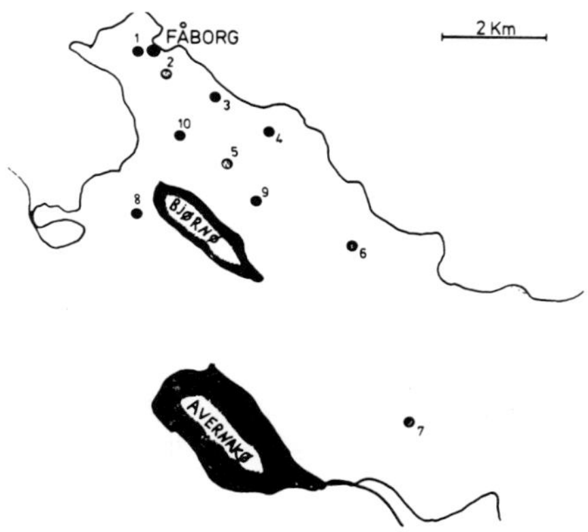

Fig. 7.1: Faaborg Fjord. The sampling stations 1-10 are shown. The point close to sampling station 1 indicates the discharge point.

7.2 MODEL DESCRIPTION

All the analysis showed that the important processes are:

1) **Settling** of the precipitated chromium(III) hydroxide and other insoluble chromium compounds.
2) **Diffusion** of the chromium, mainly as suspended matter, thorughout the firth caused by mainly tides. It implies that a eddy diffusion coefficient has to be found.
3) **Bioconcentraton** from sediment to benthic fauna.

1) and 2) can be combined in one submodel, while process 3 may be accounted for in a separate submodel.

The distribution model is based on the following simple chromium transport equation, see for instance Rich (1973):

$$\partial C / \partial t = D * \partial^2 C / \partial X^2 - Q * \partial C / \partial X - K * (C - C_o) / h \qquad (7.1)$$

where C is the concentration of total chromium in water in mg/l.

C_o is the solubility of chromium in seawater at pH = 8.1 in mg/l.
Q is the inflow to the firth = outflow by advection (m^3/ 24h).
D is the eddy diffusion coefficient considering the tide (m^2/ 24h).
X is the distance from the discharge point in m
K is the settling rate in m /24h.
h is the mean depth in m.

For a tidal firth such as Faaborg Firth with only insignificant advection Q may be set to 0.

Since the tanning plant has discharged an almost constant amount of chromium during the last two decades, we can consider the stationary situation:

$$\partial C/\partial t = 0 \tag{7.2}$$

Equation (7.2) therefore takes the form:

$$D * \partial^2 C/\partial X^2 = K *(C - C_o)/ h \tag{7.3}$$

This differential equation of second order has an analytical solution. C_u = the total discharge of chromium in g per 24 h, is known. This information is used together with F = cross sectional area (m^2) to state the boundary conditions. The following expression is obtained as an analytical solution:

$$C - C_o = (C_u / F) * \sqrt{(h/ D*K)} * \exp [-\sqrt{(K/ h*D)} * X] + C_1 \tag{7.4}$$

F is known only approximately in this equation due to the nonuniform geometry of the firth. The total discharge of chromium is known to be 22,400 kg per year. Both the consumption of chromium by the tanning factory and the analytical determinations of the waste water discharged by the factory confirm this figure. h is about 8 m in average. C_1 is an integration constant.

Equation (7.4) may be transformed to:

$$Y = K* (C - C_o) = (C_u / F) * \sqrt{(h*K/ D)} * \exp [-\sqrt{K/ h*D)} * X + K * C_1 \tag{7.5}$$

Y is, as seen, the amount of chromium (g) settled per 24h and per m^2. The equation gives Y as a function of X.

Y is, however, known from the sediment analysis. A typical chromium profile for a sediment core is shown in Figure 7.1. As we know that the increase in the chromium concentration took place 20 years ago, it is possible to find the sediment rate in mm or cm per year. As we furthermore know the concentration of chromium in the sediment, we can calculate the amount of chromium settled per year, or 24 h, and per m^2, and this is Y.

The Y-values found by this method are plotted versus X in Figure 7.2.

A non-linear regression analysis was used to fit the data to an equation of the following form:

$$Y = a^* \exp(-bX + c) \tag{7.6}$$

a, b and c are constants, which are found by the regression analysis.

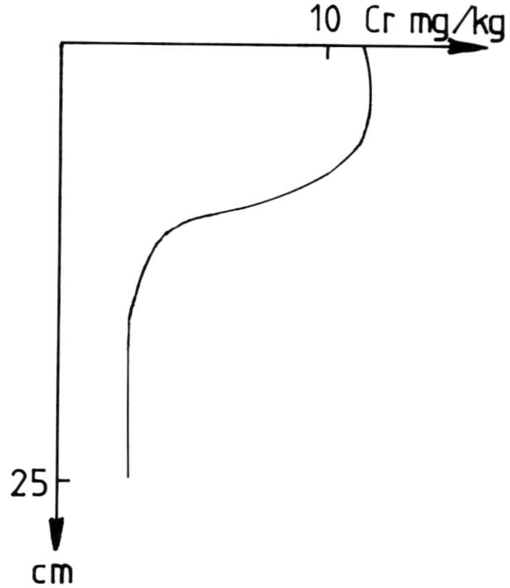

Fig. 7.2: Typical chromium profile of sediment core.

Table 7.1. shows Y = f(X). Table 7.2 gives the estimations of a, b and c found by the statistical analysis. Table 7.3 shows the result of the statistical analysis and, as seen, the model found with the values of a, b and c from Table 7.2 has a very high probability. The F-value found is 114.5, while an F-value with a probability of 0.9995 is only 30.4.

Table 7.4 translates the constants a, b and c into parameters of the model. D is found on basis of an average value for K on 1.6 m/24h. This value is found on basis of the definition of Y. Y is known as shown above. Furthermore C_o is known to be 0.2 mg/m^3, and as C is measured for all stations, K may be found from :

$$K = Y / (C - C_o) \qquad (7.7)$$

The settling rates found by this method are shown in Table 7.5.

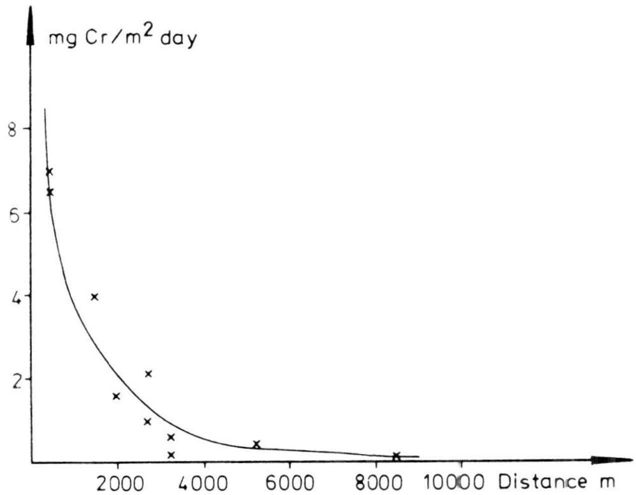

Fig. 7.3: Y, found by sediment analysis, is plotted versus X.

Table 7.1
Y versus X

Station number	g Cr/m^2 year	(Y) g Cr/m^2 day	(X) Distance from discharge point (m)
1	2.55	$7.0 * 10^{-3}$	500
2	2.39	$6.5 * 10^{-3}$	500
3	1.47	$4.0 * 10^{-3}$	1500
4	0.35	$1.0 * 10^{-3}$	2750
5	0.78	$2.1 * 10^{-3}$	2750
6	0.14	$3.8 * 10^{-4}$	5250
7	0.03	$8.2 * 10^{-5}$	8500
8	0.20	$5.5 * 10^{-4}$	3250
9	0.06	$1.6 * 10^{-4}$	3500
10	0.58	$1.6 * 10^{-3}$	2000

Table 7.2
Estimations of a, b and c

	Estimate	Asymtotic St. error
a	0.009909	0.00084
b	0.000723	0.00015
c	-0.000081	0.00045

Table 7.3
Statistical Analysis

	Degree of freedom	Sum of squares	Mean square
Model	3	0.00011337	0.00003779
Residual	6	0.00000233	0.00000033
Total	9		
	F = 114.5		

Table 7.4
Parameters

From the regression analysis we have:

$$\frac{C_u}{F} * \left(\frac{h * K_s}{D}\right)^{1/2} = 0.00990 = a$$

and

$$\left(\frac{K_s}{h * D}\right)^{1/2} = 0.000723 = b$$

which gives

$$\frac{C_u}{F} * h = \frac{a}{b} = 13.7$$

F = 35800 m^2, which seems a reasonable average value of the cross sectional area. From analysis of C_1 at stations 2, 5, 6, 7 and 8 (see. Table 7.5) we get an estimation of K_s since

$$Y = \frac{gCr}{m^2 day} = K_s\ (C_t - C_0)\ (c \sim 0)$$

C_0 is found to be 0.2 mg/m^3

Table 7.5
Settling rates

Station	mg C_r/m^2day	C_t - C_0 (mg m^{-3})	K_s (m day^{-1})
2	6.5	2.5	2.6
5	2.1	0.9	2.3
6	0.4	0.6	0.7
7	0.1	0.2	0.5
8	0.6	0.3	2.0

As seen from Table 7.4 the settling rate is approximately the same at three of the five stations. Stations 6 and 7 are given a lower value, but it should also be expected that the settling rate decreases with increasing distance from the discharge point. However, it should not be forgotten that the determination of the chromium concentration in the water is not very accurate, as it is of a rather low concentration. The value for the diffusion coefficient found on basis of the settling rate corresponds to 4.4 m^2/s - a quite reasonable value. The value for F is slightly more than the width of the inner firth, but as a weighted average for the inner and outer firth it seems a reasonable value.

Integration from o to ∞ over a half circle area gives as result 22 t of chromium or almost all the chromium discharged may be explained by the model, assuming that the distribution takes place over a half circle area.

All by all it may be concluded that the distribution model gives acceptable results. The high concentrations of chromium in the sediment gives reliable determinations, which again is the basis of the distribution model. The use of sediment analysis, as demonstrated, is, therefore, recommended to work out a distribution model.

The second submodel focuses on the chromium contamination of the benthic fauna. It may be shown (Jørgensen 1979), that under steady state conditions the relation between the concentration of a contaminant in the n'th link in the food chain and the corresponding concentration in the (n-1)'th link can be expressed by use of the following equation:

$$C_n = (MY(n) * C_{n-1} * YT(n)) / (MY(n) * YF(n) - RESP(n) + EXC(n)) = K' * C_{n-1}, \qquad (7.8)$$

where

MY(n) = the maximum growth rate for n'th link of the food chain (1/day)

C_n = the chromium concentration in the n'th link of the food chain (mg/kg)

C_{n-1} = the chromium concentration in the (n-1)'th link of the food chain (mg/kg).

YT(n) = the utility factor of chromium in the food for the n'th link of the food chain (-)

YF(N) = the utility factor of the food in the n'th link of the food chain.(-)

RESP(n) = the respiration rate of the n'th link of the food chain (1/day).

EXC(n) = the excretion rate of chromium for the n'th link of the food chain (1/day).

For some of the species present in Faaborg Firth these parameter values can be found in the literature, see for instance Jørgensen et al 1979. The mussel *Mytilus edulis* was found on almost all the stations and the following parameters are valid: (YT(n) and YF(n) are found for other species)

MY(n) = 0.03 1/day
YT(n) = 0.07
YF(n) = 0.66
RESP(n) = 0.001 1/day
EXC(n) = 0.04 1/day

The use of these values implies that K' = 0.036 for *Mytilus edulis* . In other words, the concentration of chromium in *Mytilus edulis* should be expected to be 0.036 times the concentration in the sediment.

21 mussels from Faaborg Firth have been analyzed and by statistical analysis it was found that the relation between the concentration in the sediment and in the mussels is linear:

$$C_n = C_{n-1} * K' \tag{7.9}$$

where K' was found to be 0.015 ±0.002. The discrepancy from the theoretical value is fully acceptable, when it is considered that the parameters are found in the literature and they may not be exactly the same values for all environments for all possible conditions. In general biological parameters can only be considered approximate values. The relatively low standard deviation of the observed K' value confirms, however, the relation used.

7.3 APPLICATION OF THE MODEL FOR ENVIRONMENTAL MANAGEMENT

It is proposed to use the highest K' value = 0.036, when the model is used for environmental management, as in that way the uncertainty of the K'-value is "given to the benefit of the environment".

The model was used as a management tool and the acceptable level of the chromium concentration in the sediment of the most polluted area was assessed to be 70 mg per kg dry matter. That would corresponds to a chromium concentration of 70*0.036 = 2.5 mg per kg dry biomass in mussels, or about 2 times the concentration found in uncontaminated areas of the open sea.

The distribution model was now used to assess the total discharge of chromium (kg/y), if the chromium concentration in the sediment should be reduced to 70 mg per kg dry matter in the most polluted areas (stations1 and 2). It was found that the total discharge of chromium should be reduced to 2000 kg or less per year to achieve a reduction of more than 90%.

The environmental authorities required the tanning plant to reduce its chromium discharge to 2000 kg per year, which the tanning plant has maintained from 1980.

A few samples of sediment(4) and mussels(5) taken n 1987 have been analyzed and used to validate this prognosis. The results are given in Table 7.6. Settled chromium in mg/m^2 day was found on basis of the previous determined sedimentation rate (see above). The prognosis validation was fully acceptable as the deviation between prognosis and observed average values are more or less 12%.

Table 7.6
Validation of the prognosis

Item	mg per kg dry matter Observed Value	Range	Predicted Value
Cr in sediment	65	57-81	70
Cr in mussels	2.2	1.4-4.5	2.5
mg Cr/m^2 day	0.59	0.44-0.83	0.67

REFERENCES

Jørgensen, S.E., 1979. Modelling the distribution and effect of heavy metals in an aquatic ecosystem. Ecological Modelling 6: 199-223.

Jørgensen, S.E., 1988. Fundamentals of Ecological Modelling. Elsevier, Amsterdam, Oxford, New York, Tokyo.

Jørgensen, S.E., (editor-in-chief; editorial board: M.B. Friis, J. Hendriksen, L.A. Jørgensen, S.E. Jørgensen and H.F. Mejer), 1979. Handbook of Environmental Data and Ecological Parameters. International Society of Ecological Modelling, Copenhagen.

Mogensen, B., 1978. Chromium pollution in a Danish fjord. Licentiate Thesis. Royal Danish School of Pharmacy, Copenhagen.

Mogensen, B. and S.E. Jørgensen, 1979. Modelling the distribution of chromium in a Danish firth. Proceedings of 1st International Conference on State of Art in Ecological Modelling, Copenhagen, 1978. ed. S.E. Jørgensen, Copenhagen: International Society for Ecological Modelling, pp. 367-377.

Rich, L.G., 1973. Environmental Systems Engineering, p.5. McGraw Hill, U.S.A.

CHAPTER 8

CONTAMINATION OF AGRICULTURAL PRODUCTS BY CADMIUM AND LEAD

by
S.E. Jørgensen

8.1 INTRODUCTION

Agricultural products are contaminated by lead and cadmium originating from air pollution, the application of sludge from municipal waste water plant as a soil-conditioner, and from the use of fertilizers.

The uptake of heavy metals from municipal sludge by plants has previously been modelled(see Jørgensen 1975 and 1976). This model can briefly be described as follows: Dependent on the soil composition it is possible to find for various heavy metal ions a distribution coefficient i.e. the part of the heavy metal which is dissolved in the soil-water relative to the total amount. The distribution coefficient was found by examination of the dissolved heavy metals relative to the total amount for several different types of soil. Correlation between pH, the concentration of humic substances, clay and sand in the soil on the one hand, and the distribution coefficient on the other, was also determined. The uptake of heavy metals was considered a first-order reaction to the dissolved heavy metal.

This model does, however, not take into account:

1) the direct uptake from atmospheric fall-out onto the plants.
2) the other sources of contamination such fertilizers and the long-term release of heavy metal bound to the soil and the unharvested parts of the plants.

It has been the objectives of this work to include these sources in a model for lead and cadmium, and to examine whether published data on lead and cadmium contamination in agriculture can be explained by the developed model.

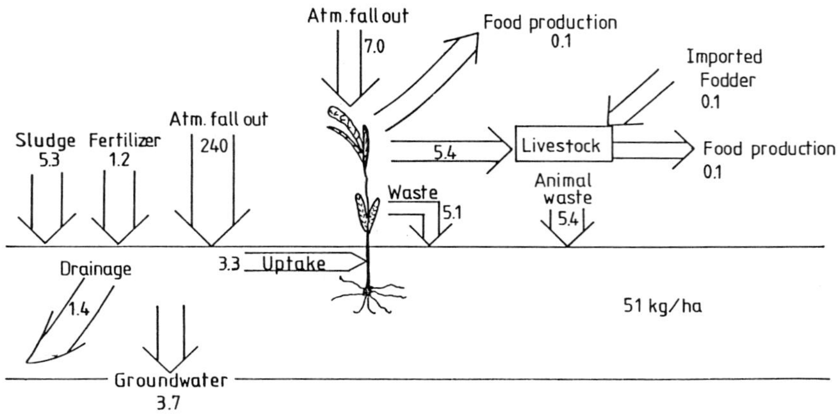

Fig. 8.1: Lead balance of average Danish agriculture land.

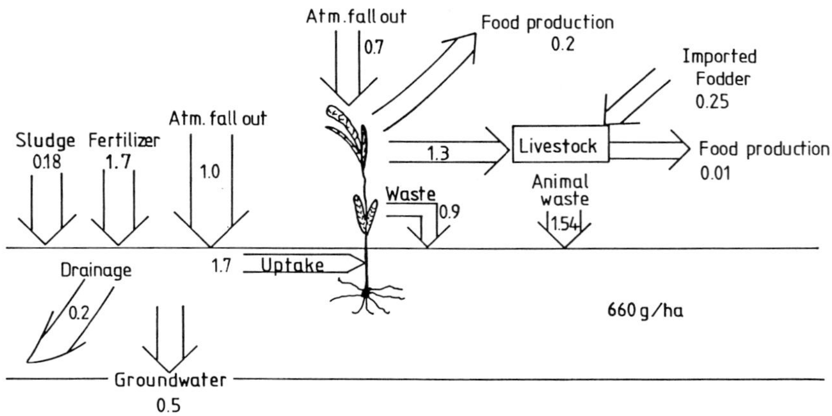

Fig. 8.2: Cadmium balance of average Danish agriculture land.

8.2 SOURCES OF CADMIUM AND LEAD CONTAMINATION IN AGRICULTURE

The basis for the model is the lead and cadmium balance for average Danish agricultural land. Figures 1 and 2 give the balances, taken from Andreasen (1985), and Knudsen and Kristensen (1987). As seen the most important source of lead contamination is the atmospheric fall-out which will gradually be reduced during the coming years when the lead concentration in gasoline is reduced, while the most important source of cadmium contamination is fertilizer. The latter can only be reduced by use of a less contaminated phosphorus ore for the production of phoshorus fertilizer.

It is seen that the amounts of lead and cadmium coming from domestic animals and plant residues after harvest are also significant contributions.

8.3 THE MODEL

Figure 8.3 shows a conceptual diagram of the model. As seen it has four state variables: Cdbound, Cdsoil, Cddetritus and Cdplant. An attempt was made to use one or two state variables for cadmium in the soil, but to get acceptable accordance between data and model output three state variables were needed. This can be explained by the presence of several soil components which bind the heavy metal differently; see Christensen (1981)(1983)(1984) and (1985), EPA, Denmark (1979), Hansen and Tjell (1981), Jensen and Tjell (1981) and Chubin and Street (1981). Cdbound covers the cadmium bound to minerals and to more or less refractory material; Cdsoil covers the cadmium bound by adsorption and ion exchange; while Cddetritus is the cadmium bound to organic material with a wide range of biodegradability.

The forcing functions are: airpoll, Cdair, Cdinput, yield and loss.

The atmospheric fall-out is known, and the allocation of this source to the soil(airpoll) and to the plants(Cdair), is in accordance with Hansen and Tjell (1981) and Jensen and Tjell (1981). Cdinput covers the heavy metal in the fertilizer and as seen from the equations Table 8.l, it comes as a pulse at day 1 and afterwards with a frequency of every 180 days. The yield corresponds to the part of the plants which is harvested. There is also a pulse function at day 180, and afterwards with a occurrence every 360 days. As seen from Table 8.1, in this case, it is 40% of the plant biomass. The loss covers transfer to the soil and groundwater below the root-zone. It is expressed as a first-order reaction with a rate coefficient dependent on the distributioncoefficient which is found from the soil composition and pH, in accordance with the correlation found by Jørgensen (1975). Furthermore the rate constant is dependent on the hydraulic conductivity of the soil. In the case in Table 1 the constant 0.01 reflects the dependence of the hydraulic conductivity.

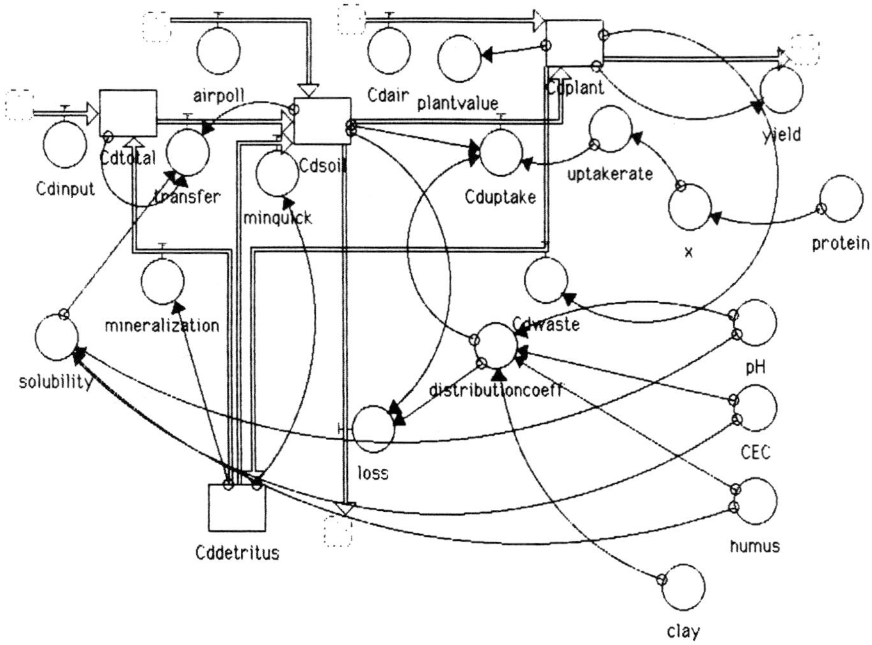

Fig. 8.3: Conceptual diagram of the model. The model has been developed on a Macintosh Plus by use of the software STELLA. Boxes indicate state variables, double line arrows give flows, circles give functions and single line arrows show feed-back mechanisms.

The transfer from Cdbound to Cdsoil indicates the slow release of cadmium due to a slow decomposition of the more or less refractory material to which cadmium is bound. The cadmium uptake by plants is expressed as a first-order reaction, where the rate is dependent on the distribution coefficent, as only dissolved cadmium can be taken up. It is furthermore dependent on tha plant species. As seen, the uptake is a step function which, in this case (grass), is 0.0005 during the growing season, and, of course, 0 after the harvest and until the next growing season starts. Cdwaste covers the transfer of plant residues to detritus after harvest. It is therefore a pulse function, which in this case is 60% of the plant biomass, as the remaining 40% has been harvested.

Table 8.1
Model equations

```
Cddetritus = Cddetritus + Cdwaste - mineralization - minquick
    INIT(Cddetritus) = 0.27
Cdplant = Cdplant + Cduptake - yield - Cdwaste + Cdair
    INIT(Cdplant) = 0.0002
Cdsoil = Cdsoil - Cduptake - loss + transfer + minquick + airpoll
    INIT(Cdsoil) = 0.08
Cdtotal = Cdtotal + Cdinput - transfer + mineralization
    INIT(Cdtotal) = 0.19
airpoll =0.0000014
Cdair = 0.0000028+STEP(-0.0000028,180)+STEP(+0.0000028,360)+STEP(-
    0.0000028,540)+STEP(+0.0000028,720)+STEP(-0.0000028,900)
Cdinput = PULSE(0.0014,1,180)
Cduptake = distributioncoeff*Cdsoil*uptakerate
Cdwaste = PULSE(0.6*Cdplant,180,360)+PULSE(0.6*Cdplant,181,360)
CEC = 33
clay = 34.4
distributioncoeff =0.0001*(80.01-6.135*pH-0.2603*clay-0.5189*humus-
    0.93*CEC)
humus = 2.1
loss = 0.01*Cdsoil*distributioncoeff
mineralization = 0.012*Cddetritus
minquick = IF TIME≤180 THEN 0.01*Cddetritus ELSE 0.0001*Cddetritus
pH = 7.5
plantvalue = 3000*Cdplant/14
protein = 47
solubility = 10^(+6.273-1.505*pH+0.00212*humus+0.002414*CEC)*112.4*
    350
transfer = IF Cdsoil<solubility THEN 0.00001*Cdtotal ELSE 0.000001*
    Cdtotal
uptakerate = x +STEP(-x,180)+STEP(x,360)+STEP(-x,540)+STEP(x,720)+
    STEP(-x,900)
x = 0.002157*(-0.3771+0.04544*protein)
yield = PULSE(0.4*Cdplant,180,360)+PULSE(0.4*Cdplant,181,360)
```

Cddetritus covers a wide range of biodegradable matter and the mineralization is therefore accounted for in the model by use of two mineralization processes: one to Cdsoil, and one to Cdbound. The first one is quick and is given a higher rate the first 180 days as the addition of municipal sludge in this case is at day 0. The second one is at about the same rate, but as the cadmium is tranferred to the Cdbound the slow release rate is taken into account by the very slow transfer from Cdbound to Cdsoil.

8.4 MODEL RESULTS

Data from Jensen and Tjell (1981) and Hansen and Tjell (1981), was used for calibration and validation of the model. It was in this phase of the modelleing procedure that it was revealed that three state variables for heavy metal in soil were needed to get acceptable results. It was particularly difficult for the right values for heavy metal concentrations the second and third year after municipal sludge had been used as a soil conditioner. This use of modelling may be called experimental mathematics, where simulation with different models are used to deduce which model structure is best. The results of experimental mathematics must of course be explained by examination of the processes involved and here can be referred to in the references given above.

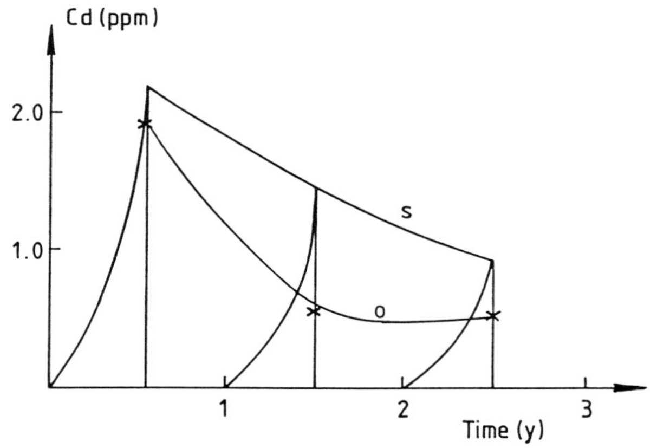

Fig. 8.4: The model was validated by use of the cadmium concentration as a function of time (y) for red clover at third and fourth harvests. x gives the observations and the curve marked s gives the corresponding model predictions.

The results of the validation phase are shown in Figures 8.4 and 8.5, and, as can be seen, the accordance between observations and model predictions is reasonable good.

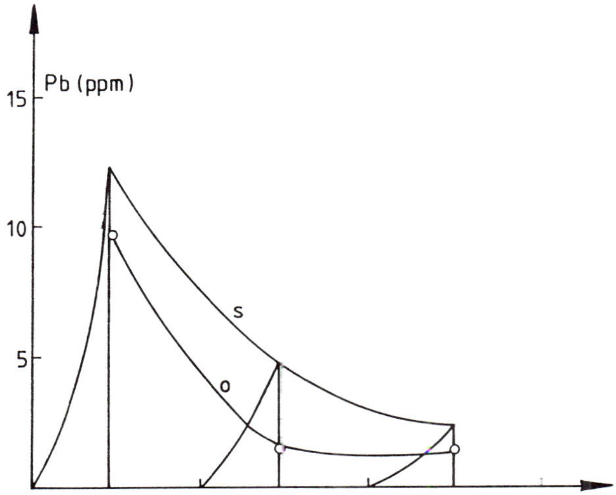

Fig. 8.5: The model was validated by use of the lead concentration as a function of time (y) for salad plants. o gives the observations and the curve marked s gives the corresponding model predictions.

8.5 DISCUSSION AND CONCLUSIONS

As seems apparent from the validation the developed model is able to explain the observations. A wider use of the model would require, however, that more data from experiments with many plant species are used to test the model.

It may, however, be concluded from these results that the model structure must account for at least three state variables for the heavy metal in soil to cover the ability of different soil components to bind the heavy metal by various processes.

The model has been calibrated and validated on basis of three years experiments and measurements and it was clear from the model exercises that the atmospheric fall-out and heavy metal in the plants residues were significant although these were not considered in the model published in 1976.

Translocation of the heavy metal to various parts of the plant was not considered in the model and this would be a natural next step to include in the model, as it is important to distinguish heavy metal concentrations in various parts of the plants.

The problem modelled is very complex and many processes are involved. On the other hand, a management model should be relatively simple and not involve too many parameters. The model can obviously be improved, but it gives at least a first rough picture of the important factors in the contamination of agricultural crops. In most cases it is not possible to get very accurate results with toxic substance models but, on the other hand, as we want to use relatively large safety factors, the need for high accuracy is not pressing.

REFERENCES.

Andreasen, I., 1985. A general ecotoxicological Model for the Transport of Lead through the System: Air-Soil(Water)-Grass-Cow-Milk. Thesis at DIA-K The Technical University of Denmark. 57 pages.

Christensen, T.H.,1981. The Application of Sludge as Soil Conditioner, Vol 3, p. 19-47. 460 pages. Polyteknisk Forlag, Copenhagen.

Christensen, T.H., 1984. Cadmium Soil Sorption at low Concentrations 1) Effect of Time , Cadmium Load, pH and Calcium. 2) Reversibility, Effect of Changes in Solute Composition, and Effect of Soil Ageing. Water, Air and Soil Pollution, Vol. 21, p. 105 - 125.

Christensen, T.H., Cadmium Soil Sorption at low Concentrations 3) Prediction and Observation of Mobility 4)Effect of Waste Leachates on Distribution Coefficients. Water , Air and Soil Pollution, Vol. 26, p. 255-274.

Chubin, R.G. and Street, J.J., 1981. Adsorption of Cadmium on Soil Constituents in the Presence of Complexing Agents. J. of Env. Qual, Vol. 10, p. 225-228.

EPA, Denmark, 1979. The Lead Contamination in Denmark. 145 pages.

Hansen, J. Aa. and Tjell, J.C., 1981. The Application of Sludge as Soil Conditioner, Vol. 2, p. 137-181. 182 pages. Polyteknisk Forlag, Copenhagen.

Jensen, K. and Tjell, J.C., 1981. The Application of Sludge as Soil Conditioner, Vol. 3, p.121-147. Polyteknisk Forlag, Copenhagen.

Jørgensen, S.E.,1975. Do heavy Metals prevent the agricultural Use of municipal Sludge? Water Research, Vol. 9, p. 163-170.

Jørgensen, S.E.,1976. An ecological Model for heavy Metal Contamination of Crops and Ground Water. Ecol. Model., Vol. 2, p. 59-67.

Møller, A., 1983. Levnedsmiddeltabeller. Statens Levnedsmiddelinstitut. Ernæringsenheden, Copenhagen.

Santillan-Medrano, J. & J.J. Jurinak, 1975. The chemistry of lead and cadmium in soil: soil phase formation. Soil. Sci. Soc. Amer. Proc. Vol. 39 p. 851-856.

CHAPTER 9

MODELLING THE RELEASE OF COPPER FROM LAKE SEDIMENT

by
S.E. Jørgensen

9.1 INTRODUCTION

Sediment is able to bind heavy metals either by adsorption or by formation of insoluble components such as carbonates, hydroxides or sulphides. The concentration found in sediment is often magnitudes greater than the concentration in water, and analysis of the metal concentration in sediment is therefore often used as indicators of a heavy metal contamination of an aquatic ecosystem. The sediment accumulates the heavy metal contamination and it is possible by determination of the concentration profile to give information on the contamination as a function of time, as also shown in the case study presented in chapter 7. In the figure 9.1 another example is shown.

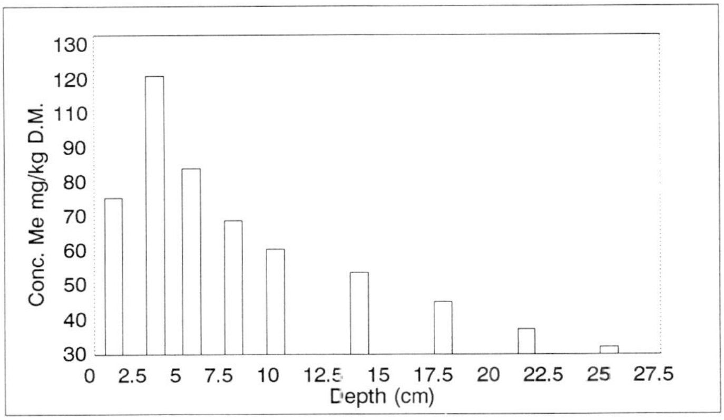

Fig. 9.1: Metal profile of lake sediment.

It is extremely difficult to determine the various forms of heavy metal in the sediment by direct analysis, although some indications can be given by use of various extraction media. However, a more or less complete analysis is very time consuming and is therefore hardly in use for environmental management where a relatively coarse determination of the most important fractions is considered sufficient in most cases.

The number of possible exchange reactions between sediment and water is very high, which renders the problem of modelling the release processes even more complex. The complexity of the problem is clearly illustrated by the following list of factors influencing the exchange processes:

1. *The temperature* obviously influences adsorption and as well as all the chemical and microbiological processes.
2. *pH* also influences all the processes. The precipitation processes are highly dependent on pH and as the charges of the interacting ions vary with pH, the adsorption will thereby be very pH-dependent.
3. *The grain size of the sediment* determines the specific surface and therefore influences the adsorption capacity The kinetics of the chemical processes are furthermore dependent on the surface area.
4. *The redox potential* will always influence the chemical and microbiological processes. It will also influence the composition of the pore water as different organic ligands are formed under aerobic and anaerobic conditions.
5. *The chemical composition of the water and the sediment* determines the concentrations of various reactive components which may precipitate or dissolve the heavy metal. The ionic strength of the water may also be of importance for the exchange processes and the solubility of heavy metal ions. All in all it may be concluded that to model the exchange processes of heavy metal between sediment and water is a very complex problem.

9.2 OVERVIEW OF THE EXCHANGE PROCESSES

Before the case-study is presented in detail an overview of the various types of exchange processes are presented below,. It is clear from the introduction to this chapter that the problem *is* complex and in most cases, it will therefore not be feasible to include all the exchange processes. As is characteristic for all modelling the case-study will therefore demonstrate, that a simplification is almost always needed and that the art of modelling is to make the right simplifications by omission of the processes and factors least important for the problem in focus. The list of processes is not exhaustive, but the most significant processes are mentioned.

1. In many cases **heavy metal precipitation** plays in many cases a major role in determination of heavy metal concentrations in the sediment. Precipitation, primarily of hydroxides, sulphides and carbonates, occurs within a water body when the corresponding solubility product is exceeded. The solubility product of an electrolyte is a special type of equilibrium constant, referred to as S. An example illustrates the concept:

$$S \text{ for } PbCl_2 = [Pb^{+2}]*[Cl^-]^2 = 1.7*10^{-5} \quad (9.1)$$

The precipitation process interferes, however, with many other processes. pH determines the concentration of anionic species, as hydrogenions are able to react with carbonates, hydroxides and sulphides. Complexes of the heavy metal ions are formed with inorganic and organic ligands as side reactions to the precipitation processes. Therefore, a quantification of a precipitation may often require calculations on several parallel processes.

2. **Ion exchange and adsorption** take place on a number of sediment-forming materials with a large surface area, such as clay minerals, iron hydroxides, organic substances and amorphous silicic acids. These processes may often be described by Freundlich's or Langmuir's adsorption isotherms. If pure ion exchange is involved an equilibrium constant may be used to describe the exchange process. The exchange capacity for some clay minerals may be as high as 100 meq/100g, for humic acids from soils even 2-5 times greater. The affinity of heavy metal ions is greater than that of alkaline earth or alkaline ions. The following affinity series gives a first approximation of the affinity (Weiss and Amstutz 1966):

   Pb> Cu> Ni> Co> Zn> Mn> Ba> Ca> Mg >K> Na.

3. **A combination of sorption and coprecipitation** takes place on hydrous oxides of aluminum, iron and manganese, particularly on the redox-sensitive iron and manganese hydroxides under oxidizing conditions.
   Even a low concentration of these components may have controlling influence on the heavy metal distribution in an aquatic system.

4. **Associations / complex formations** with organic substances have a pronounced effect on the solubilization of heavy metal. The organic matter consists of the remains of biologically produced compounds, as

well as of synthetic organic substances. They may be subdivided into different groups (Welte 1969):

1. Substances extractable with organic solvents such as benzene, ether and chloroform. They are commonly known as *bitumen.*
2. The bitumen-free or insoluble organic matter is referred to as *kerogen.*
3. *Humic acids* which are extractable with a weak alkali solution.
4. *Residual organic matter* is the component of the original mass that remains after the extraction procedures described above.
5. *Synthetic organic substances* that originated from industrial and agricultural applications such as detergent additives and pesticides.

The sequence of complex stability follows the so-called Irving-Williams (1948) series:
Pb> Cu> Ni> Co> Zn> Cd> Fe> Mn> Ca> Mg
It has been suggested that the portion of many trace metals present in natural waters as soluble organic complexes is generally greater than that present as inorganic or aqua complexes (Jenne 1976).

5. Coagulation and flocculation of metal-organic matter is caused by four mechanisms leading to destabilization of colloids or suspensions:

1. *Neutralization* of the negative charge by specific adsorption of positively charged species.
2. *Compression* of the electrical double charge surrounding the colloids or suspended particles.
3. *Inter-particle bridging* by adsorbed material.
4. *Enmeshment of clay and hydroxide particles* caused by increased number of collisions due to particle movement and higher concentrations of particulate material.

6. Coatings on clay minerals of hydroxidic or organic matter may change the adsorption and ion exchange of clay minerals radically.

7. Bacterial methylation plays an important role for the release of mercury, arsenic, lead and selenium from sediment.

A comprehensive description of these 7 processes includes the influence of the 5 factors, mentioned in section 9.1. All five factors are playing a role on the rate and the equilibrium of all the seven mentioned processes.

9.3 A CASE STUDY. FREDERIKSBORG CASTLE LAKE

K.J. Andersen (1986) has examined the release of copper from Frederiksborg Castle Lake and attempted to synthesize the results by use of a modelling approach. These results are presented in this section.

This case study has been selected, because the sediment of Frederiksborg Castle Lake is contaminated by copper mainly through wash-out of copper from the castle copper roof.

The lake has an area of 21,000 m^2 and a volume of 640,000 m^3. The average depth is 3.1 m. The primary production of the lake is high due to a high internal loading of nutrients.

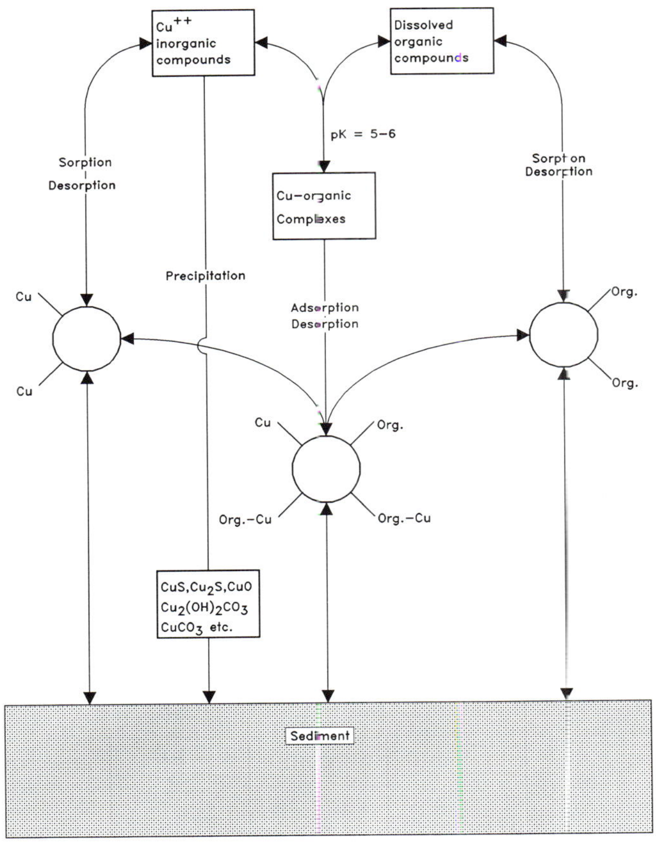

Fig. 9.2: Sediment-water interactions for Copper. Circles represent suspended matter.

Figure 9.2 presents a conceptual diagram after Davies and Leckie (1978) of the exchange processes for copper between sediment and water. This conceptual model has been used as a basis for model development. As seen from this diagram the most important processes are precipitation of insoluble copper compounds such as copper sulphide, copper hydroxide, copper oxide and copper carbonate. Copper can be dissolved as organic complexes, which however have a high capacity for adsorption on suspended matter. The most important processes seem to be precipitation (which is very dependent on pH), complex formation and adsorption (sorption and desorption).

The precipitation may be described quantatively by use of the mass equation - the solubility product - the adsorption by use of Langmuir's or Freundlich's adsorption isotherm, while it is more difficult to describe the complex formations.

Hart (1981) has introduced the concept; 'complex capacity', which is defined as the total amount of cations - in this case copper ions - which can be complex-bound by organic ligands. The capacity may furthermore be found as μmol/mg DOC (DOC= dissolved organic carbon).

Fig. 9.3 shows the conceptual diagram of the model applied in this case study. As can be seen it is only a modification of the model presented in Fig. 9.2.

The formation of copper compounds may be described by use of mass equations. The possible processes and their mass constants are shown in Table 9.2. Two conditions may be considered: aerobic or anaerobic. The solubility of copper-hydroxo-carbonate at $pH < 7.0$ is the determining process in the first case, while the solubility of copper oxide is the determining process at $pH > 7.0$. The solubility of copper sulphide (CuS) is the determining process under anaerobic conditions, while Cu_2S may be dominant under certain conditions.

The formation constant for humic acid complexes is determined in this case study to be $10^{5.0}$ by pH = 4.4 and $10^{5.9}$ by pH = 7.0. In each case-study it is necessary to determine this constant because the variability of humic compounds is very high from case to case.

An equilibrium model is developed by the use of mass equations for all the processes shown in Table 9.1. The concentrations of the actual ions are determined by chemical analysis. The concentration of humic compounds are determined as DOC, and the application of a constant ratio between humic compounds and DOC.

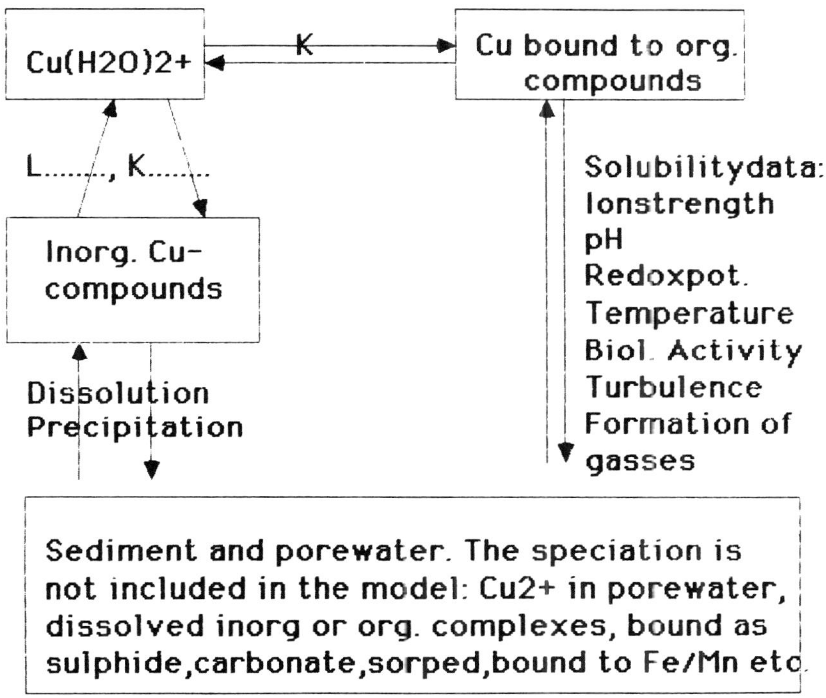

Fig. 9.3: Conceptual diagram of model.

The results of the equilibrium model used in the case of Frederiksborg Castle Lake are given in Figures 9.4 and 9.5.

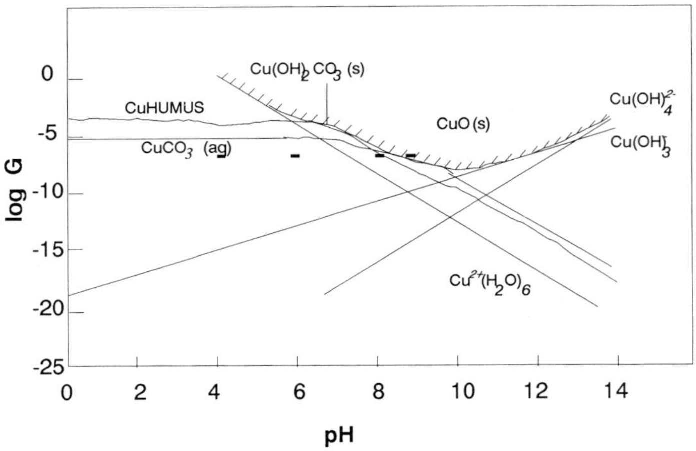

Fig. 9.4: Solubility diagram for Cu(II) based on this case study under aerobic conditions.

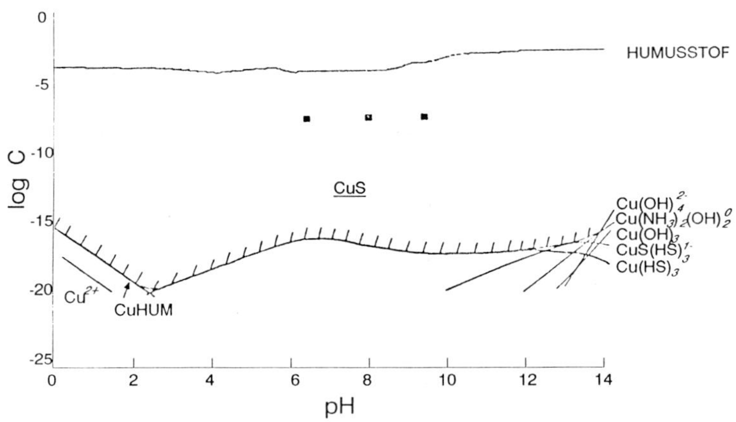

Fig. 9.5: Solubility diagram for Cu(II) based on this case study under anaerobic conditions.

Figure 9.4 shows the solubility of copper under aerobic conditions. Total carbonate is determined to be 0.001 M. The concentration of humic compounds in the water phase is pH-dependent and the analytical results are given in the computer program as a table function. CuHUM indicates the concentration of organic copper compounds. • shows the measured values which may be diretly compared with the theoretical values.

Figure 9.5 gives the results of the solubility as a function of pH for the anearobic case found by use of the equilibrium model. The total concentration of sulphide (the sulphide ion + the hydrogen sulphide ion + dihydrogen sulphide) is 0.0001M. Again • indicates the measured values and, as can be seen, in this case they are significantly different from the theoretical values found by use of the equilibrium model.

9.4 CONCLUSIONS AND FURTHER RESEARCH NEEDS

The model presented illustrates a typical model in environmental chemistry. It may be concluded that the model has not given predictions of the copper solubility with sufficient accuracy under anaerobic conditions, while the values found by aerobic conditions may be considered acceptable. The model is clearly too simple, particularly under anaerobic conditions. The assumptions, which may be questioned and therefore will require further research, are:

1) It is not certain that *an equilibrium model* can be used under all possible conditions.
2) *The sorption* is not considered in the model and in many cases may play a major role.
3) *Chemical reactions between copper ions and suspended matter* included suspended matter coated with organic compounds which may have formed surface complexes with copper ions are not included in the equilibrium model.
4) *Only 1:1 reactions between copper ions and humic compounds* are considered, while humic compounds have the possibility to bind two or more copper ions per molecule.

The case study demonstrates, however, how it is possible to synthesize present knowledge of a complex system and how the discrepancy between model and measurements may be used to set up a list of the questions that need an answer by further development of the model.

Table 9.1
Cu(II)- Equilibrium

	Equilibrium			log K (25°C, I=0)	
1	$Cu^{2+} + H_2O$	=	$CuOH^+$	-8	a
2	$2Cu^{2+} + 2H_2O$	=	$Cu(OH)_2^{2+} + 2H^+$	-10.95	a
3	$Cu^{2+} + 3H_2O$	=	$Cu(OH)_3^- + 3H^+$	-26.3	a
4	$Cu^{2+} + 4H_2O$	=	$Cu(OH)_4^{2-} + 4H^+$	-39.4	a
5	$Cu^{2+} + CO_3^{2-}$	=	$CuCO_3$ (aq)	6.75	b
6	$Cu^{2+} + 2CO_3^{2-}$	=	$Cu(CO_3)_2^{2-}$	10.01	b
7	$CuO + 2H^+$	=	$Cu^{2+} + H_2O$	7.65	a
8	$Cu_2(OH)_2CO_3$ (s) $+ 4H^+$	=	$2Cu^{2+} + 3H_2O + CO_2$ (g)	14.16	a
9	$Cu_3(OH)_2(CO_3)_2$ (s) $+ 6H^+$	=	$3Cu^{2+} + 4H_2O + 2CO_2$ (g)	21.24	a
10	Cu^{2+} + HUM	=	CuHUMUS	5.5	d
11	$Cu^{2+} + NH_3$	=	$Cu(NH_3)^{2+}$	5.8	c
12	$Cu^{2+} + 2NH_3$	=	$Cu(NH_3)_2^{2+}$	10.7	c
13	$Cu^{2+} + 3NH_3$	=	$Cu(NH_3)_3^{2+}$	14.7	c
14	$Cu^{2+} + 4NH_3$	=	$Cu(NH_3)_4^{2+}$	17.6	c
15	$Cu^{2+} + 2NH_3 + 2OH^-$	=	$Cu(NH_3)_2(OH)_2$	22.5	c
16	$Cu^{2+} + 3NH_3 + OH^-$	=	$Cu(NH_3)_3(OH)^+$	21.1	c
17	$Cu^{2+} + 3HS^-$	=	$Cu(HS)_3^-$	26.5	c
18	$Cu^{2+} + 4HS^-$	=	$CuS(HS)_3^{3-} + H^+$	18.12	c
19	CuS	=	$S^{2-} + Cu^{2-}$	-36.1	b

a P. Schindler (1967)
b Smith and Martell (1976)
c Sillén and Martell (1964, 1971)
d K.J. Andersen (19)

REFERENCES

Andersen, K.J., 1986. Exchange of Copper between Freshwater-sediment and Water. Ph.D. Thesis, Danmarks Farmaceutiske Højskole, Copenhagen.

Hart, B.T., 1981. Trace Metal Complexing Capacity of Natural Waters: A Review. Environ. Technol. Lett. 2: 95-110.

Schindler, P., 1967. Heterogenous Equilibria Involving Oxides, Hydroxides, Carbonates and Hydroxide Carbonates in Equilibrium Concepts in Natural Water Systems, Advances in Chemistry Series, No. 67, American Chemical Society, Washington, D.C.

Sillén, L.G. and A.E. Martell 1964, 1971. Stability Constants of Metal-Ion Complexes. Chem. Soc. Spec. Publ. 17 and 25, London.

Smith, R.M. and A.E. Martell, 1976. Critical Stability Constants Vol. 4: Inorganic Complexes. Plenum Press.

CHAPTER 10

A MERCURY MODEL FOR MEX BAY, ALEXANDRIA

by

S.E. Jørgensen, Aboul Dahab and U. Grydersen

10.1 DESCRIPTION OF THE MERCURY PROBLEM IN MEX BAY

Mex Bay is located west of Alexandria and is suffering from serious pollution problems due to discharge of waste water from many heavy industries, such as a cement plant, tanneries, an oil refinery and a chlorine alkali plant.

The most serious pollution problem of the bay is probably the mercury contamination of fish. The concentration of mercury in most fish caught in the bay exceed the limit for human food set by WHO (1 ppm).

Figure 10.I shows a map of Mex Bay. The surface area is 29 km^2 and the mean depth is 10 m.

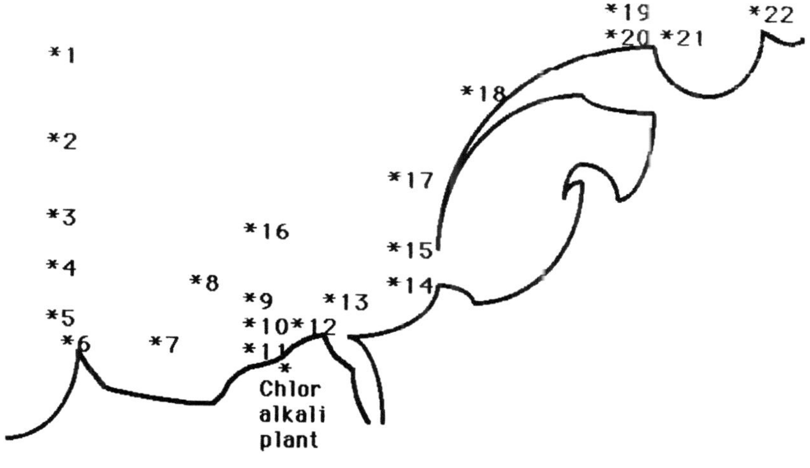

Fig. 10.1: Map of Mex Bay

A comprehensive investigation of the mercury pollution of the bay has been carried out at Alexandria University. The stations used for these examinations are shown on the map and the distances from the municipal waste water outlet are given in Table 10.1. The results on which the development of the model is based are published in the following references: Aboul Dahab (1985), Aboul Dahab et al (1984), El-Gindy et al (1985), El-Rayis et al (1984) and Halim et al (1984).

Table 10.1
State variables

	State variable	Unit	Comments
1	Salinity	%	Measured for all stations, and in different depths
2	Hg - inorganic	µg/l	Hg - inorganic
3	Hg - organic	µg/l	Hg - organic
4	Hg - total dissolved	µg/l	Hg - total dissolved is the sum of Hg - inorganic and Hg - organic
5	Hg - particulate	µg/l	Hg - particulate
6	Hg - total	µg/l	Hg - total is the sum of Hg - total dissolved and Hg - particulate
7	Inorganic Hg plankton	µg/kg WW	Hg inorg. in plankton
8	Total Hg - plankton	µg/kg WW	Hg total in plankton
9	Inorganic Hg - Pelagic fish	µg/kg WW	5 different forms of Pelagic fish were examined
10	Total Hg - Pelagic fish	µg/kg WW	Is measured in all the five species in muscle (flesh)
11	Inorganic Hg - benthic fish	µg/kg WW	Two species of benthic fish were examined
12	Total Hg - Benthic fish	µg/kg WW	
13	Inorganic Hg - Filter feed fish	µg/kg WW	Two species of filter feed fish were examined
14	Total Hg - Filter feed fish	µg/kg WW	Is measured in muscle (flesh)
15	Inorganic Hg - Conv. fish	µg/kg WW	Two species of Conv. fish were examined
16	Total Hg - Conv. fish	µg/kg WW	Is measured in muscle
17	Inorganic Hg - Benthic invertebrates	µg/kg WW	Two species of benthic invertebrates were examined
18	Total Hg - Benthic invertebrates	µg/kg WW	
19	Suspended matter	µg/l	
20	Suspended matter	µg C/l	Use of equation
21	Leachable Hg - sediment	µg/g DM	
22	Organic Hg - sediment	µg/g DM	
23	Total Hg - sediment	µg/g DM	
24	Hg f(weight(time)) in Pelagic fish		
25	Hg f(weight(time)) in filter feed fish		
26	Hg f(weight(time)) in Benthic fish		
27	Hg f(weight(time)) in Conv. fish		
28	Hg f(weight(time)) in Benthic invertebrates fish		

10.2 DESCRIPTION OF THE MODEL

A static model is used to describe the mercury contamination of the bay. The model is based on a mass balance for the bay. The principles are given in Figure 10.2, where the following processes are indicated:

1) Discharge of municipal and industrial waste water.
2) Atmospheric fall-out - dry and wet deposition.
3) Volatilization.
4) Exchange with the open sea.
5) Sedimentation.
6) Release from the sediment.
7) Fishery.

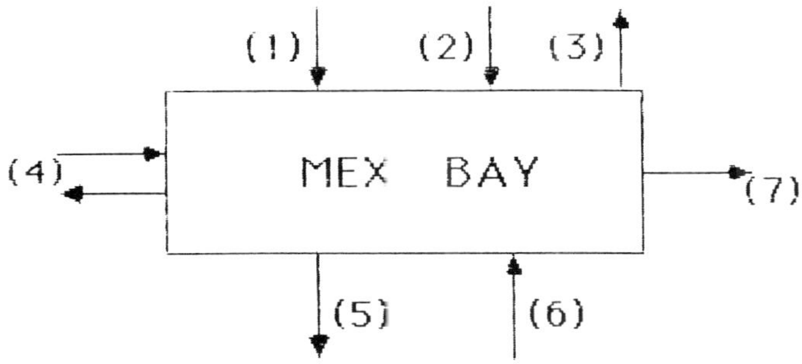

Fig. 10.2: Mass balance principle.

The model consists of five submodels which are interrelated as shown in the conceptual diagram Figure 10.3.

Submodel I deals with the mercury concentration in water. It describes the mercury concentration as a function of the distance from the outlet, see also Figure 10.2. The change in mercury concentration with time is the result of: (dispersion - advection - settling + methylation). As the discharge of mercury has been almost constant for several years, we are able to transform the partial differential equation to a differential equation. We are furthermore not interested in dayly fluctuations but in the general pollution picture. We get:

$$\frac{\partial Hgt}{\partial t} = 0 = D\frac{\partial^2 Hgt}{\partial x^2} - (\frac{Q}{AE})\frac{\partial Hgt}{\partial x} - (\frac{SR}{Depth}) Hgt + (M*MC) * Hgts$$

This equation needs to be in another form, so that the computer language, CSMP can handle it:

$$D\frac{d^2 Hgt}{dx^2} - (\frac{Q}{AE})\frac{dHgt}{dx} + (\frac{SR}{Depth}) Hgt - (M*MC) * Hgts =$$

$$\frac{d^2 Hgt}{dx^2} - (\frac{Q}{D*AE})\frac{dHgt}{dx} + (\frac{SR}{D*Depth}) Hgt - (\frac{M*MC}{D}) Hgts$$

where

Q	is the flow of water from The Umum Drain = $7.6*10^6$ (m^3/day)
AE	is the width of the bay multiplied by the depth = BBxDepth (m^2)
Depth	the mean depth of the bay = 10 m
SR	settling rate in water is calculated (m/day)
M	methylation rate = 0.01 (day^{-1}) or less
MC	Modification coefficient is the amount of organic carbon divided by the highest value of organic carbon in the bay
D	diffusion coefficient = 10^9 (m^2/day) or less

Submodel 2 considers the concentration of suspended matter in water. It describes the concentration of suspended matter as function of the distance from the outlet. It is again possible to tranform a partial differential equation to a differential equation, as the discharge of suspended matter has been constant for a longer period. We are furthermore not interested in the changes on a day-to-day basis, but on the general pollution picture of the Mex Bay. The concentration of suspended matter is the result of: (dispersion - advection - settling):

$$\frac{\partial TSM}{\partial t} = 0 = D\frac{\partial^2 TSM}{\partial x^2} - (\frac{Q}{AE})\frac{\partial TSM}{\partial x} - (\frac{SR}{Depth}) TSM$$

This equation needs to be in another form, so that the computer language, CSMP can handle it:

$$D\frac{d^2 TSM}{dx^2} - (\frac{Q}{AE})\frac{dTSM}{dx} + (\frac{SR}{Depth}) TSM =$$

$$\frac{d^2 TSM}{dx^2} - (\frac{Q}{D*AE})\frac{dTSM}{dx} + (\frac{SR}{D*Depth}) TSM$$

where

Q	is the flow of water from The Umum Drain = $7.6*10^6$ (m^3/day)
AE	is the width of the bay multiplied by the depth = BBxDepth (m^2)
Depth	the mean depth of the bay = 10 m
SR	settling rate in water is calculated (m/day)
D	diffusion coefficient = 10^9 (m^2/day) or less
TSM	the amount of Total Suspended Matter

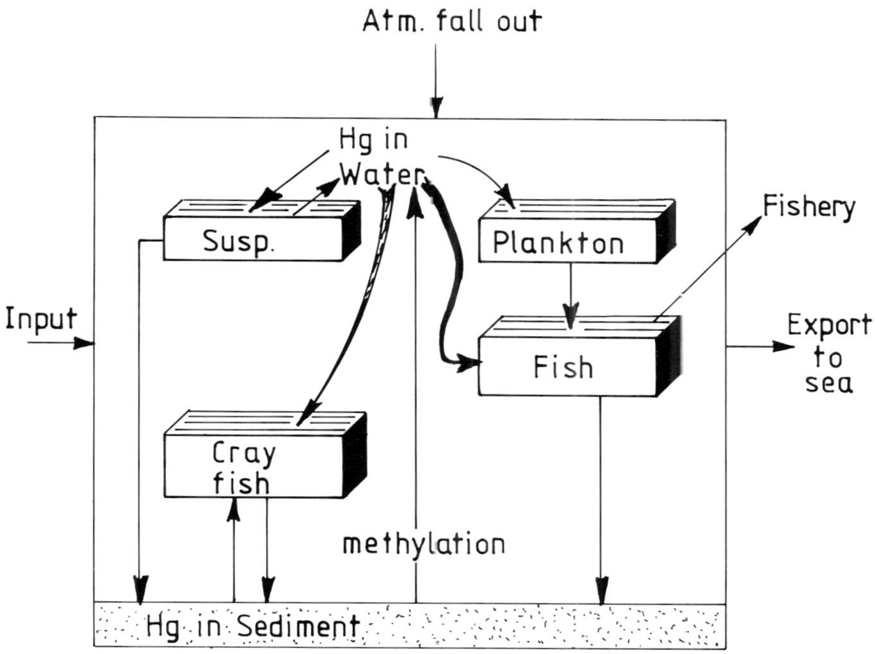

Fig. 10.3: The total model.

Submodel 3 describes the concentration of mercury in phytoplankton. The model distinguishes between organic mercury and inorganic mercury in phytoplankton. They are both described simply as a concentration factor * the concentration in water.

Submodel 4 deals with mercury in the sediment. The concentration in the sediment is a result of the settling (from submodel 1) and the methylation (also described in submodel 1). As the mercury concentration in the sediment is a function of these two processes, which are considered constant with the time (see again submodel 1), the concentration in the sediment is considered a constant at a given station - it is only dependent on the distance from the outlet and the depth.

Submodel 5 considers the mercury in fish and distinguishes between inorganic and organic mercury. The mercury concentration of fish, HgF is determined by:

1) the uptake from water: $= CF^{*}$ Hg in water* dw/dt
2) the uptake from food $a^{*}w^{b*}$ Hg in food* eff
3) the excretion: excretion coefficient* Hg in fish

where CF is the concentration factor (fish/water), w is the weight of the fish, which implies that dw/dt is the growth of the fish, a and b are characteristic constant describing the food uptake by the fish, eff is the efficiency of the mercury uptake from the food (it is different for inorganic and organic mercury). The change in mercury concentration of the fish is determined by:

d HgF/ dt = uptake from water + uptake from food - excretion.

The growth of the fish is found by:

$$dw/dt = a^{*}w^{b} - r^{*}w^{c}$$

where a and b are constants mentioned above, while r and c are other constants. In accordance with several investigations, b = 0.68 and c = 0.8.

Two possible food chains are included in the model, see Figure 10.4. For each station the mercury concentration of the sediment and the phytoplankton is dertermined by use of submodels 3 and 4. A probability generator determines in which of the station the "average" filter feeders *(Sardina pilchardus)*, the "average benthic invertebrates *(Penaeus kerathurus)* and the "average" Pelagic fish *(Boops boops)* are at a given day.

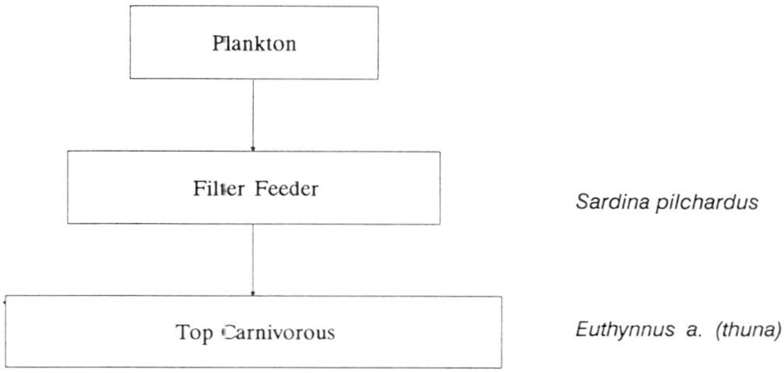

Fig. 10.4A: Food chain A

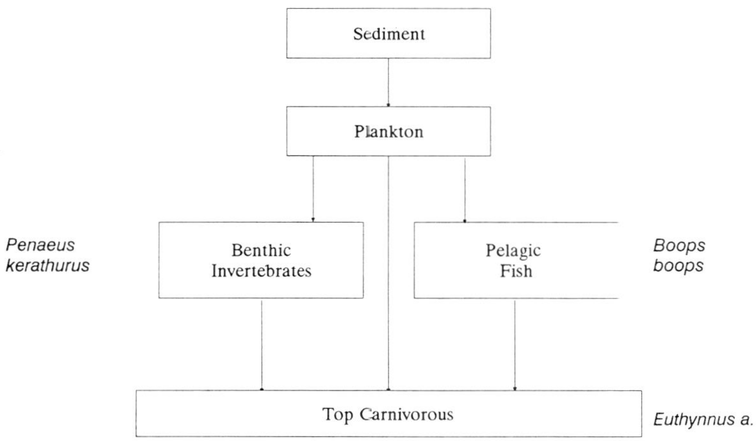

Fig. 10.4B: Food chain B.

The station determines the mercury concentration of the food for these three species. Their concentrations are currently determined by use of the above mentioned equations and the concentration of the carnivorous predators is determined by use of the same set of equations, but now using the mercury concentration of their average food sources. The ratio of the three species, that comprise the food, is determined by use of analysis of the stomach contents of the fish compared with general knowledge of the species preferred food items.

The state variables and forcing functions of the model are listed in Tables 10.1 and 10.2.

Table 10.2
Forcing functions

	Forcing function	Unit	Comments
1	Wind	km/hr	Monthly mean scalar wind speed, is measured as an average over 20 years. Alexandria Meteorological station
2	Effluent 1	m^3/day	Umum Drain has a flow of $7*10^6$ m^3/day
3	Effluent 2	m^3/day	Industrial waste water from Chlor Alkali Plant has a flow of $35*10^3$ m^3/day (O. Aboul Dahab, 1985)
4	Effluent 1 Hg - inorganic	µg/l	The inorganic Hg of Umum Drain is measured as dissolved reactive mercury
5	Effluent 1 Hg - organic	µg/l	The organic Hg of Umum Drain is measured as dissolved organic
6	Effluent 1 Hg - particulate	µg/l	The particulate Hg of Umum Drain is measured as particulate
7	Effluent 1 suspended matter	µg/l	The suspended matter of Umum Drain is measured
8	Effluent 2 Hg - inorganic	µg/l	The inorganic Hg of the Chlor Alkali effluent is measured as dissolved reactive
9	Effluent 2 Hg - organic	µg/l	The organic Hg of the Chlor Alkali effluent is measured as dissolved organic
10	Effluent 2 Hg - particulate	µg/l	The particulate Hg of the Chlor Alkali effluent is measured
11	Effluent 2 suspended matter	µg/l	The suspended matter of the Chlor Alkali effluent is measured as particulate
12	Sea water temperature	°C	Water temperature is measured at different depths
13	Atmospheric fall-out f(wind)	µg Hg/m^2 day	
14	Sediment composition	% sand, % mud	Sediment composition is measured
15	Aerobic/anaerobic conditions in sediment		Is determined
16	Open sea Hg	µg/l	Station 1 = open sea
17	Open sea salinity	%	Station 1 = open sea
18	Open sea suspended matter	µg/l	Station 1 = open sea
19	Settling rate (net)	mm/year	Is determined by means of sediment analysis
20	Density f(salinity, temp)	kg/l	Salinity and temperature are measured. Density is f(salinity, temp)
21	Air temperature	°C	Air temperature is measured
22	Depth	m	Depths are measured for all stations
23	Precipitation		Tables available

10.3 MODEL RESULTS

The model has at this stage only been calibrated. Submodel 1 and 2 are second-order differential equations and the the concentration of mercury Hgt and suspended matter TSM at x = 0 and d Hgt/dx and dTSM/dx at x = 0 have therefore been included in the calibration. Values based upon the measurements are used as initial guesses. The initial guesses of the settling rates are found on the basis of sediment analysi according to the method used in chapter 7. The Chlor Alkali plant started its production in 1950 and the settling rates found on the basis of the sediment profiles - Figure 7.2 illustrates an example - are shown in Table 10.3.

Table 10.3
Settling rate (1960)

Station	SR (cm/year)	SR (g/m^2 day)
7	0.81	24.4
8	-	-
9	0.93	25.8
10	0.93	25.8
11	0.93	25.8
12	0.81	24.4
13	0.81	24.4
14	0.81	24.4
15	0.70	19.4
16	0.81	24.4
17	0.93	25.8
18	0.93	25.8
		24.6

Figures 10.5 and 10.6 show the model results compared with the measured values for submodels 1 and 2.

Figures 10.7 and 10.8 give the results of the mercury content of the fish species *Euthynnus alletteratus and Boops boops.* The accordance between model results and measured values is acceptable. As seen, the tuna fish will exceed a mercury concentration of l mg/kg at a weight of 350 g. Results of a simulation for the tuna fish is shown in Figure 10.9. The mercury concentration versus the weight is shown on the assumption of a 90% reduction of the mercury discharge. The result shows that this reduction gives a satisfactory low mercury concentration in the tuna fish and these results should be used in environmental management.

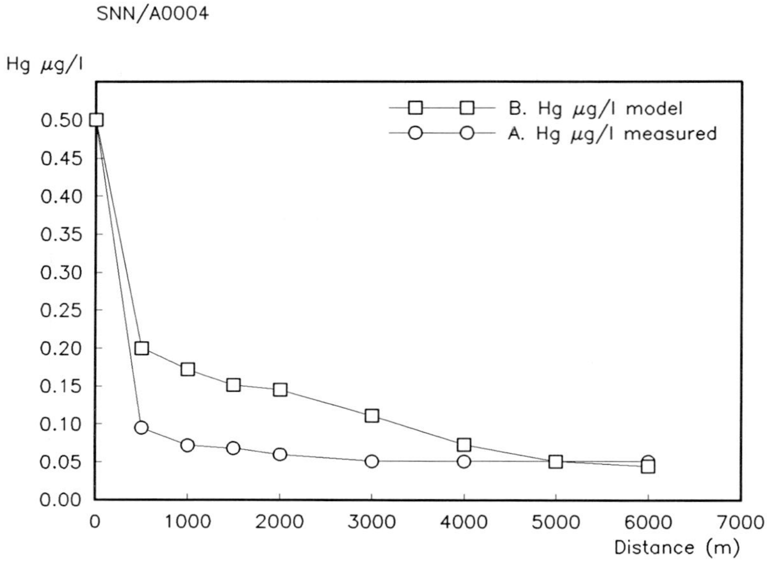

Fig. 10.5: Results of submodel 1. Hg as function of distance. A: measured values. B: model results.

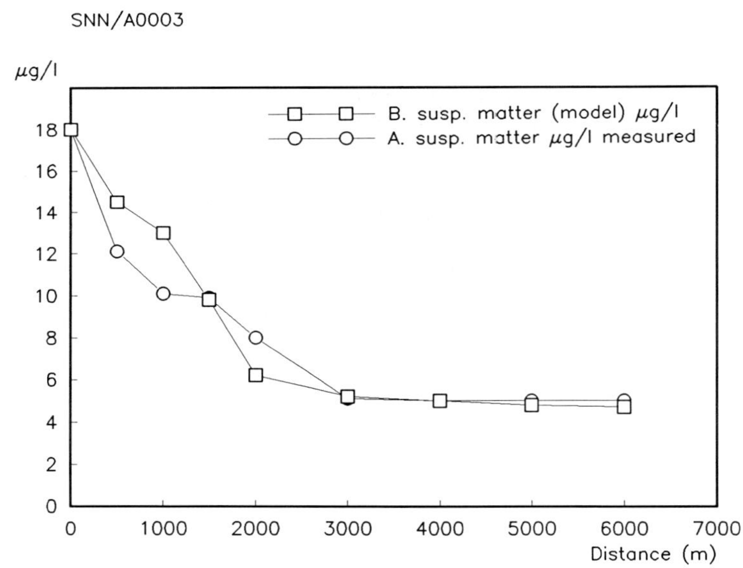

Fig. 10.6: Results of submodel 2. Total suspended matter as function of distance. A: measured values. B: model results

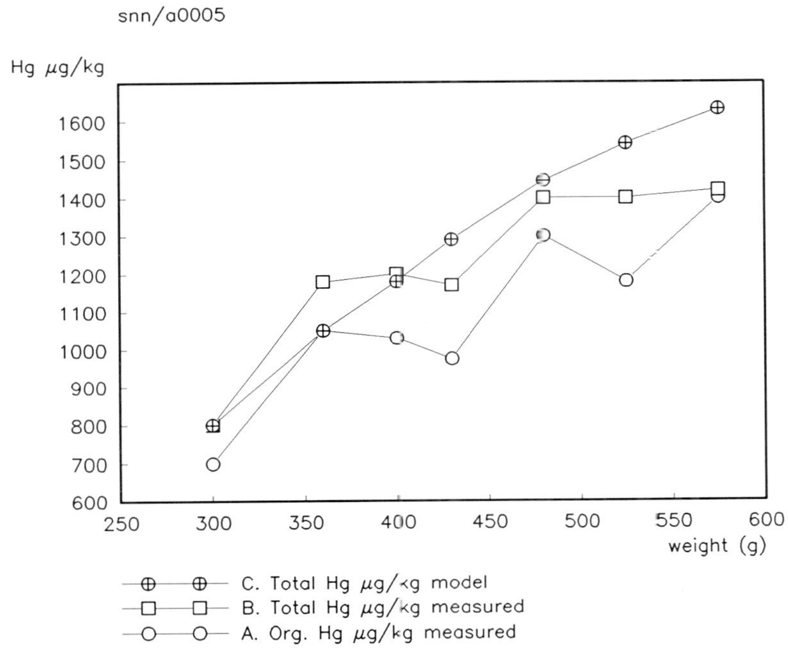

Fig. 10.7: Mercury concentration in *Euthynnus alletteratus* (μg/kg) as function of weight. A gives organic Hg in μg/kg (measured), B gives total Hg in μg/kg (measured) C shows the model results of total Hg in μg/kg.

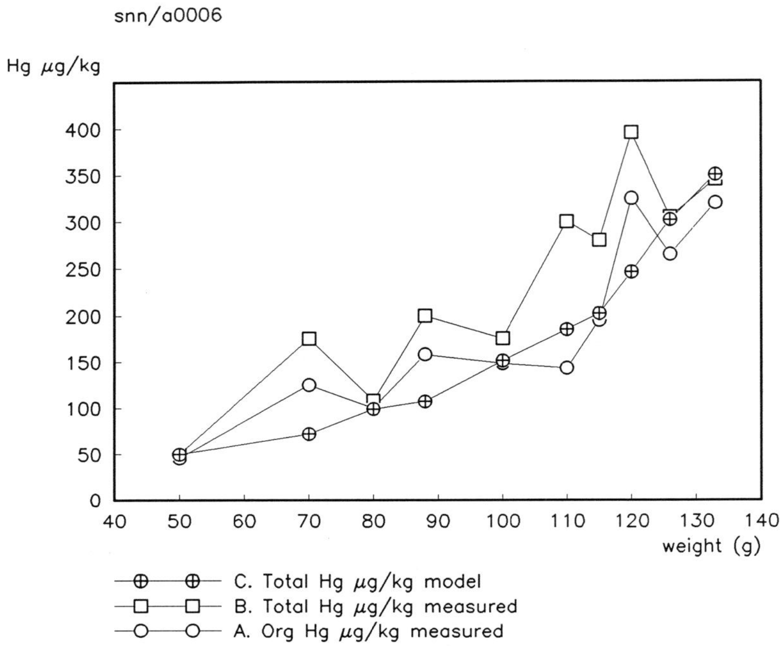

Fig. 10.8: Mercury concentration in *Boops boops* (μg/kg) as function of weigth. A gives organic Hg in μg/kg (measured) B gives total Hg in μg/kg (measured) C shows the model results of total Hg in μg/kg.

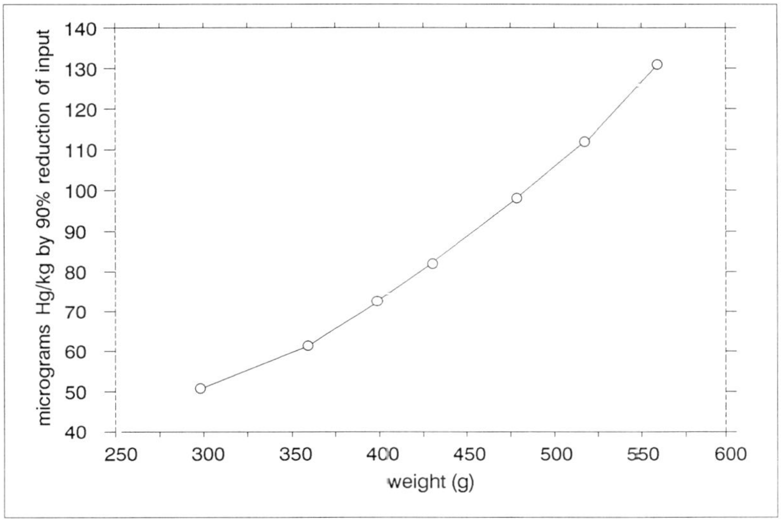

Fig. 10.9: Mercury concentration in Tuna (μg/kg) as function of weigth. A gives organic Hg in μg/kg (measured) B gives total Hg in μg/kg (measured) C shows the model results of total Hg in μg/kg.

10.4 CONCLUSIONS AND FURTHER RESEARCH NEEDS

The model applied in this case study is rather simple compared with the complex biological and hydrodynamic processes responsible for the mercury concentrations of the fish species, which are the most central state variables. However, an acceptable accordance between measured values and model values is found, although submodel 1 does not give an acceptable fit for the relationship between mercury concentration and the distance from the outlet probably due to a too simple description of the hydrodynamics.

The model is an illustrative example of what can be achieved by use of a simple model, resulting from considerations of where simplification can be made and what are the most essential processes and state variables. If the experience gained by development of this model and the chromium model presented in chapter 7 is used to set up a procedure for development of a management model for the control of heavy metal pollution in aquatic ecosystems, we would recommend the following procedure:

1. A relationship between the concentration in water and/or in sediment is developed by use of Fick's second law. This submodel has probably the lowest accuracy of the included submodels and if higher accuracy is required further research on the hydrodynamics of the system should be implemented for the improvement of this submodel.
2. The parameters in this relationship are found by use of determinations of the heavy metal concentrations in water and sediment. The concentration profiles of heavy metal in the sediment are of particular interest, as they may be used to determine the net annual sedimentation and the settling rate.
3. The concentration of heavy metal in the species with high level contamination can be determined by use of concentration factors and description of bioaccumulation. If a description of the increasing concentration with the growth is of importance, growth equations are included in this submodel.

REFERENCES

Aboul Dahab, O.M.T., 1985. Chemical cycle of Inorganic Pollutants in the Ecosystem West of Alexandria between Anfoushy and Agamy. Thesis for O.M.T. Aboul Dahab, University of Alexandria.

Aboul Dahab, O.M.T., Halim, Y., El-Rayis, O.A., 1984. Mercury Species in Coastal Marine Organisms from different Trophic Levels West of Alexandria. FAO Fisheries Report No. 325 Supplement. pp. 1-7.

El-Gindy, A., Aboul Dahab, O.M.T., Halim, Y., 1985. Preliminary Estimates of Water and Trace Metal Balances in Mex Bay West of Alexandria.

El-Rayis, O.A., Halim, Y., Aboul-Dahab, O.M.T., 1984 Total Mercury in the Coastal Marine Ecosystem West of Alexandria. FAO Fisheries Report No. 325 Supplement pp. 58-72.

Halim, Y., Aboul Dahab, O.M.T., El-Rayis, O.A., 1984. Chemical Forms of Mercury in Flesh, Gills and Liver from Fish Species of different Habits from two Localities West of Alexandria. FAO Fisheries Report No. 325 Supplement pp. 99-103.

CHAPTER 11

MODELLING POLLUTANT EXCHANGE BETWEEN PLANT AND ENVIRONMENT: UPTAKE AND METABOLISM OF SULFUR DIOXIDE BY DIFFERENT LEAF CELL COMPARTMENTS

by
K. Siebke, F.W. Badeck, G.H. Kohlmaier, M. Plöchl & C. Wientzek

11.1 INTRODUCTION

The computer model we present here describes the diffusive transport of sulfur dioxide into plant leaves and its subsequent solvation, dissociation, and metabolization. The need to elaborate such a model arose from the attempt to simulate the effects of gaseous air pollutants on plant metabolism - a potential cause of forest die-back (Waldsterben). For more than a hundred years it has been a well known fact that sulfur dioxide in high concentrations may cause the death of the vegetation near industrial sites such as ore smelter or power plants. The contribution of SO_2, and other sulfur oxides as well as sulfuric acid to the formation of acid rain is well known, too. But up to now the specific cause-effect relationships of air pollution and forest die-back, especially in areas of relatively low pollutant concentrations are not well understood. The interest therefore focuses on synergisms of several pollutants, interactions between organs of the living plant and the soil, as well as changes in the morphological and physiological state of the organism. In order to investigate these phenomena it is necessary to determine given pollutant concentrations at the site of their physiological action. The aim of the current model is to provide information on that topic. The model structure may also be adapted to describe any gas exchange processes of plant leaves and their gaseous environments.

An important progress in **mathematical modelling of gas exchange processes** has been made by Gaastra (1959), who presented a model to calculate the relation of carbon dioxide uptake and diffusion resistance of photosynthesizing leaves. As with the following models he described the gas transport according to Fick's laws of diffusion. The model structure implies a simplification of leaf anatomy; the boundary layer, epidermis, intercellular air spaces, and mesophyll are regarded as stratified layers which exhibit specific resistances to the diffusion of CO_2.

Subsequently several models have been developed which comprised descriptions of further details. For example, Lommen et al. (1971) took into account the release of carbon dioxide by the mitochondrial respiration. Parkhurst (1977) presented a model reflecting the three-dimensional structure of the mesophyll. Nobel (1974) distinguished the resistance of cellular and intracellular membranes. Bennet et al. (1973) adopted the model structure for the **simulation of gaseous pollutant sorption by leaves**.

Until present modelling of air pollutant uptake did not consider the several components of mesophyll resistance. Membrane resistances and the effect of intracellular gas concentration on gas flux have been summarized as 'mesophyll resistance'. In the current model these components are considered separately: the short distance transport across plasmalemma, chloroplast envelope and tonoplast, as well as, dissociation and metabolization of SO_2. A similar model has been elaborated by Laisk et al. (1988a,b).

11.2 MODEL DESCRIPTION

The mathematical model consists of a set of coupled linear differential equations. When *changes in pH* are incorporated as well, a quadratic differential equation is incorporated. The gas transport between free air space surrounding the leaf and the leaf sink sites is determined by the diffusion potential of the gas (gas concentration gradient) and the resistance of the pathway to the gas transfer. The specific SO_2 flux (J) is proportional to the gas concentration gradient (c_0-c_i), and inversely proportional to the resistance to transfer (R) as it is described by Fick's law:

$$J = 1/R\,(c_0 - c_i) \qquad (11.1)$$

The main path for the uptake occurs via the stomates. Uptake of gaseous SO_2 across the cuticle plays a minor role because the resistance of the cuticle is about 10^4 times that of the stomatal resistance. Lendzian reported a cuticular resistance to SO_2 of $8.2 \cdot 10^3$ s/cm for the tomato, *Lycopersicon esculentum* (1984). Besides the stomatal resistance (R_3) the total resistance for the gaseous diffusion of SO_2 into plant leaves is made up of the resistance of the leaf boundary layer (R_2) and of the intercellular air space (R_4), see Fig. 11.1.

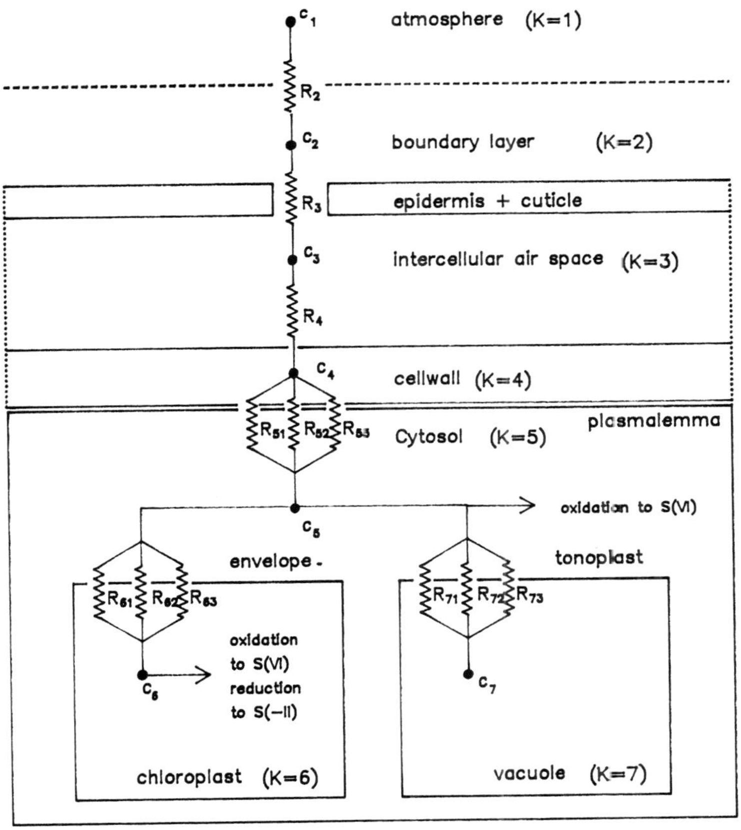

c_K: SO_2 bzw. S(IV)–concentration in compartment K
R_{Ki} : resistance between compartment K–1 (K–2 respectively) and K for the molecule species i

i=1: $SO_2 \cdot H_2O$ i=2: HSO_3^- i=3: SO_3^{--}

Fig. 11.1: Simplified leaf anatomy and S(IV) flux model.

On entering the cell walls from the intercellular air space SO_2 dissolves in their liquid phase. The equilibrium concentration of sulfur dioxide is described by Henry's law. The concentration in the liquid phase (c_l) is proportional to the partial pressure of the gas in the atmosphere (p_g), with H being Henry's constant:

$$c_l = H * p_g \tag{11.2a}$$

or equivalently by introducing the dimensionless partition coefficient $H_v = H^* V_m * P_{atm}$ (V_m = molar volume, P_{atm} = total atmospheric pressure):

$$C_l = H_v * C_g \tag{11.2b}$$

In the liquid phase sulfur dioxide not only dissolves, but also dissociates, depending on the pH, into HSO_3^-, SO_3^{2-} and protons, respectively. At given a pH of 5.7 for the liquid phase of the cell walls, stabilized by a buffer, the bulk of the imported sulfur(IV) will occur as hydrogen sulfite HSO_3^- and sulfite SO_3^{2-}. But the part of S(IV) which is reflected in the concentration gradient between the sulfur dioxide in the gaseous phase and the cell walls is made up only by the physically dissolved SO_2. Therefore the equation for the specific flux per leaf area (J) can be written similar to equation (11.1), whereby the S(IV) concentration in the liquid phase of the cell wall (c_4) needs to be divided through the coefficient of solubility (Hv) and multiplied with the mole fraction of SO_2 physically dissolved ($X_{4\ 1}$):

$$J = (c_0 - X_{4\ 1} * Hv^{-1} * c_4) / R \tag{11.3}$$

S(IV) is transported from the cell walls into the cytosol across the plasmalemma. This transport is assumed to be a diffusion process whereby the resistance of membranes for the diffusion of the anions exceeds the resistance for the diffusion of the undissociated species manifold. Following Laisk et al. (1988a) it is assumed that there will be a transfer of protons parallel to the diffusion of the ions in order to compensate for their electric charge. The diffusion across the plasmalemma has to be calculated separately for every molecular species (i) of S(IV) taking into account its mole fraction in the liquid phase of the cell wall (X_{4i}) and in the cytosol ($X_{5\ i}$) and its resistance across the plasmalemma ($R_{5\ i}$). Now the equation for the concentration change in the cell wall per time can be written as follows:

$$\frac{dc_4}{dt} = \frac{1}{d_4} \left[\frac{c_1 - X_{4\ 1} * Hv^{-1} * c_4}{R_2 + R_3 + R_4} - \left[c_4 \sum_{i=1}^{3} \frac{X_{4\ i}}{R_{5\ i}} - c_5 \sum_{i=1}^{3} \frac{X_{5\ i}}{R_{5\ i}} \right] * \frac{A_5}{A} \right] \tag{11.4}$$

where i=1 stands for SO_2, i=2 for HSO_3^-, i=3 for SO_3^{2-},

d_4 thickness of the layer representing the cell wall and A_5/A is the area of the plasmalemma per leaf area.

The changes of the concentration of S(IV) in cytosol chloroplasts, and vacuole are calculated similarly assuming diffusive transport across the chloroplast membranes and tonoplast (Fig. 11.1). The respective degrees of dissociation of S(IV) in each compartment are mediated by the compartmental proton concentration (acidity). The compartmental buffer systems are simulated as one single buffer which stabilizes the pH-values at 7.4 in the cytosol, 8.0 in the chloroplasts, and 5.8 in the vacuole. The simplest variation of the model can be run by assuming an ideal buffer with an infinite buffer capacity maintaining constant pH-values.

Aside from these processes governing the uptake of SO_2, the overall intracellular concentrations of the S(IV) species are influenced by the rates of the metabolizing reactions. The oxidation of S(IV) to S(VI), i.e. sulfate - mediated by a radical chain mechanism - in the cytosol and chloroplasts is modelled as a simple first order reaction with the oxidation coefficient ko_5 in the cytosol and ko_6 in the chloroplasts. In contrast the reduction of S(IV) and S(VI) to S(-II) in the chloroplasts through a series of enzyme catalyzed reactions is reflected in the model by a simple Michaelis-Menten term. Since there are some enzymes which are required for both, S(IV)- and S(VI)-reduction, we assume a competitive inhibition between the two species. Sulfur of the oxidation state S(-II) is a component of amino acid synthesis, or can be alternatively emitted as H_2S or $(CH_3)_2S$ up to a certain degree (Rennenberg, 1984).

Under these conditions the following equations describe the changes of the S(IV)-concentration in the cytosol (c_5), the chloroplasts (c_6) and the vacuole (c_7):

$$\frac{dc_5}{dt} = \frac{1}{d_5}\left[\frac{A_5}{A}\sum_{i=1}^{3}\frac{X_{4\,i} * c_4 - X_{5\,i} * c_5}{R_{5\,i}} - \frac{A_6}{A}\sum_{i=1}^{3}\frac{X_{5\,i} * c_5 - X_{6\,i} * c_6}{R_{6\,i}} - \frac{A_7}{A}\sum_{i=1}^{3}\frac{X_{5\,i} * c_5 - X_{7\,i} * c_7}{R_{7\,i}}\right] - ko_5 * c_5 \tag{11.5}$$

$$\frac{dc_6}{dt} = \frac{A_6}{d_6 * A}\sum_{i=1}^{3}\frac{X_{5\,i} * c_5 - X_{6\,i} * c_6}{R_{6\,i}} - ko_6 * c_6 - \frac{Red_6 * c_6}{Kmr_6 + c_6 + \frac{Kmr_6}{Kmra_6} * [SO_4^{2-}]} \tag{11.6}$$

$$\frac{dc_7}{dt} = \frac{A_7}{d_7 * A} \sum_{i=1}^{3} \frac{X_{5\,i} * c_5 - X_{7\,i} * c_7}{R_{7\,i}} \tag{11.7}$$

A_6 : Area of the chloroplast envelope
A_7 : Area of the tonoplast
d_5 : Thickness of the layer representing the cytosol
d_6 : Thickness of the layer representing the chloroplasts
d_7 : Thickness of the layer representing the vacuoles
Kmr_6 : Km-value (Michaelis-Menten) for the reduction of S(IV)
$Kmra_6$: Km-value (Michaelis-Menten) for the reduction of S(VI)
$R_{6\,i}$: Resistance of the chloroplast envelope for the molecule species i
$R_{7\,i}$: Resistance of the tonoplast for the molecule species i
$X_{6\,i}$: Mole fraction for the molecule species i in the chloroplasts
$X_{7\,i}$: Mole fraction for the molecule species i in the vacuole
Red_6 : Maximal reduction rate for S(IV) in the chloroplasts

The system of equations is simplified under the assumption of a very fast equilibration between the outside SO_2 concentration and the liquid phase of the cell wall (steady state assumption $dc_k/dt = 0$ for $k = 1,2,3,4$). Under that assumption the S(IV) concentration in the cytosol (c_5) becomes directly dependent on the concentration of sulfur dioxide in the surrounding air space:

$$\frac{dc_5}{dt} = \frac{1}{d_5} \left[\frac{C_1 - g * c_5}{R_{eff}} - \frac{A_6}{A} \sum_{i=1}^{3} \frac{X_{5\,i} * c_5 - X_{6\,i} * c_6}{R_{6\,i}} - \frac{A_7}{A} \sum_{i=1}^{3} \frac{X_{5\,i} * c_5 - X_{7\,i} * c_7}{R_{7\,i}} \right] - ko_5 * c_5 \tag{11.8}$$

with:
$$g = X_{4\,1} * Hv^{-1} * \sum_{i=1}^{3} \frac{X_{5\,i}}{R_{5\,i}} * \left[\sum_{i=1}^{3} \frac{X_{4\,i}}{R_{5\,i}} \right]^{-1}$$

and
$$R_{eff} = R_2 + R_3 + R_4 + X_{4\,1} * Hv^{-1} * \left[\sum_{i=1}^{3} \frac{X_{4\,i}}{R_{5\,i}} * \frac{A_5}{A} \right]^{-1}$$

The transport of SO_2 into the cytosol is governed by an effective resistance R_{eff}. The high plasmalemma resistance of 80 s/cm (Pfanz et al., 1987a), being a component of R_{eff} is nearly negligible both because the surface area of the plasmalemma is about 11 times greater than the leaf surface area and the concentration of dissolved S(IV) species in the cell wall is much higher than the corresponding outside concentration of SO_2. When considering the concentration gradient between gaseous SO_2 of the surroundings (c_1) and the concentration of total dissolved S(IV) species in the cytosol (c_5), we need to multiply c_5 by the factor g reflecting the accumulation of S(IV) in the liquid phase as a result of solubility and dissociation.

As stated above, the simplest way to calculate the model is to assume an ideal buffer with an infinite buffer capacity maintaining a constant pH-value. Laisk et al. (1987a) developed a method to calculate the acidification of cell compartments caused by the dissociation of the sulfurous and sulfuric acid taking into account the cellular buffer systems. The feed back between pH-value and dissociation is reflected in the model by calculating alternately the pH-values, the mole fractions and the transport rates. The law of mass action delivers the equations for the calculation of the mole fractions ($X_{k\,i}$):

$$X_{k\,1} = \left[1 + \frac{K_1}{[H^+]} + \frac{K_1 * K_2}{[H^+]^2} \right]^{-1} \tag{11.9}$$

$$X_{k\,2} = \left[\frac{[H^+]}{K_1} + 1 + \frac{K_2}{[H^+]} \right]^{-1} \tag{11.10}$$

$$X_{k\,3} = \left[\frac{[H^+]^2}{K_1 * K_2} + \frac{[H^+]}{K_2} + 1 \right]^{-1} \tag{11.11}$$

K_1, K_2: Equilibrium constants of the first and second dissociation step of the sulfurous acid.

According to the Henderson-Hasselbalch equation the proton concentration in a buffered solution can be calculated as follows:

$$[H^+] = Eb\,[HA] / [A^-] \tag{11.12}$$

Eb : Equilibrium constant of the buffer
A^- : Buffer anion
HA : Protonated buffer

Since every HSO_3^- ion liberates one proton, every SO_3^{2-} and SO_4^{2-} two protons which protonate buffer anions, their presence causes a shift in the buffer equilibrium increasing the concentration of the protonated buffer and decreasing the concentration of free buffer anions. The proton concentration is therefore given by:

$$[H^+]_K = Eb_K * \frac{[HA]_K + X_{K\,2} * c_K + 2 * X_{K\,3} * c_K + 2 * [SO_4^{2-}]_K}{[BF]_K - ([HA]_K + X_{K\,2} * c_K + 2 * X_{K\,3} * c_K + 2 * [SO_4^{2-}]_K)} \qquad (11.13)$$

with $[BF] = [HA] + [A^-]$
$[BF]_k$: Buffer concentration in compartment k.

Additionally we have to consider the fate of sulfate generated by S(IV) oxidation. Sulfate can be transported actively into the vacuole (Cram 1983). Parallel to this transport sulfate can pass the tonoplast by diffusion, but the permeability is very low (Cram, 1983). The sulfate concentration in the vacuole ($[SO_4^{2-}]_7$) therefore is given by:

$$\frac{d[SO_4^{2-}]_7}{dt} = \frac{1}{d_7} \left[\frac{[SO_4^{2-}]_5 - [SO_4^{2-}]_7}{R_{74}} * \frac{A_7}{A} + \frac{Vt_7 * [SO_4^{2-}]_5}{Kmt_7 + [SO_4^{2-}]_5} * \frac{A_7}{A} \right] \qquad (11.14)$$

Kmt_7 : Km value for the sulfate transport into the vacuole
R_{74} : Resistance of the tonoplast against sulfate diffusion
Vt_7 : Maximal transport rate

The transport of sulfate between the chloroplast and the cytosol has been investigated by Mourioux & Douce (1979), and Hampp & Ziegler (1977). It is transported via a translocator. Several concepts about the nature of this translocator are under discussion. For the simulation of the sulfate transport across the chloroplast envelope we therefore use the simplification of a passive diffusion. As in the case of HSO_3^- and SO_3^{2-} transport here it is assumed that together with every sulfate ion two protons are transported too. The changes of the sulfate concentration in the cytosol and the chloroplasts are given by:

$$\frac{d[SO_4^{2-}]_5}{dt} = ko_5 * c_5 - \frac{1}{d_5} \left[\frac{[SO_4^{2-}]_5 - [SO_4^{2-}]_7}{R_{74}} * \frac{A_7}{A} + \frac{Vt_7 * [SO_4^{2-}]_5}{Kmt_7 + [SO_4^{2-}]_5} * \frac{A_7}{A} + \frac{[SO_4^{2-}]_5 - [SO_4^{2-}]_6}{R_{64}} * \frac{A_6}{A} \right] \qquad (11.15)$$

$$\frac{[SO_4^{2-}]_6}{dt} = ko_6 * c_6 - \frac{Reda_6 * [SO_4^{2-}]_6}{Kmra_6 + \frac{Kmra_6}{Kmr_6} * c_6 + [SO_4^{2-}]_6} - \frac{[SO_4^{2-}]_5 - [SO_4^{2-}]_6}{d_6 * Ra_6} * \frac{A_6}{A} \qquad (11.16)$$

R_{64} : Resistance of the chloroplast envelope against sulfate diffusion.
$Reda_6$: Maximal rate of reduction of SO_4^{2-} in the chloroplasts.

The parameters used in the model (cell compartment dimensions, resistances, dissociation constants, buffer concentrations, oxidation and reduction rates) have been adopted from Garsed & Read (1977), Pfanz et al. (1987a,b), Laisk et al. (1988a,b) and Seel (1965). A list of the parameters is given in Table 11.1.

It is a well known fact, that the factors influencing uptake, distribution, and metabolization of S(IV) are changing during the course of the day. For instance, the stomata of C_3-plants are open at day-time and closed during the night. The acidity of the chloroplast stroma changes, dependent on light intensity (Werdan et al., 1975). The rate of S(IV) and S(VI) reduction and the rate of S(IV) oxidation also changes with the light intensity (Wilson et al., 1978; Hällgren & Frederiksson, 1982; Rothermel & Alscher, 1985).

Changes of light intensity in the course of the day are modelled according to Richter (1985). Energy of the radiation per area and time is calculated as follows:

$$\sin(lh) = \sin(ll) * \sin(ld) + \cos(ll) * \cos(ld) * \cos\left[\frac{2\pi}{24} * (th+12)\right] \qquad (11.17)$$

$$ld = -0.408 * \cos\left[\frac{2\pi(td+10)}{365}\right] \qquad (11.18)$$

$$li = 640 * \sin(lh) * e^{[-cloud/\sin(lh)]} \qquad (11.19)$$

lh : Height of the sun
ld : Declination of the sun
ll : Geographical latitude
th : Time in hours
td : Day of the year
li : Light intensity (energy/area/time)
cloud : Parameter describing absorption by the atmosphere

Table 11.1
Standard parameter set

Parameter	Value	Comments
c_1	0.5 $\mu mol*m^{-3}$	SO_2 concentration in the ambient air
d_5	$4.95*10^{-4}$ cm	Effective thickness of the cytosol
d_6	$8.25*10^{-4}$ cm	Effective thickness of the chloroplasts
d_7	$4.95*10^{-3}$ cm	Effective thickness of the vacuole
A_5	11.35 cm^2	Surface area of the plasmalemma per cm^2 leaf area
A_6	13.6 cm^2	Surface area of the chloroplast envelope per cm^2 leaf area
A_7	9.7 cm^2	Surface area of the tonoplast per cm^2 leaf area
R_{2+4}	0.5 $s*cm^{-1}$	Resistance of boundary layer and intracellular air space to SO_2xH_2O
R_3	0.5 $s*cm^{-1}$	Minimal stomatal resistance to SO_2xH_2O
$R_{5\,1}$	80 $s*cm^{-1}$	Plasmalemma resistance to SO_2xH_2O
$R_{5\,2}$	$6*10^6$ $s*cm^{-1}$	Plasmalemma resistance to HSO_3^-
$R_{5\,3}$	$1*10^9$ $s*cm^{-1}$	Plasmalemma resistance to SO_3^{2-}
$R_{6\,1}$	90 $s*cm^{-1}$	Resistance of chloroplast envelope to SO_2xH_2O
$R_{6\,2}$	$1*10^7$ $s*cm^{-1}$	Resistance of chloroplast envelope to HSO_3^-
$R_{6\,3}$	$1*10^9$ $s*cm^{-1}$	Resistance of chloroplast envelope to SO_3^{2-}
$R_{6\,4}$	$1*10^9$ $s*cm^{-1}$	Resistance of chloroplast envelope to SO_4^{2-}
$R_{7\,1}$	80 $s*cm^{-1}$	Tonoplast resistance to SO_2xH_2O
$R_{7\,2}$	$6*10^6$ $s*cm^{-1}$	Tonoplast resistance to HSO_3^-
$R_{7\,3}$	$1*10^9$ $s*cm^{-1}$	Tonoplast resistance to SO_3^{2-}
$R_{7\,4}$	$1*10^9$ $s*cm^{-1}$	Tonoplast resistance to SO_4^{2-}
pH_4	5.7	Acidity of the cell walls
pH_5	7.4	Acidity of the cytosol
pH_6	8.0	Acidity of the chloroplasts
pH_7	5.8	Acidity of the vacuole
BF_5	$6.2*10^7$ $pmol*l^{-1}$	Buffer capacity of the cytosol
BF_6	$6.2*10^7$ $pmol*l^{-1}$	Buffer capacity of the chloroplast
BF_7	$3.0*10^7$ $pmol*l^{-1}$	Buffer capacity of the vacuole
Eb_5	$1.3*10^2$ $pmol*l^{-1}$	Equilibrium contant of the buffer within the cytosol
Eb_6	$3.2*10^1$ $pmol*l^{-1}$	Equilibrium contant of the buffer within the chloroplast
Eb_7	$5.0*10^3$ $pmol*l^{-1}$	Equilibrium contant of the buffer within the vacuole
K_1	$1.2*10^7$ $pmol*ml^{-1}$	First dissociation contant for SO_2xH_2O = HSO_3^- + H^+
K_2	$1*10^2$ $pmol*ml^{-1}$	Second dissociation constant for HSO_3^- = SO_3^{2-} + H^+
ko_5	$5*10^{-5}$ s^{-1}	Oxidation coefficient for S(IV) in the cytosol
ko_6 (lko_6)	$5*10^{-5}$ s^{-1}	Oxidation coefficient for S(IV) in the chloroplasts
$IKmox_6$	150 $W*ms^{-2}*s^{-1}$	MM-constant for the light dependant oxidation in the chloroplasts
Red_6	$5*10^{-2}$ pmol/ml*ms	Maximal reduction rate for S(IV) in the chloroplasts
$IKmr_6$	60 $W*ms^{-2}*s^{-1}$	MM-constant for light dependant S(IV) reduction in the chloroplasts
Kmr_6	$5*10^{+5}$ $pmol*ml^{-1}$	MM-constant for S(IV) reduction in the chloroplasts
$Reda_6$	$5*10^{-2}$ pmol/ml*ms	Maximal reduction rate for Sulfate in the chloroplasts
$IKmra_6$	60 ms^{-1}	MM-constant for light dependantSulfatereduction in the chloplasts
$Kmra_6$	$5*10^{+5}$ $pmol*ml^{-1}$	MM-constant for Sulfate reduction in the chloroplasts
Vt_7	$5*10^{-4}$ $pmol/cm^2*ms$	Maximal velocity of sulfate transport into the vacuole
Kmt_7	$1*10^4$ $pmol*ml^{-1}$	MM-constant for the sulfate transport into the vacuole
II	0.873	Geographical latitude (50°)
cloud	$0.1 \leq cloud \leq 0.18$	Parameter describing absorption by the atmosphere

Since the exact correlation of pH in chloroplasts of intact leaves and light intensity is unknown, it is approximated by a Michaelis-Menten-term:

$$[H^+]_6 = [H^+]_{6d} - \frac{IvH * li}{IKmH + li} \qquad (11.20)$$

$[H^+]_6$: Proton concentration in the chloroplasts
$[H^+]_{6d}$: Proton concentration in the chloroplasts during the night
IvH : Maximal reduction of proton concentration in the chloroplasts
IKmH : Light intensity with half maximal reduction of proton concentration

Similarly the change in rate of reduction of S(IV) and S(VI) has been modelled to change with light

$$rate[(S(IV) - S(-II)] = IRed_6 * \frac{c_6}{Kmr_6 + c_6 + \frac{Kmr_6}{Kmra_6}[SO_4^{2-}]} * \frac{li}{IKmr_6 + li} \qquad (11.21)$$

$$rate[(S(VI) - S(-II)] = IRed_6 * \frac{[SO_4^{2-}]_6}{Kmra_6 + [SO_4^{2-}]_6 + \frac{Kmr_6}{Kmra_6} c_6} * \frac{li}{IKmr_6 + li} \qquad (11.22)$$

$IRed_6$: Maximal rate of reduction of S(IV)
$IKmr_6$: Light intensity with half maximal reduction rate for S(IV)
IRedL6 : Maximal rate of reduction of S(VI)
$IKmra_6$: Light intensity with half maximal reduction rate for S(VI)

The oxidation rate in the cytosol is assumed to be constant, while it is changing in the chloroplasts following a Michaelis-Menten-term:

$$Rate\ [\ S(IV) - S(VI)\ ] = Iko_6 * \frac{li}{IKmox_6 + li} * c_6 \qquad (11.23)$$

Iko_6 : Maximal oxidation coefficient
$Ikmox_6$: Light intensity with half maximal oxidation coefficient

The formula describing the changes in the aperture of stomates has been adopted from Lommen et al. (1971):

$$\frac{1}{R_3} = \frac{li}{li + IKmR_3} * \frac{1}{IR_3} \qquad (11.24)$$

IR_3 : Resistance of fully opened stomates

$IKmR_3$: Light intensithy at which the resistance of stomates is twice the resistance for fully opened stomates

The daily changes in light intensity, pH of the chloroplasts, oxidation coefficient, and the maximal rate of reduction are shown in Fig. 11.2.

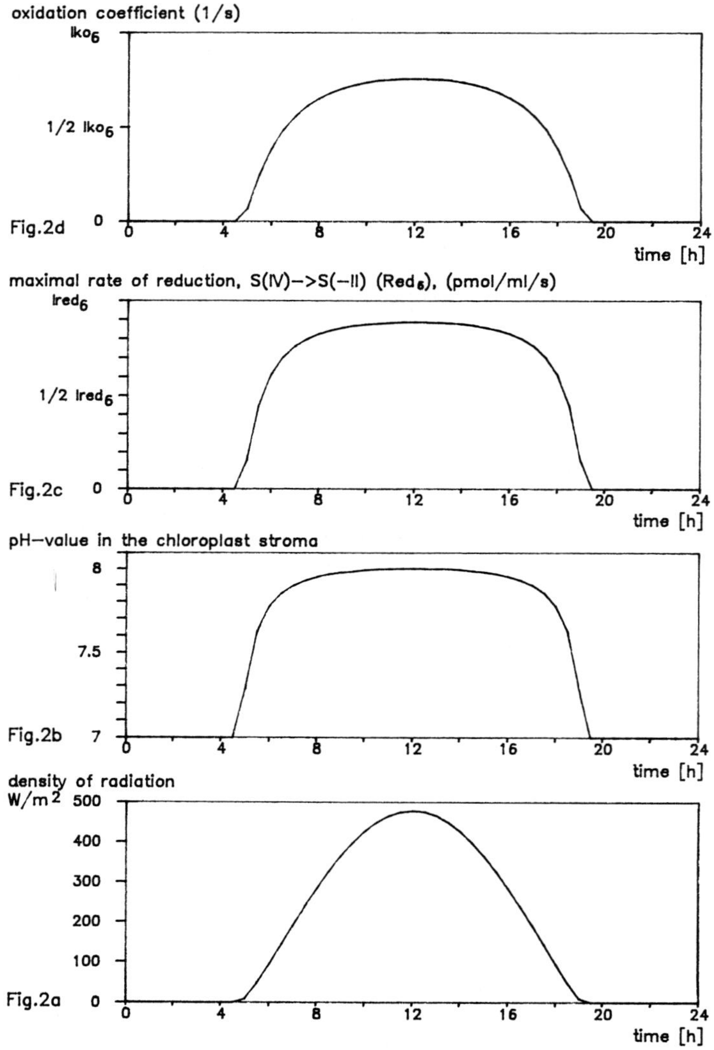

Fig. 11.2: Density of radiation for the 130th day of the year (a); variation of the pH in the chloroplast (b); the maximal rate of reduction (c); and the oxidation coefficient (d)

11.3 MODEL RESULTS

Starting with a hypothetical system comprising two gaseous compartments of different SO_2 concentrations, we will consider some factors determining the equilibrium concentration inside a system (= compartment 2) which is subjected to a pollutant, SO_2, abundant in the surrounding atmosphere (= compartment 1). Assuming, at first, that both compartments can only accomodate gaseous species the flux of the pollutant into the system will be determined by the concentration gradient and the resistance for the pollutant's diffusion, respectively, according to Fick's second law. Assuming that the concentration of the pollutant in the environment is constant in time, the concentration in compartment 2 will rise until thermodynamic equilibrium is established. Under these circumstances the concentration inside compartment 2 will be equal to the **environmental pollutant concentration**.

Assuming in a second step that compartment 2 is constituted by a liquid phase some additional factors arise which govern the level of the internal equilibrium concentration. The pollutant's solubility and dissociation must be taken into consideration. Because of the high solubility of SO_2 in water when thermodynamic equilibrium is established, SO_2xH_2O in solution will be 40 times the concentration of SO_2 in the gaseous phase. Additionally, the total S(IV) concentration will be higher than $[SO_2xH_2O]$ because, according to Henry's Law, only the dissolved gas is directly equilibrated with the pollutant in the gas phase but not the hydrogen sulfite and sulfite species. Any pollutant molecule transported into compartment 2 will be subjected to dissociation. The equilibrium of the different S(IV) species will be determined by the actual pH. On the other hand any dissociation step will release a proton. In the consequence the pH decreases and the dissociation equilibrium is changed in favour of the undissociated species. Overall S(IV) is again accumulated in compartment 2: 40 fold because of the solubility and another 300 fold because of dissociation.

In a third step the results of computer simulations of SO_2 uptake need to be evaluated. In these experiments diffusion of the pollutant has been simulated applying the model equations described above. Metabolization of S(IV) is not yet taken into account. Unlike the system constituted of one gaseous and one liquid phase described above, we now deal with several internal compartments in which the acidity is controlled by buffer solutions. Figs. 11.3, 11.4 and Table 11.2 present the results of two experiments:

a) assuming *an ideal buffer* (model run 1.1)
b) assuming limited buffer concentration of $6.2*10^{-2}$ M (model run 1.2)

Under these conditions a thermodynamic equilibrium is established too.

With a pollutant concentration of $0.5*10^{-6}$ mM in the ambient air the equilibrium concentration of S(IV) in the cytosol reaches 4.50 mM with unlimited buffer concentration, and 3.12 mM with limited buffer concentration. That means S(IV) is enriched about 6 million times. A high level of a potentially poisonous compound is built up inside the plant cells.

As can be concluded from these results the action of cellular buffers bears conflicting results: On the one hand, the buffer protects the plant from running out of the physiological range of pH-values in which the enzymes are stable and can operate optimally. On the other hand, the stabilization of cellular pH hinders the ratio of undissociated to dissociated S(IV) species to be increased with decreasing pH. In consequence the equilibrium concentration for intracellular S(IV) is much higher than it would be in an unbuffered liquid phase.

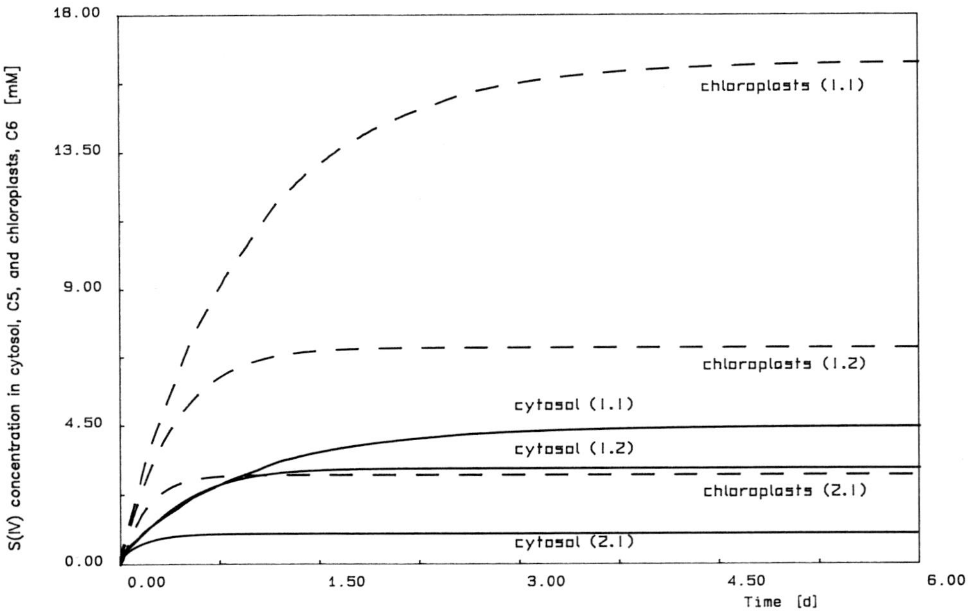

Fig. 11.3: S(IV) concentrations for the model runs 1.1, 1.2 and 2.1.

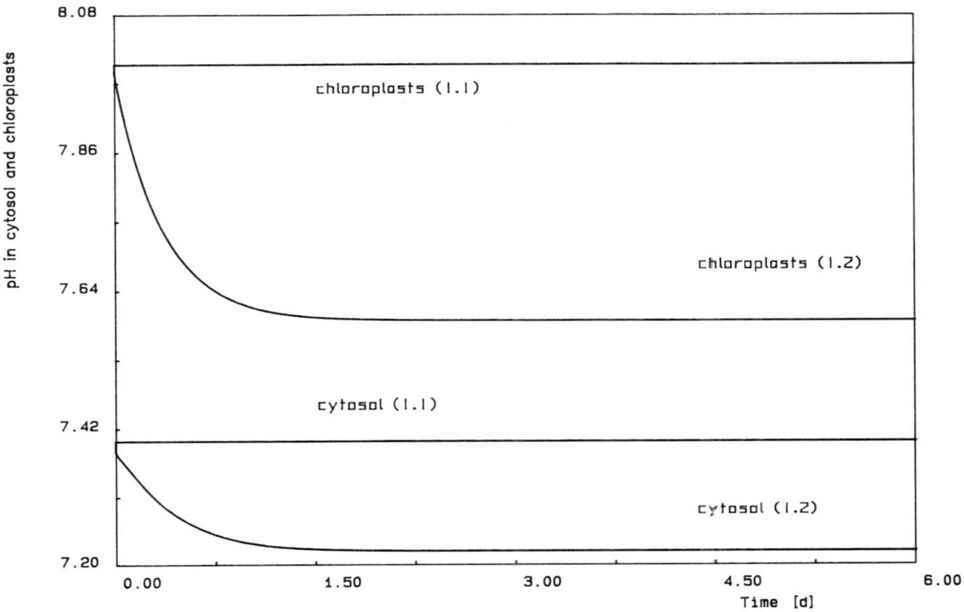

Fig. 11.4: pH in cytosol and chloroplast, model run 1.1 and 1.2.

Table 11.2
Results of the model runs 1.1, 1.2 and 2.1, 2.2 2.3

model run	S(IV) concentration in equilibrium or steady-state-equilibrium C5 [mM]	C6 [mM]	C7 [mM]	Time required to equilibrate [days]
1.1	4.50	16.48	0.158	20.0
1.2	3.12	7.09	0.154	8.2
2.1	1.02	2.91	0.0358	4.4
2.2	0.273	0.263	0.00962	0.5
2.3	0.0618	0.0078	0.00217	0.125

The increase in equilibrium concentration by buffering can be demonstrated by comparing the data for the systems with infinite and limited buffer concentrations, respectively (see. Figs. 11.3 and 11.4). The ideal buffer stabilizes the pH-value in the cytosol at 7.40. With a buffer concentration of $6.2*10^{-2}$ M the pH-value in the cytosol is lowered to 7.22, leading to a reduction in the equilibrium concentration of S(IV) by 30.7%.

In cell compartments with different buffer systems the equilibrium concentration of S(IV) is altered correspondingly. In the more alkaline compartments (e.g. chloroplasts, pH 8.0) the mole fraction of SO_2xH_2O is lower and therefore the total S(IV) concentration higher than in the more acidic compartments. Consequently the equilibrium concentration of S(IV) in the chloroplasts is higher than in the cytosol (16.48 mM vs. 4.50 mM) and much lower than in the more acidic vacuoles (0.158 mM at pH 5.7). Proton stress and potential toxic effects brought about by S(IV) alter with the alkalinity of the respective compartment.

In another computer experiment we proceed one step closer to the real situation inside plants. In addition to diffusion, solution, dissociation and effects towards acidity now elimination of S(IV) is simulated. Oxidation of S(IV) and transport of sulfate ions, together with two protons from chloroplasts and cytosol into the vacuole, are modelled.

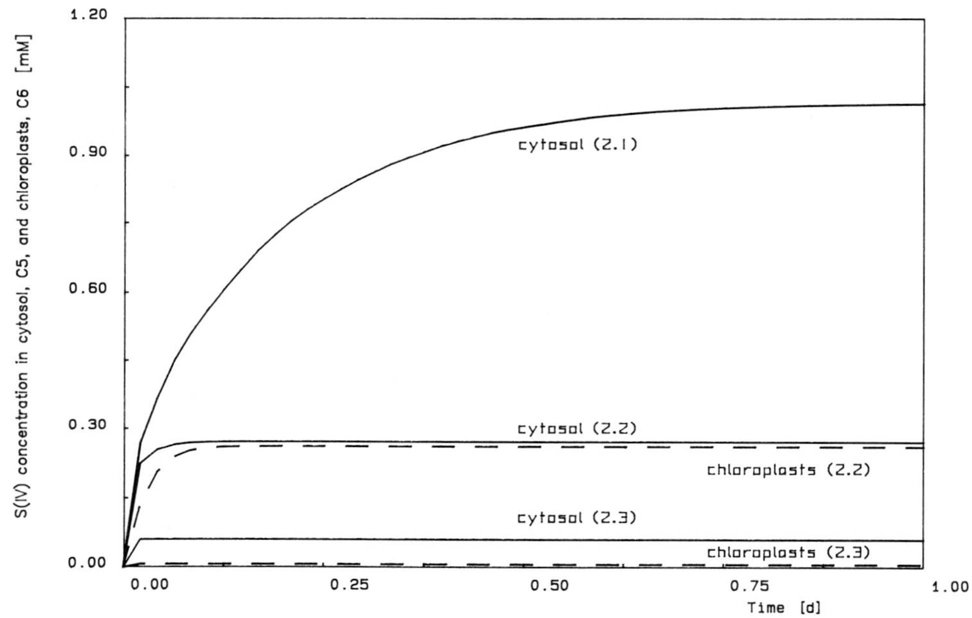

Fig. 11.5: S(IV) concentrations for the model runs 2.1, 2.2 and 2.3. Oxidation coefficients: $1*10^{-5}$ in (2.1), $1*10^{-4}$ in (2.2), $1*10^{-3}$ in (2.3).

As can be seen from the results presented in Figs. 11.3, 11.5 and Table 11.2, the equilibrium concentrations of S(IV) in the cellular compartments differ significantly from the concentrations given for a system without oxidation. High oxidation rates result in low equilibrium concentrations for S(IV). It was 4.5 mM for the cytosol in the model run 1.1 and decreases to 1.02 mM with an oxidation coefficiemt of $2*10^{-5}$ s^{-1} (model run 2.1), 0.27 mM with an oxidation coefficient of $2*10^{-4}$ s^{-1} (model run 2.2), and 0.06 mM with an oxidation coefficient of $2*10^{-3}$ s^{-1} (model run 2.3). Compared with the previous experiment the enrichement of S(IV) in the cytosol is reduced. Without oxidation it was enriched 9 million times; now the enrichment is 2 million times in model run 2.1, $5.4*10^5$ times in model run 2.2, and $1.2*10^5$ times in model run 2.3.

In contrast with the previous experiments the thermodynamic equilibrium for the S(IV) concentration in the current experiment is no longer established. The S(IV) concentration is now governed by a steady state equilibrium. The influx of SO_2 does not come to an end when the equilibrium is established. With a pollutant concentration of 0.5 $\mu mol*m^{-3}$, a total resistance for the gaseous diffusion ($R_2 + R_3 + R_4$) of 1 $s*cm^{-1}$ and an oxidation coefficient of $5*10^{-5}$ s^{-1} the influx of SO_2 in the steady state equilibrium is 0.145 pmol/(cm^2 leaf area*s).

Besides the factors discussed so far the relation of influx (diffusion) and efflux (elimination reactions) now also affects the level of the equilibrium concentration for S(IV). Alterations of the resistances for diffusion and the reaction constants of the elimination reactions will cause alterations in the steady state equilibria.

In further model runs reduction of S(IV) has been simulated as well as oxidation. The parameters were chosen according to the standard parameter set (see Table 11.1) pH was kept constant. Oxidation and reduction rates have not been altered, while the concentration of SO_2 in the ambient air has been assumed to be 0.05, 0.1, 0.5, 1.0, 5.0 and 10.0 $\mu mol*m^{-3}$ respectively. After 72 hours a steady state equilibrium is established in all of the model runs. The concentrations of S(IV) in the cytosol and chloroplasts, as well as the total amount of S(VI) and S(-II) produced by oxidation and reduction at this time, are given in Table 11.3.

The total amount of reduced S(-II) is rising with increasing pollutant concentration. But with a pollutant concentration of 0.5 $\mu mol*m^{-3}$ and more, the amount of reduced sulfur stays constant, i.e. the enzymes of the reduction pathway are saturated. In contrast the oxidation rate does not change with rising substrate concentrations. Therefore the ratio of oxidized to reduced sulfur species rises with an increase in the pollutant concentration.

For model run 4 the same parameters as for the runs 3.1 through to 3.6 were used. The SO_2 concentration in the ambient air was assumed to be 13.7 $\mu mol*m^{-3}$. The resulting uptake rate is shown in Fig. 11.6.

Table 11.3
Results of the model runs 3.1 - 3.6

Model run	C1 [μmol/m^3]	C5 [mM]	C6 [mM]	S(VI) [mmol]	S(-II) [mmol]	S(VI)/S(-II)
3.1	0.05	$6.78*10^{-2}$	$1.47*10^{-1}$	$8.18*10^{-4}$	$3.49*10^{-3}$	$2.35*10^{-1}$
3.2	0.1	$1.50*10^{-1}$	$3.60*10^{-1}$	$2.35*10^{-3}$	$6.52*10^{-3}$	$3.60*10^{-1}$
3.3	0.5	$9.70*10^{-1}$	$2.70*10^{0}$	$3.01*10^{-2}$	$1.11*10^{-2}$	$2.71*10^{0}$
3.4	1.0	$1.99*10^{0}$	$5.63*10^{0}$	$6.56*10^{-2}$	$1.09*10^{-2}$	$6.01*10^{0}$
3.5	5.0	$1.01*10^{1}$	$2.89*10^{1}$	$3.48*10^{-1}$	$1.08*10^{-2}$	$3.24*10^{1}$
3.6	10.0	$2.03*10^{1}$	$5.81*10^{1}$	$7.52*10^{-1}$	$1.15*10^{-2}$	$6.56*10^{1}$

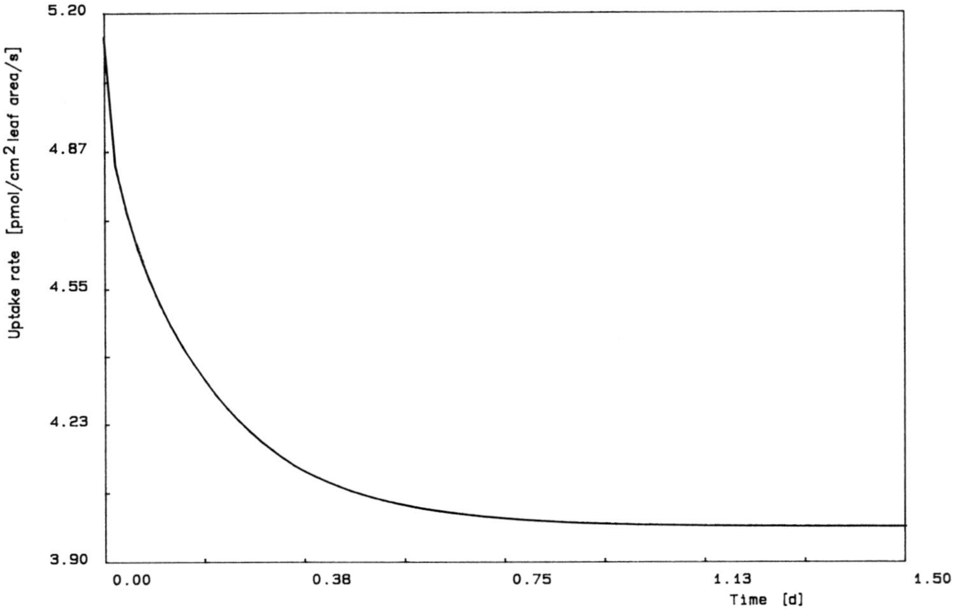

Fig. 11.6: Uptake rate for model run 4. SO_2 concentration in the ambient air: 13.7 μmol/m^3.

In Figs. 11.7 to 11.9 the results are illustrated for a simulation of diurnal changes in the acidity of the chloroplast-stroma, the rate of S(IV) reduction, the rate of oxidation in the chloroplast, and the stomatal resistance (model run 5).

Fig. 11.7 shows the accumulation of total S taken up and metabolized by the leaf. It is increasing during day-time when the stomata are open. In the night, when gas exchange is almost interrupted, it stays unchanged. The uptake rate (Fig. 11.9) exhibits a striking daily characteristic, with a maximum in the midday hours when the stomata are opened widely and the oxidation rate in the chloroplast is maximal.

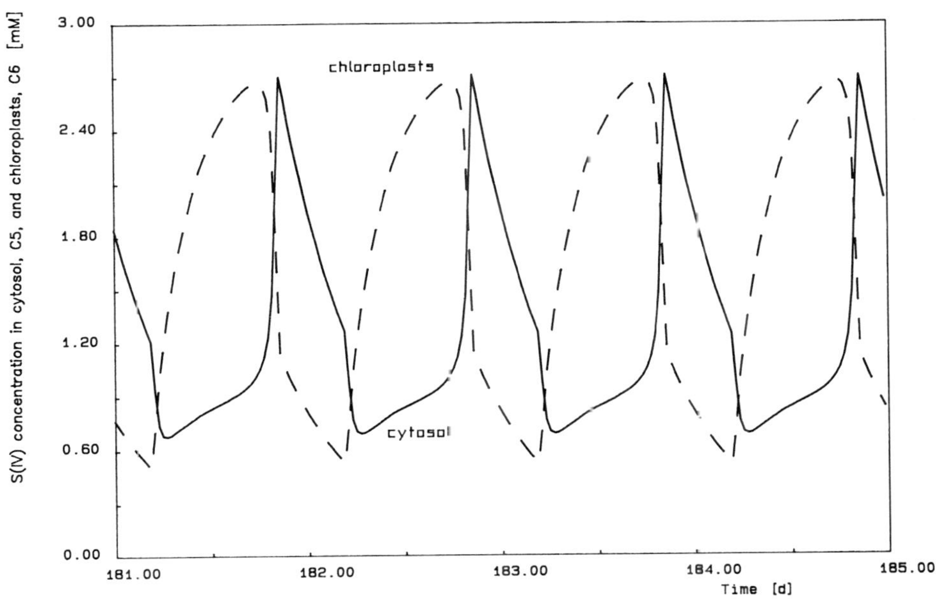

Fig. 11.7: S(IV) concentration for the model run 5, including simulation of diurnal changes in pH, oxidation and reduction rates and stomatal conductance.

The S(IV) concentration in the chloroplasts and cytosol change with a diurnal rhythm which is induced by the diurnal changes in the pH of the chloroplasts. At sunset when the acidity of the chloroplast stroma increases (from pH 8.0 to pH 7.0) SO_2xH_2O is streaming out into the cytosol causing a rapid increase of the S(IV) concentration in the cytosol and a rapid decrease in the chloroplasts. During the night the S(IV) concentration decreases in all cell compartments because of the light-independent oxidation in the cytosol. At dawn when the acidity in the chloroplast stroma starts to decrease again evening S(IV) fluxes are reversed, but on a lower level of total S(IV).

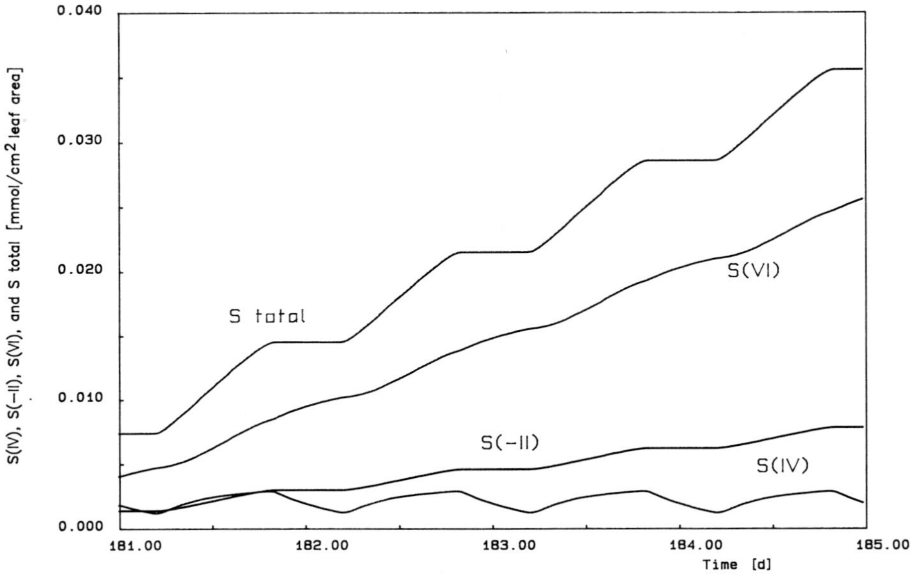

Fig. 11.8: Accumulation of several sulfur-species in model run 5.

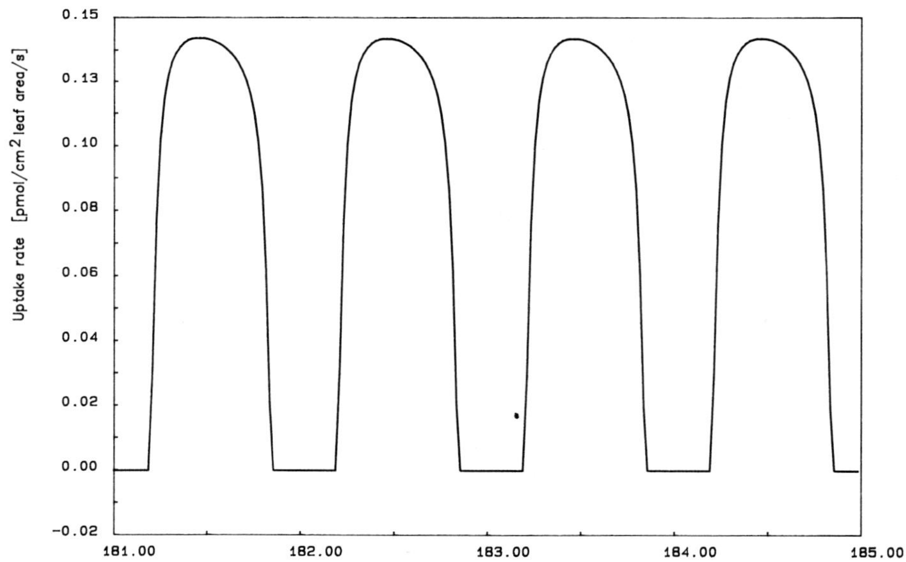

Fig. 11.9: Uptake rate for model run 5. SO_2 concentration in the ambient air: 0.5 µmol/m^3.

11.4 DISCUSSION

1. The model structure reproduces and explains the ratios of the resistances for the pathway of gaseous diffusion and "mesophyll" resistance determined experimentally. Although the membrane resistances for the S(IV) species are 10 to 10^9 times (for SO_2xH_2O and SO_3^{2-}) the resistances for the gaseous diffusion pathway, from measured SO_2 fluxes into leaves very small "mesophyll resistances" have been calculated. From the formula for the effective resistance (R_{eff} see equation (11.8)) it can be seen that the effective resistance for the uptake across the plasmalemma (r_5) is modified by three factors:

   First, the surface area of plasmalemma is 11.35 times greater than leaf surface area. Membrane resistances computed in relation to leaf surface area are therefore diminished by a factor of 11.35 (= A_5/A).

   Furthermore the enrichment of S(IV) in the liquid phase of the cell walls because of the equilibrium of solubility (40 times)
   and pH-dependant dissociation (about 6000 times at pH 5.7 in cell wall liquids) needs to be taken into account.
   With the parameters used in the model runs the effective resistance for the plasmalemma is 0.163 s^*cm^{-1}. This result is in good accordance with the mesophyll resistances calculated from experimental data. Black & Unsworth (1979) reported a mesophyll resistance of 0.0 ± 0.2 s^*cm^{-1} for *Vicia faba.*

2. Experiments of Pfanz (1987) demonstrated a reduction of pH in isolated protoplasts of *Hordeum vulgare* when they were bathed in a solution (pH 5) containing 297 nM $^{35}SO_2$. After 46 minutes the pH in the cytoplasm had decreased by 1.65 units (7.00 to 5.45).
   This phenomenon, registered for short exposure times and isolated protoplasts, is being simulated in the computer model; furthermore the accumulation of the acids H_2SO_4 and H_2SO_3 as described by the model, leads to a long term increase in acidity. This does not represent a real physiological event. There is even experimental evidence for an increase in pH of leaf homogenates of intact plants after longer exposure times. Jurat (pers. comm.) fumigated leaves of *Helianthus annuus* from the first to the 33rd day with 10 ppb SO_2. In the soluble fraction of leave tissue homogenates she measured a significantly increased pH-value.
   The decrease in pH computed in the model runs can be regarded as a kind of "virtually accumulated" acid load; that means it can be regarded as an indicator of the trend towards the acidification the plant has to cope with.

The effect of acids on the acidity and thereby on plant metabolism, is not only mediated by a given amount of buffer substances. The metabolism of the constituents of cellular buffer systems, as well as additional biochemical and biophysical mechanisms participating in the regulation of cellular pH, need to be taken into account (Smith & Raven, 1979).

3. The results of the model runs demonstrate high concentrations of potentially poisenous and acidifying S(IV) species in intracellular fluids (up to 10^7 times of the pollutant concentration in the ambient air) caused by the enrichment of the pollutant because of solubility and dissociation in the course of passive diffusion. Cellular buffers protect the plant from the trend to decrease pH but lead to higher concentrations of total S(IV). Metabolization of S(IV) serves the plant as a mean to diminish the internal concentration of the pollutant. In the case of the oxidation of S(IV) to S(VI), a decrease of the S(IV) concentration is accompanied by an increase in proton-stress, because the oxidation product, sulphuric acid (H_2SO_4), is a stronger acid than H_2SO_3. Only reduction of S(IV) to S(-II) reduces the trend to increase the proton-stress as well as the concentration of potentially poisonous S(IV). The capacity of the plant to use this preferable pathway of detoxification is limited by rate limiting enzymatic processes and concentration of amino acids serving as acceptor-molecules for S(-II). The model runs 3.1 to 3.6 show a relative increase of oxidized sulphur species, S(VI); and a relative decrease of reduced sulphur, S(-II) with increasing concentration of the pollutant. Garsed (1985) reviewing several fumigation experiments shows the same trend in the ratio of oxidation vs. reduction of S(IV) with increasing pollutant concentrations. With the concentrations calculated by the model we evaluated the importance of the inhibition of Ribulose Bisphosphate Carboxylase by S(IV) (Plöchl et al., 1989).
In any case, augmentation of metabolization of S(IV) leads to a higher influx of SO_2. Higher rates of metabolizing processes lead to lower intracellular S(IV) concentrations and thereby to constantly high effective concentration differences between ambient air and cellular fluids. Under these circumstances this concentration difference, being the driving force for the SO_2-influx will lead to an increase in uptake rates.

4. The uptake rates for SO_2 calculated in the simulation runs are of the same order of magnitude as the rates determined experimentally by Taylor & Tingey (1983). Fumigating *Geranium carolinianum L.* in gas exchange chambers with 0.3-0.8 $\mu l*l^{-1}$ SO_2 (13.7-36.6 $\mu mol*m^{-3}$) for four hours they calculated stomatal resistances between 1.0 and 3.9

s*cm^{-1} and an SO_2-uptake through the stomates of 18-44 nmol*cm^{-2} leaf area*h^{-1} (5-12 pmol*cm^{-2} leaf area*s^{-1}).
In a simulation (model run 4) uptake rates between 5.14 (at the beginning of the simulation) and 4.37 (after 4 hours) pmol*cm^{-2} leaf area*s^{-1} have been calculated with a stomatal resistance of 0.5 s*cm^{-1} and a pollutant concentration of 13.7 μmol*m^{-3}. Using an extended version of the model presented here we calculated the total uptake of Western German forests to amount to about 40,000 t S yr^{-1} (Badeck et al., 1989).
Up to now the model presented here does not cover possible reactions of S(IV) outside the cells, particulary in the cell wall, and long distance transport of the several sulfur species. Also several regulatory mechanisms involved in the stabilization of cellular pH already mentioned above, are not yet incorporated. In order to reflect the concrete reciprocations of sulfur- and proton-stresses and regulation of metabolism additional modelling of the balances of ions, organic acids and organic bases is required. But even on the current level of model development the results presented above prove **that the effective concentration of pollutants in plant cells and cell compartments are dependent on the internal state of the plant**. The dependence of active pollutant concentrations in different cellular compartments on processes regulating cellular pH, on elimination reactions and on the aperture of stomates has been demonstrated.

REFERENCES

Anderson, J.W., 1980. Assimilation of Inorganic Sulfate into Cysteine. In: Biochemistry of Plants a comprehensive treatise (Stumpf, P.K.; Conn, E.E.; eds.), Vol. 5: Amino acids and derivatives. (Miflin, B.J., ed. of vol.): 203-223.

Badeck, F.-W., G.H. Kohlmaier, M. Plöchl, 1989. Modelling the uptake of SO_2 into leaves of forest canopies. In: Mechanisms and Effects of Pullutant-Transfer into Forests (H.-W. Georgii: ed.) Dordrecht, Kluwer Academic Publishers: 177-184.

Black, V.J. & M.H. Unsworth, 1979. Resistance analysis of sulphur dioxide fluxes to *Vicia faba.* Nature 282: 68-69.

Cram, J., 1983. Characteristics of sulfate transport across plasmalemma and tonoplast of carrot root cells. Plant Physiol. 72: 204-211.

Gaastra, P., 1959. Photosynthesis of crop plants as influenced by light, carbon dioxide, temperature and stomatal diffusion resistance. Medelingen van de Landbouwhogeschool te Wageningen, Nederland 59(13): 1-68.

Garsed, S.G., 1985. SO_2 uptake and transport. In: Sulphur dioxide and Vegetation: Physiology, ecology and policy issues (W.E. Winner, H.A. Mooney, R.A. Goldstein eds.) Stanford, Stanford University Press: 75-95.

Garsed, S.G. & D.J. Read, 1977. Sulfur dioxide metabolism in soy-beans, *Glycine max var. biloxi,* Part II. New Phytol. 99: 583-592.

Hamp, R. & I. Ziegler, 1977. Sulfate and sulfite translocation via the phosphate translocator of the inner envelope membrane of chloroplasts. Planta 137: 309-312.

Hällgren, J.E. & S.A. Frederiksson, 1982. Emission of hydrogen sulfide from sulfur dioxide-fumigated pine trees. Plant Physiol. 70: 456-459.

Laisk, A., H. Pfanz, M.J. Schramm & U. Heber, 1988a. SO_2 fluxes into different compartments of leaves photosynthesizing in a polluted atmosphere, I. Computer Analysis. Planta 173: 230-240.

Laisk, A., H. Pfanz & U. Heber, 1988b. SO_2 fluxes into different compartments of leaves photosynthesizing in a polluted atmosphere, II. Consequences of SO_2 uptake as revealed by computer analysis. Planta 173: 241-252.

Lendzian, K.J., 1984. Permeability of plant cuticles to gaseous air pollutants. In: Gaseous air pollutants and plant metabolism (M.J. Koziol & F.R. Whatley: eds.) Butterworths, London: 77-81.

Lommen, P.W., C.R. Schwintzer, C.S. Yocum & D.M. Gates, 1971. A model describing photosynthesis in terms of gas diffusion and enzyme kinetics. Planta 98: 195-220.

Mourioux, G. & R. Douce, 1979. Transport du sulfate à travers la double membrane limitante, ou enveloppe, des chloroplastes d'épinard. Biochimie 61: 1283-1292.

Nobel, P.S., 1974. Introduction to biophysical plant physiology. San Francisco, W.H. Freeman & Co.

Parckhurst, F., 1977. A three-dimensional model for CO_2 uptake by continuously distributed mesophyll in leaves. J. theor. Biol. 67: 471-488.

Pfanz, H., 1987. Aufnahme und Verteilung von Schwefeldioxid in pflanzlichen Zellen und Organellen. Auswirkungen auf den Stoffwechsel. Dissertation, Würzburg: 134 p.

Pfanz, H., E. Martinoia, O.-L. Lange & U. Heber, 1987a. Mesophyll resistances to SO_2 fluxes into leaves. Plant Physiol. 85: 922-927.

Pfanz, H.E., E. Martinoia, O.-L. Lange & U. Heber, 1987b. Flux of SO_2 into leaf cells and cellular acidification by SO_2. Plant Physiol. 85: 928-933.

Plöchl, M., F.-W. Badeck, G.H. Kohlmaier, 1989. Modelling the effect of sulfur dioxide, hydrogen sulfite and sulfite on the metabolism of plants. In: Mechanisms and Effects of Pollutant-Transfer into Forests (H.-W. Georgii: ed.). Dordrecht, Kluwer Academic Publishers: 185-192.

Rennenberg, H., 1984. The fate of excess sulfur in higher plants. Ann. Rev. Plant Phys. 35: 121-153.

Richter, O., 1985. Simulation des Verhaltens ökologischer Systeme. Weinheim, Verlag Chemie.

Rothermel, B. & R. Alscher, 1985. A light-enhanced metabolism of sulfite in cells of *Cucumis sativus L. cotyledons.* Planta 166: 105-110.

Schiff, J.A., 1983. Reduction and other metabolic reactions of sulfate. In: Encyclopedia of plant physiology (eds.: A. Pirson & M.H. Zimmermann), vol. 15A: Inorganic plant nutrition (eds. of vol.: A. Läuchli & R.L. Bieleski). Berlin, Springer Verlag. 401-421.

Seel, F., 1965. Grundlagen der analytischen Chemie unter besonderer Berücksichtigung der Chemie in wässrigen Systemen. 4. Auflage. Weinheim: Verlag Chemie.

Smith, F.A. & J.A. Raven, 1976. H^+ transport and regulation of cell pH. In: Encyclopedia of plant physiology. New Series, vol. 2A: Transport in plants II (eds.: U. Lüttge & M.G. Pitman). Berlin, Heidelberg, New York, Springer. 317-346.

Taylor, G.E., D.T. Tingey, 1983. Sulfur Dioxide Flux into Leaves of *Geranium carolinianum L.* Plant Physiol. 72: 237-244.

Werdan, K., H.W. Heldt & M. Milanovancev, 1975. The role of pH in the regulation of carbon fixation in the chloroplast stroma: Studies on CO_2 fixation in the light and dark. Biochim. Biophys. Acta 396: 276-292.

CHAPTER 12

ATMOSPHERIC TRANSPORT OF SULPHUR DIOXIDE ON A LOCAL SCALE: A CASE STUDY

by

Mario C. Cirillo, Giancarlo Clerici & Donatella Manzi

12.1 INTRODUCTION

This chapter deals with an assessment of the local impact of the fossil fuel energy system of the Piacenza province on the surrounding environment, with special emphasis on air pollution from SO_2.

The study is part of a broader project having the following main objectives:

1. The analysis and assessment of the impact from energy systems on the natural and developed environment, including the social, economic, territorial and institutional situations.

2. The identification of possible measures of intervention, prevention and control, aimed at minimizing or even eliminating undesired effects (Pinchera et al., 1984).

The area considered is the province of Piacenza, in nothern Italy (Fig. 12.1).

The total population of the province, which covers 2,590 km^2, was 280,127 in 1981, the reference year of the analysis. The main field activities in the province are agriculture and industry, including electricity generation, metalworking, chemical and cement factories. The industrial installations are concentrated in areas which have become significant sources of pollution: the area of the town of Piacenza, which includes two oil-fired electric power stations and two cement factories; the area of Castel San Giovanni - Sarmato, with the 1280 MWe oil-fired power plant of La Casella and a large sugar factory; the area of Lugagnano Val d'Arda, with a large cement factory.

Around 77% of the energy flow which enters the province in the form of conventional fuels serves the generation of electric energy, of which about 7% is used locally.

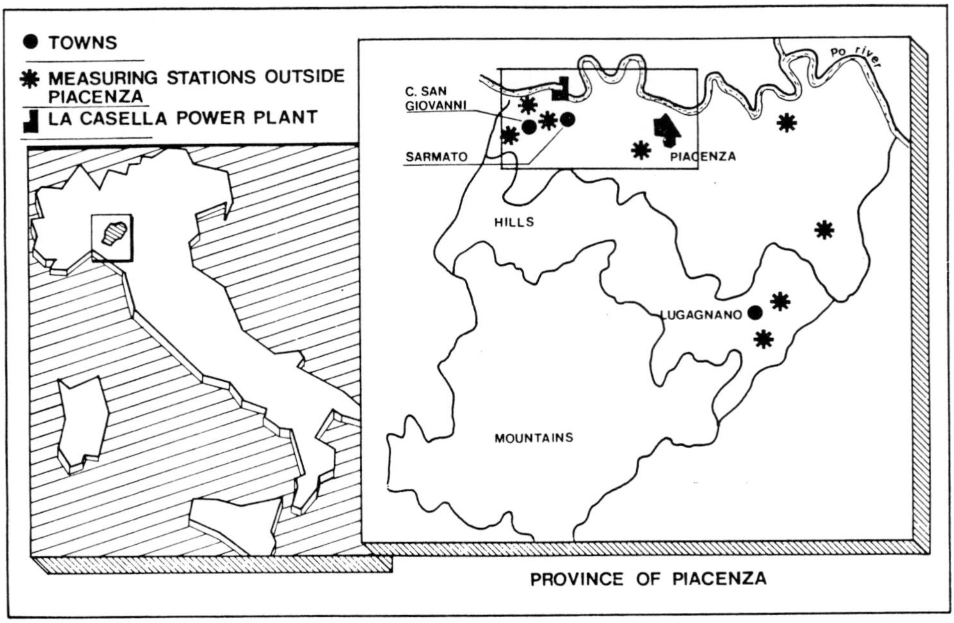

fig. 12.1: Sketch of the area considered in the study.

The data collection and organization, and the technologiocal characterization were carried out by means of a critical appraisal of all information already available, followed by a specific field investigation to obtain the missing elements. The study seeks to use, as much as possible, specific data on the plants included in the unit of analysis.

Regarding energy characterization and air pollutant emission estimates, the whole province was considered, while as far as pollutant dispersion is concerned, we concentrated the study in the rectangle (325 km^2) that includes the town of Piacenza on the east side and the municipality of Castel S. Giovanni on the west (Fig. 12.1); it encompasses most of the local industrial installations, and in particular the three electric power plants using mainly fuel oil; the area has the highest population density and the highest concentration of local agricultural resources.

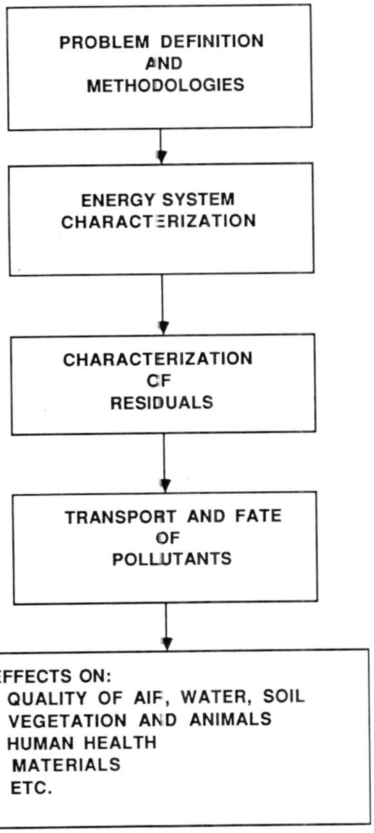

Fig. 12.2: Sketch of the various steps of the whole project.

Fig. 12.2 shows the various steps of the work, from energy characterization of the unit of analysis to the estimate of negative effects. Emphasis is given here to air pollution from SO_2. The quantitative estimate of crop damage due to air pollution is reported elsewhere (Cirillo et al., 1987).

An attempt is made particularly to characterize the pollution load distribution among the various emitting sources in terms of both emissions and air concentrations of SO_2 in the area examined. The analysis provides decision-makers with several elements for planning possible action for intervention, prevention and control.

12.2 ENERGY CHARACTERIZATION OF THE UNIT OF ANALYSIS AND AIRBORNE EMISSIONS

Consumption of various fuels by sectors in 1981 are shown in Table 12.1. Note the high quota (77.2%) of fuel destined to the power stations (operated by the Italian National Electricity Generating Board, ENEL) and the pre-eminence (76.5%) of fuel oil (mostly burned for the generation of electricity) with respect to other fuels.

Oil-derived fuels that are consumed in the province of Piacenza for a total of about 2520 ktoe (thousands of tons of oil equivalent) correspond to 2.68% of the national consumption. About 83% of the total provincial consumption of petroleum products are assigned to the generation of electricity.

In 1981, the Emilia plant (140 MWe, initial year of operation: 1953) and the Levante plant (680 MWe, 1965-67), situated in the town of Piacenza, were in operation, plus the La Casella plant (1280 MWe, 1971-73), situated in the Castel San Giovanni municipality.

The spatial distribution of the industrial plants are concentrated in the town of Piacenza, where two cement factories (Unicem and Rossi) and two power stations (Levante and Emilia) are located; in the area of Castel San Giovanni - Sarmato, dominated by the presence of the La Casella thermal power plant and by a large sugar refinery; in Lugagnano, in the mid Val d'Arda, where there is a large cement factory, plus several industries for the production of tiles.

The three oil-fired power plants in Piacenza burn mainly residual fuel oil with a sulfur content of 2.8% medium and 3% maximum and a maximum ash content of 0.25%. Natural gas was used only in the Levante plant for an overall consumption of $228*10^6$ m^3, equivalent to 188 ktoe (thousands of tons of oil equivalent). No systems are installed to control air pollution, except for the addition of magnesium oxide to reduce acid smuts and, more broadly, sulphur trioxide.

The estimate of the overall emissions of sulfur dioxide, nitrogen oxides and dusts was made on the basis of the fuel consumption reported by a field survey using the emission factors given by the U.S. Environmental Protection Agency (1985). This estimate was found to be relatively easy as regards the electricity generation, residential-services, transportation and agricultural sectors. Estimated emissions in the industrial sector are, however, far more uncertain, since the methodology used, which is based on the simple disaggregation of fuel consumption and on the use of homogeneous emission factors per type of fuel, does not take into account the different potentialities and characteristics of furnaces and kilns. With regard to cement factories, only emissions by rotating ovens are assumed as attributable to petroleum consumption.

Table 12.1
Fuel consumption (ktoe) by source and sector in Piacenza provincie during 1981

Fuel/sector	Power plants	Industry	Domestic	Trans- port	Agri- culture	Total	%
Fuel oil	2897.30	153.29	- -	-	- -	2258.59	76.46
Natural gas	188.21	100.90	104.20	5.00	- -	398.31	13.54
Distillate/ diesel oil	1.82	10.98	45.10	107.00	19.30	184.50	6.27
Gasoline	- -	- -	- -	77.27	- -	77.25	2.63
Kerosene	- -	- -	2.16	- -	- -	2.16	0.07
Coal	- -	17.08	- -	- -	- -	17.08	0.58
Wood	- -	- -	4.30	- -	- -	4.30	0.15
Other	- -	4.09	4.26	- -	- -	8.35	0.28
Total	2287.33	286.34	160.02	189.55	19.30	2942.54	100.00
%	77.23	9.73	5.44	6.44	0.66	100.00	

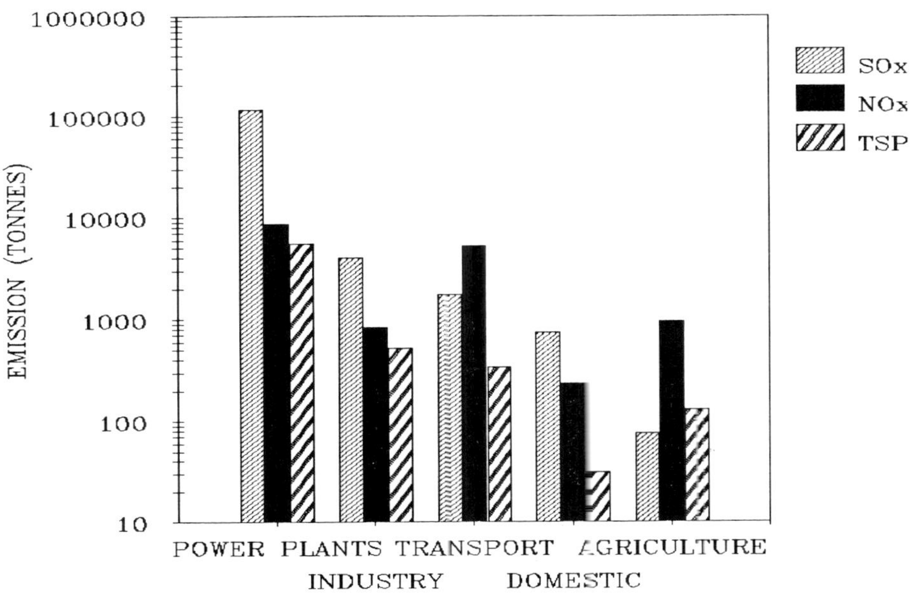

Fig. 12.3: Air pollutant emission estimate by sector in the Piacenza province (1981).

A general overview of the estimated emissions of SO_2 (sulphur dioxide), NO_x (nitrogen oxides) and TSP(total suspended particulates) generated during 1981 by the various sectors in the Piacenza province is given in Fig. 12.3. Monthly emissions were also estimated for each municipality in the province and for the major point sources.

12.3 TRANSPORT AND FATE OF SULPHUR DIOXIDE ON A LOCAL SCALE

The DIMULA model

The study of the transport and fate of air pollutants is a part of the general problem of describing how the different polluting agents generated by various sources are transformed and distributed in the environment (air, water, and soil).

The pathways of pollutants are described by means of mathematical models which allow their space and time distribution to be calculated. Recent developments in the study of turbulent diffusion and the growing public interest in pollution control have led to a great number of model families; however, the approaches used in developing these models are rather limited. Basically, it is possible to distinguish between semi-empirical models, based on physical conservation laws (both in deterministic and in stochastic form), and statistical models that relate the measured concentrations to some "concomitant variables" (atmospheric conditions) by means of regression analysis.

Other classifications are possible and useful, according to the space and time scale covered by the model, as well as the possibility of accounting for the spatial and temporal variability of the atmospheric field, for territorial heterogeneity, dry and wet deposition, chemical transformation, and so on.

A multisource diffusion model was developed to calculate the concentration of pollutants in the air at ground level. This model is called **DIMULA** (Modello di Diffusione Multisorgente Atmosferica) and was developed to provide a description of a situation of environmental pollution generated by a number of sources, concentrated as well as distributed, over distances of some tens of kilometres. It has been designed to be a simple and easy-to-use tool, but is at the same time sufficiently accurate to provide an adequate and realistic description of the situation examined, in particular, to describe the rather frequent situations of calm and foggy weather in the Po Valley.

Other significant characteristics of the model are as follows:

1. The appropriately processed **Gaussian "plume" and "puff"** formulae provide a series of models that simulate the various possible situations, both concerning the characteristics of the sources as well as meteorological situations.
2. The calculation of **the effective height of the plume** is provided by means of the Briggs (1975) formulae.
3. The phenomenon of **dry deposition** can be simulated on request.

The calculation field is a rectangular domain with rectangular mesh-elements, at the vertices of which the concentration is ca culated. Some of the limitations inherent in the Gaussian dispersion models are also applicable to the multisource model. In particular, the use of the plume formulae presumes stationary and homogeneous atmospheric and territorial conditions. The space scale of the model comprises some tens of kilometers; the time scale can measure concentrations averaged over periods, ranging from a few ten-minute intervals to over several hours, up to a day or more if the input data are given sequentially, e.g. on an hourly basis. Another version of the **DIMULA** accepts the input meteorological data in the form of joint frequency functions; this allows the calculation of concentrations averaged over longer periods (a month, a season, a year).

There are several factors influencing the response of the model. In general, this `response depends on its structure and on the input data. With the **DIMULA**, some problems correspond to the well-known drawbacks presented by Gaussian models. In particular, the estimate of dispersion coefficients, especially critical in cases of low wind-speed conditions, is a crucial problem. A further sensitive parameter of the model is the estimate of the actual height of the plume. The commonly used Briggs formulae (Briggs, 1975), chosen also for this model, gives acceptable results in most cases. However, these formulae do not supply good results under conditions of low wind-speed and convective turbulence (instability); according to some experimental data (Frank and Röber, 1983), with a wind-spped of less than 1.5 m/sec, the Briggs formulae significantly overestimate the actual rise of the plume. Thus it follows that calm weather conditions are the most critical for simulation. On the other hand, these are of great interest, considering the specific meteorological characteristics of the area examined.

The most significant uncertainties are those relating to the quantity of the various emissions and their distribution in time, the thermal and dynamic characteristics of flue gases, the height of the inversion layer in the atmosphere, and its evolution during the day.

Table 12.2 shows the basic equations used in the **DIMULA model**. Further details can be found in Cirillo et al., 1986.

Table 12.2
Basic equations adopted in the DIMULA model for describing the dispersion of gaseous nonreactive air pollutants

---

Depending on the type of emission source, the meteorological conditions and the available information on wind direction, DIMULA uses different equations for estimating the pollutant concentration. These equations can be classified following this scheme:

Type of emission source: point or areal
Anemological conditions: presence of wind or calm
Inversion layer: modelling of an elevated inversion layer or not
Information on wind direction: in degrees or by sector

Here we give the basic equations for a point source, assuming the wind direction parallel to the X axis: the other models (areal sources, wind direction given by sector) are derived on the basis of the following considerations:

- for area sources we applied the superposition principle, i.e. an area source is considered the sum of infinitesimal point sources;
- if the wind direction is given by sector, we considered a uniform crosswind pollutant distribution inside the sector and zero outside (see e.g. Yang 1979).

For the full list of all equations used in the **DIMULA** model, see Cirillo et al., 1986.

<u>A. Plume model without inversion at elevation</u> <u>(Pasquill and Smith, 1983)</u>

$$C(x,y,z) = \frac{Q}{2\pi u \partial_y \partial_z} \exp\left(\frac{-y^2}{2\partial_y^2}\right) \left\{\exp\left[\frac{-(z-H)^2}{2\partial_z^2}\right] + \exp\left[\frac{-(z+H)^2}{2\partial_z^2}\right]\right\}$$

C(x,y,z): pollutant air concentration (e*y*μg/m^3) at point (x,y,z)
Q : pollutant emission per unit time (e.g. μg/sec), assumed to be constant
u : mean wind velocity at height H (m/sec), assumed to be parallel to the X axis; the crosswind direction is along Y axis
∂_y, ∂_z : statistical measures of crosswind and vertical dispersions (m); both are function of atmospheric stability and downwind distance
H : effective height of the plume (m), given by the sum of the stack height and the height of rise due to the efflux velocity and temperature (for hot plumes).

The underlying assumptions are those of a simple plume geometry, with convenient gaussian crosswind and vertical distributions and perfect "reflection" from the ground surface.

<u>B. Plume model with inversion at elevation</u> <u>(Schnelle, 1979)</u>

$$C(x,y,z) = \frac{Q}{2\pi u \partial_y \partial_z} \exp\left(\frac{-y^2}{2\partial_y^2}\right) \left\{\exp\left[\frac{-(z-H)^2}{2\partial_z^2}\right] + \exp\left[\frac{-(z+H)^2}{2\partial_z^2}\right] + E_T\right\}$$

where

$$E_T = \sum_{i=1}^{n} \left\{\exp\left[\frac{-(z-H-2iL)^2}{2\partial_z^2}\right] + \exp\left[\frac{-(z+H-2iL)^2}{2\partial_z^2}\right] + \exp\left[\frac{-(z-H+2iL)^2}{2\partial_z^2}\right] + \exp\left[\frac{-(z+H+2iL)^2}{2\partial_z^2}\right]\right\}$$

L : height of the bottom of the inversion layer (m)
n : number of reflections considered.

Table 12.2 (continued)

The model developed to account mixing under a stable layer employs the idea of a complete reflection from the diffusion lid as in the case for the plume contacting the ground. The number of reflections is given by n where n = 4 is sufficient to include all the important contributions.

C. Diffusion in calm conditions (Cagnetti and Ferrara, 1982)

$$C(x,y,z) = \frac{Q}{4\pi r}\left[\frac{1}{\sqrt{(u_d K_z r + u_d^2\ (z-H)^2)}} - \frac{1}{\sqrt{(u_d K_z r + u_d^2\ (z+H)^2)}}\right]$$

r : receptor-source distance in the X-Y plane (m); $r = \sqrt{(x^2+y^2)}$
u_d : diffusion rate (m/sec)
K_z : vertical diffusion coefficient (m^2/sec).

The starting point is the gaussian puff model integrated in time between 0 and ∞ (stationary conditions).

D. Diffusion in calm condition with inversion at elevation

$$C(x,y,z) = \frac{Q}{2\sqrt{(2\pi)}\ arL}$$

a : constant analogous to a diffusion rate (m/sec).

This model supposes a vertical uniform distribution of pollutants below the inversion. The formula can be derived by integrating the diffusion equation imposing an uniform distribution along Z and assuming that K_H = ar, where K_H is the horizontal diffusion coefficient (m^2/sec).

The case study

The province of Piacenza can be subdivided into two main areas: the mountainous area in the south covering a surface of approximately 950 km^2, and the northern plain territory with a surface area of about 700 km^2.

Most of the industrial installations, and particularly the electric power plants using fuel oil, are located in the plain area, as is the majority of the population. For this reason, the northern plain of the province was examined with particular care, especially the rectangle that includes the city of Piacenza on the eastern side and Castel San Giovanni on the west.

Local wind circulation can be considered similar to the characteristic regional circulation of the middle Po Valley. The prevailing wind direction in this area is therefore parallel to the main axis of the valley, due to deviation of the synoptic wind by the Apennine mountain chain. During the winter, the prevailing wind direction is West-North West, but during the summer it is East-South East (Fig. 12.4).

Furthermore, there is a high percentage of instrument-calm weather (wind speed less than approximately 0.5 m/sec), which during the year covers about 50% of the cases and is more frequent in winter months. Fog is present during the winter and practically absent throughout the summer, when the atmosphere tends more towards unstable conditions. The air is generally humid. The average annual rainfall in the area is not abundant. Maximum rainfalls occur in May (50 mm) and in September-October (70 mm). Heavy precipitations are rare. Storms are usually limited to the summer months.

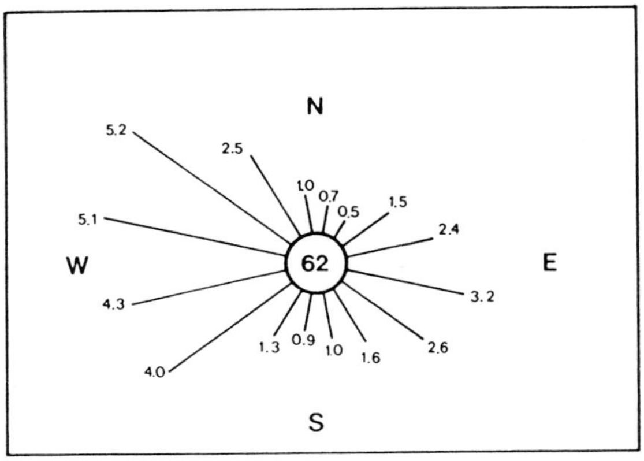

Fig. 12.4A: Wind rose: winter data (ENEL/AM, 1981). Frequencies are expressed in %. The number in the circle is the percentage of calm conditions.

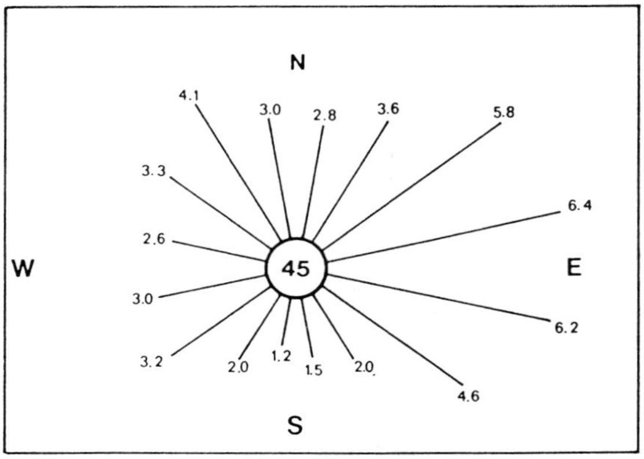

Fig. 12.4B: Wind rose: summer data (ENEL/AM, 1981). Frequencies are expressed in %. The number in the circle is the percentage of calm conditions.

The **DIMULA model** was applied to the short-term (diurnal, 12-h average) as well as long-term (annual) dispersion of sulphur dioxides.

The input data needed to run the model can be subdivided into two classes:

1. **Information on emission sources**: type (point or area), location, stack characteristics (height, diameter), effluent characteristics (temperature, velocity), magnitude of SO_2 emissions. Only pollution sources within the rectangle, or external sources for which a significant contribution in terms of air concentration in the rectangle is envisaged, are considered; pollution souruces external to the province are also disregarded. We considered 27 point sources and 22 area sources that simulate both industrial and domestic SO_2 emissions. Sulphur dioxide emissions from mobile sources are not accounted for.

2. **Meteorological information,** stability class[(1)], wind velocity and direction, presence of elevated inversion and its characteristics[(2)].

   The first type of information was collected by means of a critical appraisal of all the information already available, followed by a specific field investigation to obtain the missing elements.

   The data from the local meteorological monitoring network were used for the short-term wind direction and velocity (the measured direction is not precise, since the entire circle is divided into only 8 sectors): two 10-m high stations (Casten San Giovanni and Gossolengo), and one 100-m high meteorological tower located in Piacenza. We decided to relate the wind measurement taken on the tower to the high sources of pollution. For lower sources the wind data for Gossolengo or Castel S. Giovanni were considered, depending on which of the two stations was nearest to the source in question.

   The joint frequency function of stability class, wind direction and velocity reported in ENEL/AM (1981) publication, were used to estimate annual concentrations.

1. *The dispersion capacity of the lower atmosphere,* especially the phenomena correlated with vertical mixing, are characterized via the concept of "atmospheric stability". Dispersion is linked to the turbulent motions of air masses, which depend on a variety of factors, among which the vertical temperature gradient is worthy of note. When unstable conditions occur, upward and downward motions of the air, and of the pollutants contained in it are amplified; in contrast stable conditions inhibit vertical mixing (an air parcel, if vertically displaced, tends to return); an intermediate condition is neutrality, corresponding to a situation of indifferent equilibrium. Neutrality is, by definition, characterized by a negative vertical temperature gradient which equals the dry adiabatic lapse rate (0.98°C fall per 100 m rise) - the rate at which a parcel of dry air changes its temperature adiabatically as a result of the change of pressure associated with vertical displacement. Thus, neutral stability is associated with a

near-zero gradient, while unstable conditions are associated with lapse (decrease of temperature with height) and stable conditions with inversion (increase of temperature with height). Usually in the applications 7 stability classes are considered, from A (very unstable) to F (very stable); D is neutrality.

2. When an **inversion** occurs at a certain height, with neutral or unstable conditions below, we call it an "elevated inversion layer". Since the inversion inhibits vertical motions, an elevated inversion acts like a "cork" which traps the plume below, causing significant impact at ground level. If a plume is emitted by a tall stack with sufficient buoyancy and momentum, it is possible that perforates the inversion layer: in this case the same layer prevents a downward dispersion of the plume.

Sulphur dioxide concentration is measured in 6 monitors, 3 located in Piacenza, 3 in open country around the La Casella power plant.

Short-term (diurnal, 12-h average) concentration estimate of SO_2

We considered the diurnal SO_2 concentration, because daytime meteorological conditions generally differ from those occuring at night; in point of fact, due to the comparatively more unstable conditions, the tall emitting sources do also cause a significant impact near the point of emissions, i.e. on a local scale, during the day; furthermore, some emission sources (e.g. those from the domestic sector and some industries) are only active in daytime.

We considered 7 days, trying to cover some of the possible interesting situations:

A. Days characterized by calm wind

June 12, 1981 The prevailing wind direction is from the East. The meteorological tower (100 m height, located in Piacenza) reports a 12% calm during the diurnal period; this percentage is obviously higher at 10 m.

In *this* condition, La Casella has no significant impact on the city of Piacenza, while calculation estimates a contribution of about 20% from the Levante power plant (stack height 120 m) and of about 50% from the Emilia power plant (stack height 47 m) to the mean SO_2 concentration of the city (Fig. 12.5). This and the following graphs show the estimated SO_2 concentration field in the rectangle which appears in Fig. 12.1. Note that La Casella power plant is located at about 4.5 km East and 10 km North from the lower left corner, while the Levante and Emilia power plants are about 22.5 km East and 6 km North from the same corner. The city of Piacenza covers an area of about 12 km^2, centred at about 21.5 km East and 5 km North from the lower left corner.

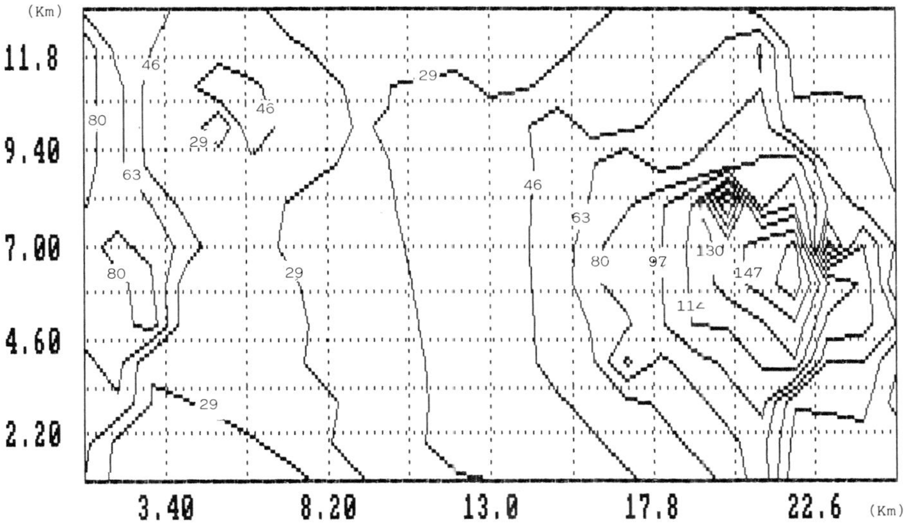

Fig. 12.5: Diurnal 12-h average concentrations expressed in $\mu g/m^3$ - June, 12, 1981. Calculated SO_2 concentrations in the rectangle including Piacenza (right side) and Castel S. Giovanni (left side).

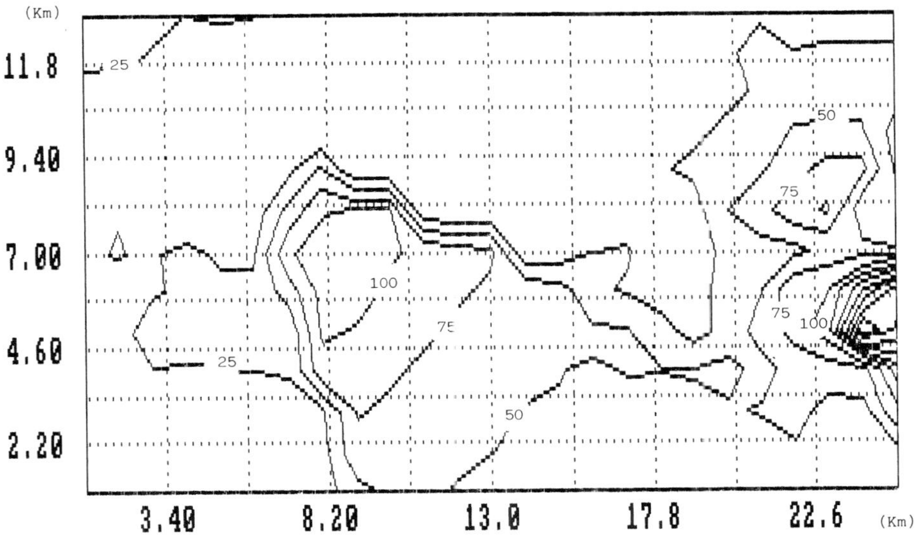

Fig. 12.6: Diurnal 12-h average concentrations expressed in $\mu g/m^3$ - October, 5, 1981. Calculated SO_2 concentrations in the rectangle including Piacenza (right side) and Castel S. Giovanni (left side).

October 5, 1981 The prevailing wind direction is from the North-West. The meteorological tower (100 m) reports a 5% calm wind during the day; the percentage is higher at 10 m. There is no appreciable impact on the city from La Casella and Levante, while the estimated contribution from Emilia is significant (Fig. 12.6).

November 10, 1981 The main wind direction is from the West-North West. The meteorological tower (100 m) reports a 20% calm wind. Calculation shows a contribution from La Casella to the Piacenza SO_2 concentration of the order of 15%. With this condition there is no appreciable contribution to the town by the Levante power plant, while the contribution from Emilia is around 10% (Fig. 12.7).

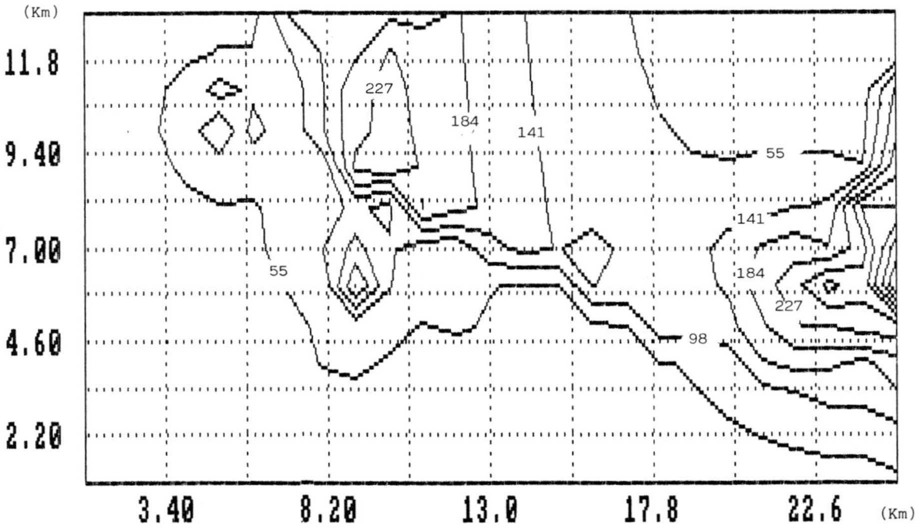

Fig. 12.7: Diurnal 12-h average concentrations expressed in µg/m^3 - November 10, 1981. Calculated SO_2 concentrations in the rectangle including Piacenza (right side) and Castel S. Giovanni (left side).

B. Days with fog

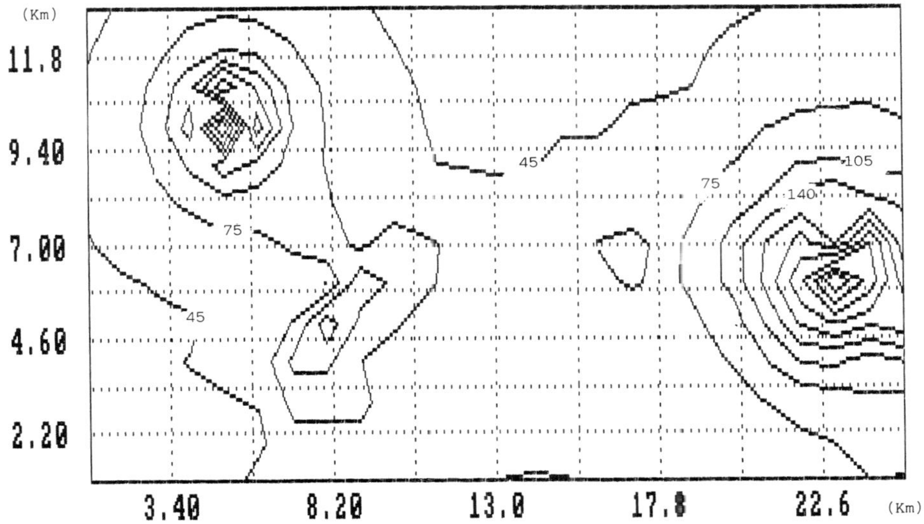

Fig. 12.8: Diurnal 12-h average concentrations expressed in µg/m^3 - November 20, 1981. Calculated SO_2 concentrations in the rectangle including Piacenza (right side) and Castel S. Giovanni (left side).

November 20, 1981 Wind is from the West-North West. The meteorological tower reports a 33% calm The estimated impact of La Casella on the city is very low, while those of Levante and Emilia power plants are of the same order (about 18% each), Fig. 12.8.

November 23, 1981 Wind is from the West, but with a very high percentage of calm (54% from meteorological tower, 100 m height). The contribution from La Casella to the city is again low, while those from the Levante and Emilia power plants are estimated to be significant (about 25% each), Fig. 12.9.

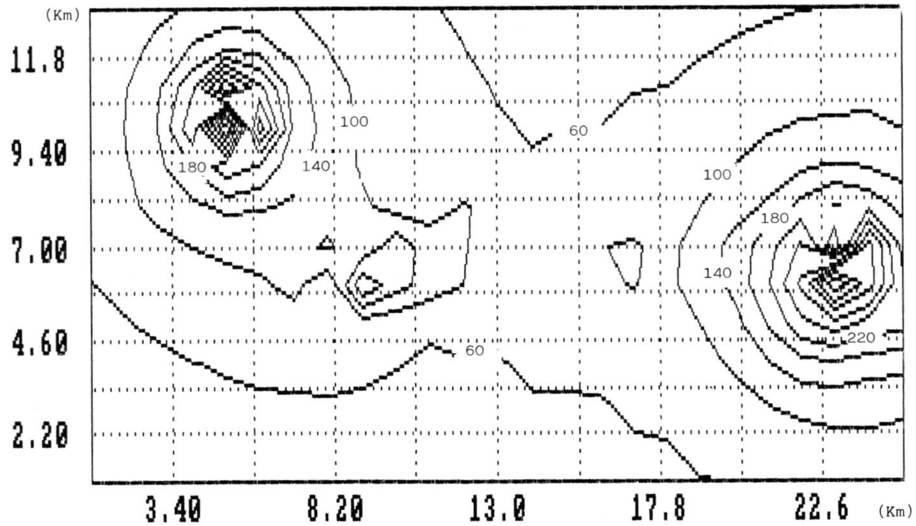

Fig. 12.9: Diurnal 12-h average concentrations expressed in $\mu g/m^3$ - November 23, 1981. Calculated SO_2 concentrations in the rectangle including Piacenza (right side) and Castel S. Giovanni (left side).

C. Days characterized by presence of wind

August 5, 1981 Wind from the East has a velocity of 4.5 - 6.5 m/sec at 100 m height (meteorological tower). On this day there is obviously no contribution from the domestic sector. The Levante power plant also has no emission, because it is not in operation. The SO_2 concentration is very low over the whole area examined (Fig. 12.10).

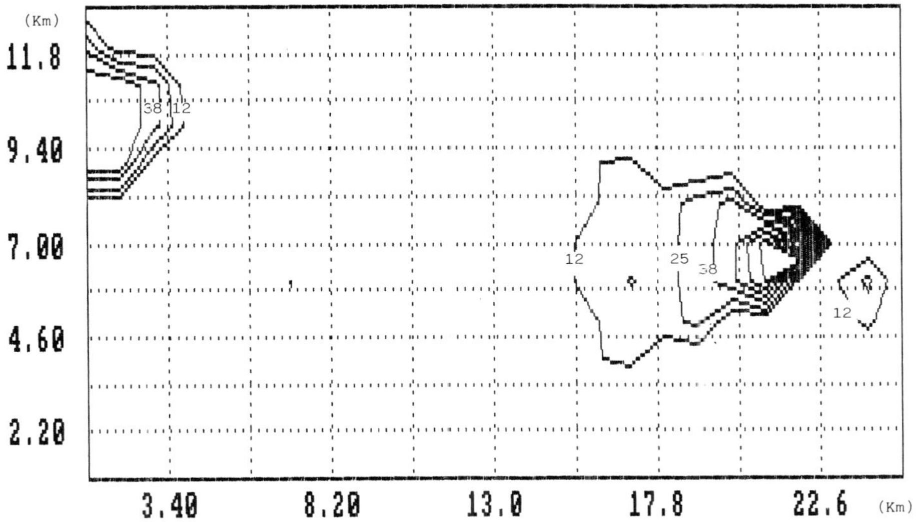

Fig. 12.10: Diurnal 12-h average concentrations expressed in $\mu g/m^3$ - August 5, 1981. Calculated SO_2 concentrations in the rectangle including Piacenza (right side) and Castel S. Giovanni (left side).

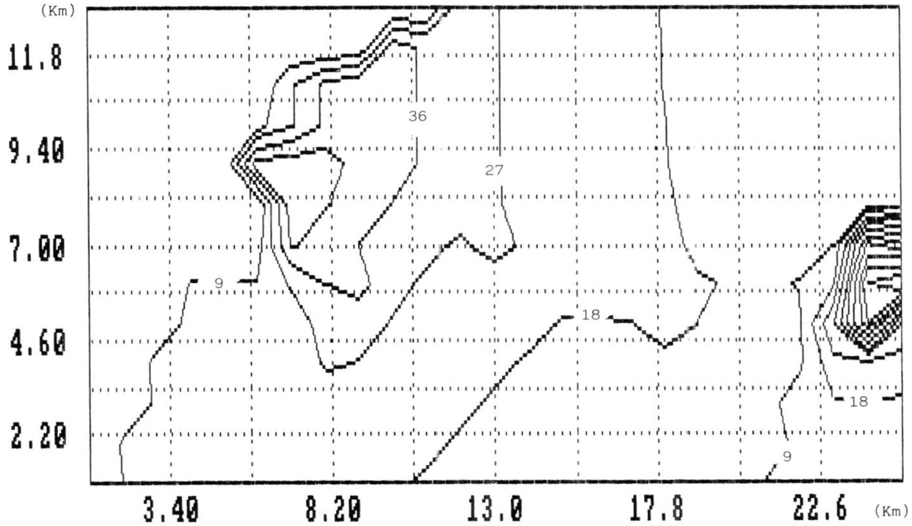

Fig. 12.11: Diurnal 12-h average concentrations expressed in $\mu g/m^3$ - August 28, 1981. Calculated SO_2 concentrations in the rectangle including Piacenza (right side) and Castel S. Giovanni (left side).

August 28, 1981 Wind direction is variable from the West, North, North-West. Wind speed reaches 7 m/sec at 100 m height. Note the emission from the domestic sector and from Levante power plant are still zero. The SO_2 concentrations are low over the whole area (Fig. 12.11).

Figure 12.12 shows a comparison of 12-h average computed and observed SO_2 concentrations. The open country monitors, located at Castel S. Giovanni, Sarmato and Pievetta, are considered individually, while for Piacenza the mean concentration in the city as calculated by **DIMULA** is compared with the measured concentration averaged over the three measuring stations located in the town; this choice was made to take into account the limited spatial resolution of the model and of the present input emission and meteorological data, which do not allow the calculation of a detailed field concentration in the town of Piacenza.

Fig. 12.12 shows that the **DIMULA model** tends to overestimate the urban concentrations. To a lesser extent, open country SO_2 measurements are underestimated.

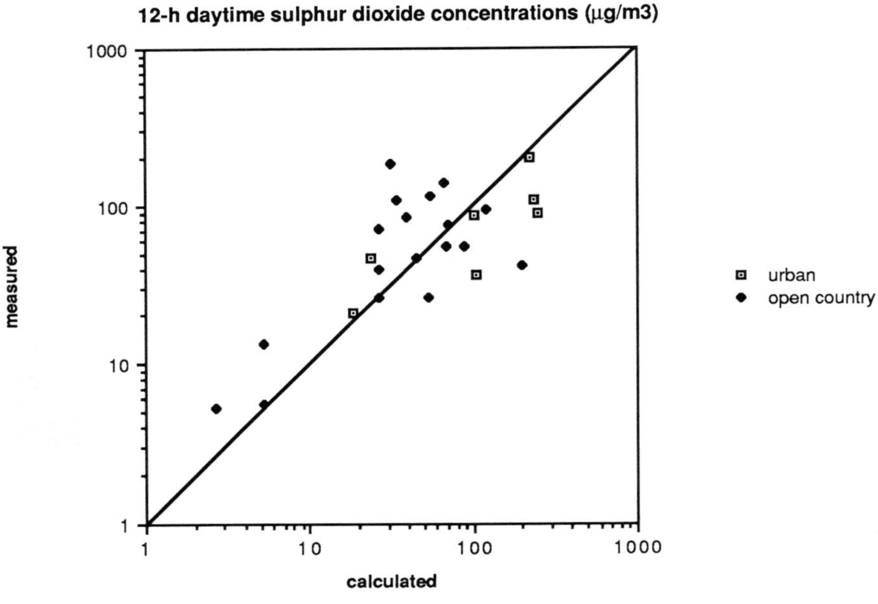

fig. 12.12: Measured versus calculated SO_2-concentrations.

Annual concentration estimate of SO_2

Fig. 12.13 shows the annual SO_2 concentration field in the rectangle, as calculated by DIMULA. The maximum SO_2 values are located in the city of Piacenza and in Pievetta.

The model tends to overestimate the actual concentrations measured by the local pollution monitoring network; the ratio between calculated and measured sulphur dioxide concentrations is 0.98 for Castel S. Giovanni, 1.3 for Sarmato, 1.9 for Pievetta and 1.4 for Piacenza.

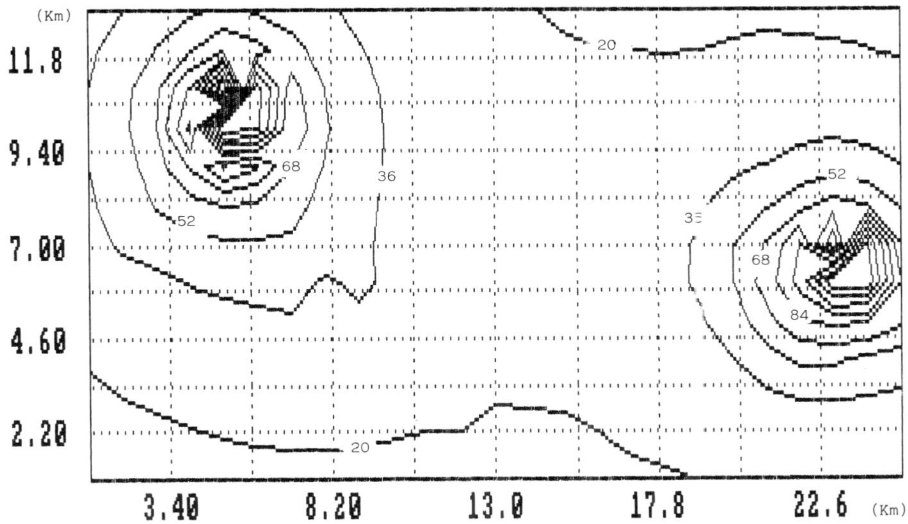

Fig. 12.13: Calculated annual SO_2 concentrations expressed in $\mu g/m^3$ in the rectangle including Piacenza (right side) and Castel S. Giovanni (left side). Year: 1981.

The spatially averaged SO_2 annual concentration was considered as an indicator of the relative weight of the impact of each source in the considered area. It is easily seen that this indicator is strongly influenced by location (relative to the emission sources), and the extent and shape of the spatial domain considered, so that the observations presented below are greatly dependent on these choices.

Fig. 12.14 gives the percentage contribution from the various source categories to both total SO_2 emission and SO_2 concentration (expressed as spatially averaged annual concentration). The figures suggest at least two considerations:

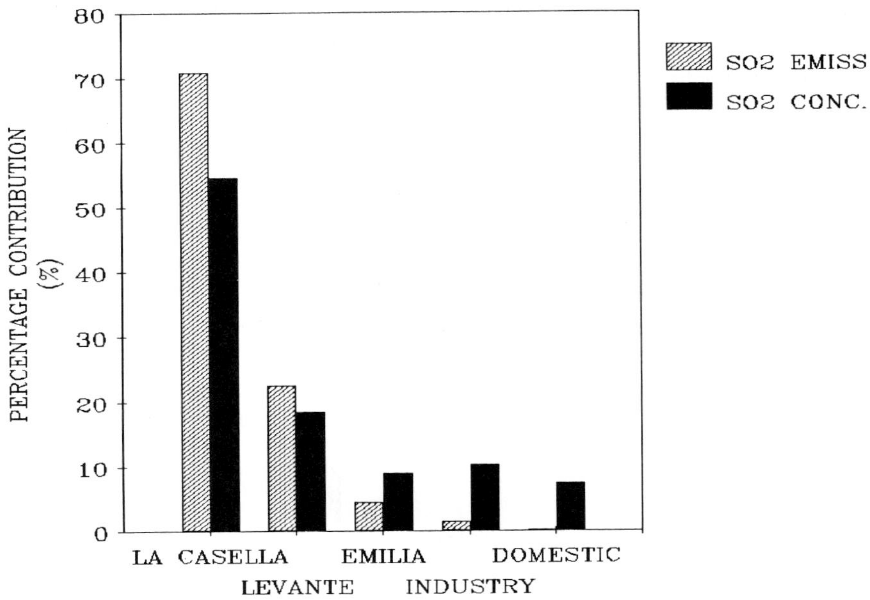

Fig. 12.14: Pollution load distribution of sulphur dioxide in terms of emission and of estimated air concentration in 1981 (only sources simulated by DIMULA are considered).

1. At a local level the relative impact of domestic and industrial sources is far greater than that of the power plants equipped with tall stacks. Thus, the increase of 1% with respect to the total SO_2 emission in the domestic sector would cause an increase in the concentration of the order of 25%; similarly, a 1% increase of the total SO_2 emission in industry would result in an increase of about 7% in the concentration; finally, if the power plants increase their emission of 1% of the total SO_2 emitted, there is a marginal impact in terms of concentrations of 0.7% only; this is mainly due to the fact that an elevated source has a minor relative impact on the immediately surrounding environment.

2. The absolute estimated contribution of the three power plants in terms of concentration (about 80%) is far from being negligible. This large contribution is mainly due to dispersion under instrument-calm conditions (about 50% of the cases over the year); these are the most critical conditions as far as the reliability of the dispersion models is concerned, so the uncertainty connected with this estimate must not be

disregarded. Notwithstanding this, the model suggests that attention must be paid to the local impact (in addition to the long-range impact) of elevated emission sources, where particular meteorological conditions occur, as is the case of the Po Valley.

If we restrict our attention to the town of Piacenza, the influence of La Casella power plant in terms of annual SO_2 concentration is about 9%, (55% in the whole rectangle); Levante and Emilia power plants account for about 35% and 19% respectively (18% and 9% in the whole rectangle); industry and domestic sectors give a contribution of about 16% and 21% respectively (10% and 8% in the whole rectangle). In the city of Piacenza there is a comparatively minor impact from La Casella, while greater impact comces from the Piacenza and Emilia power plants. Industry, and especially the domestic sector, have a far greater impact on the town. Compared with the average situation in the rectangle, it was reasonable to expect all these things.

12.4 DISCUSSION AND CONCLUSIONS

To conclude this presentation we wish to discuss briefly all those elements, that in our opinion, are affected by *major* uncertainty, and hence have a negative influence on the reliability of the results either because the simulation is worse than in the case with better data or (more dangerously) because a "good" result was obtained randomly.

Here we are discussing only those problems arising from uncertainty inherent in input data and model parameters. We do not discuss the goodness of the model structure in the face of possible alternatives.

The critical points of the analysis were found to be the emissions, stability classes, characteristics of thermal inversion, wind direction.

SO_2 emissions are estimated by means of emission factors, taking into account the main chemical and physical characteristics of the fuel utilized, the combustion process, and presence and efficiency of air pollution abatement systems. The more detailed this information, the more reliable the choice or the definition of an appropriate emission factor (understood as the quantity of pollutant produced per unit of fuel utilized), and consequently the more precise the emission estimate. The available information allowed us to estimate emissions on a monthly basis, a time scale which is very different from that considered in our short-term estimates (several hours).

It is obvious that emissions can vary greatly throughout different days of the week and hours of the day. Hence the emission information is much less detailed than would be necessary for studying short-term SO_2 dispersion.

The wind speed and insolation data, together with general information on the meteorology of the area on the specific day (cloudiness, fog etc.) are used to determine the stability classes. The method applied is basically as proposed by Pasquill in 1961. The aim was to determine approximately the atmospheric dispersion characteristics via routine meteorological observations. A correct determination of the stablity class is strongly affected by quantity and quality of available meteorological information, especially those about insolation.

Regarding the elevated inversion layer, that in persistent form occurs in the Po Valley during the presence of fog, it is important to predict whether the plume can penetrate the layer (with zero impact at ground level), or remains below the inversion (plume trapping), in which case the impact at ground level is significant and can also be very strong. Briggs' model enables one to estimate (with some degree of uncertainty) if a plume penetrates, or not, an elevated inversion; to do this, the model needs some information, i.e. the height, the thickness and the temperature drop between the top and bottom of the inversion layer. As for the SO_2 emissions, the best should be to have their characteristics in the days considered, while these data were hypothesized on a basis of statistics about thermal atmospheric inversion in the Po Valley (Bocci, 1973).

The local meteorological network considers 8 sectors only (each of 45°) for defining wind direction. This causes a significant uncertainty in the plume trajectory, increasing at greater distances from the emission point. For long-term simulations the difference in considering either 8 or 16 sectors tends to disappear, but for short-time simulation results may be quite different: it is preferable to have the wind direction distributed over 16 sectors, or even the exact mean direction of the wind for each time interval considered. The latter information has the advantage of allowing the use of a gaussian crosswind distribution of the pollutant concentration, that is a smooth function (instead of a uniform crosswind distribution inside the sector and zero elsewhere), which is by far more realistic. Unfortunately the local meteorological network, used for our short-term simulations, considers only 8 sectors for wind direction: this source of uncertainty must not be disregarded.

In summary, we stress two points:

1. **Better information of the source term** (the SO_2 emission data) is a preliminary condition for more reliable simulation, no matter the type and sophistication of the model adopted.

2. We need **a deeper understanding of the (possible) links between some key parameters of the diffusion models and routine meteorological data**, to extend the reliable application of these models to situations where "ad hoc" measurements are not available.

12.5 EPILOGUE

At present, of the two power plants that operated in the city of Piacenza in 1981, one (Emilia, 140 MWe, stack height 47 m) is no longer in operation; the other (Levante, 680 MWe, stack height 120 m) burns less residual fuel oil (112,000 t in 1986, 485,000 in 1981) and more natural gas ($479*10^6$ m^3 in 1986, $228*10^6$ m^3 in 1981) which is practically sulphur free.

REFERENCES

Bocci, R., 1973. Andamento delle inversioni termiche nel periodo 1967-1972 (Semestre freddo). CRIAL Lombardia.

Briggs, G.A., 1975. Plume Rise Prediction. In: "Lectures on air pollution and environmental impact analysis." Workshop proceedings, Boston, Mass., Sept. 29-Oct. 3, 1975, pp. 59-111. American Meteorological Society, Boston, Mass.

Cagnetti, P. & V. Ferrara, 1982. Two possible simplified diffusion models for very low wind-speed. Rivista di Meteorologia Aeronautica, Vol. 42, pp. 399-404.

Cirillo, M.C., G.C. Clerici & D. Manzi, 1986. Manuale d'uso del codice DIMULA - Versione 1.0. ENEA report RT/STUDI/86/2.

Cirillo, M.C., W. Bocola & L. Triolo, 1988. Environmental Impact of Air Pollutants with Emphasis on Crops: the Piacenza Case Study. In: "Advances in Environmental Modelling", A. Marani (ed.), Elsevier, pp. 293-312.

ENEL/AM, 1981. Caratteristiche diffusive dei bassi strati dell'atmosfera. Ente Nazionale per l'Energia Elettrica e Servizio Meteorologico Aeronautica Militare, Rome.

Frank, J. & K.R. Röber, 1983. Acoustic sounder (SODAR) investigations for the measurements of plume rise of wet and dry plumes up to stack height of about 300 m. In: "Proceedings of the VIth World Congress on Air Quality, Vol. 3, pp. 167-173. International Union of Air Pollution Prevention Associations, Paris.

Pasquill, F. & F.B. Smith, 1983. Atmospheric diffusion, 3rd edn. Ellis Horwood Ltd., Chichester.

Pinchera, G.C., W. Bocola, M.C. Cirillo, M. Martinelli & L. Triolo, 1984. An assessment of the environmental effects of energy systems, with emphasis on oil. Environmental International, Vol. 10, pp. 396-417.

Schnelle, K.B., 1979. The engineers guide to air pollution meteorology.

U.S. Environmental Protection Agency, 1985. Compilation of air pollutant emission factors. Fourth edition.

Yang, B.T., 1979. Atmospheric dilution. In: "Handbook of environmental engineering, Vol. 1." Edited by L.K. Wang and N.C. Pereira, the Humana press, Clifton, New Jersey, pp. 199-270.

CHAPTER 13

SIMULATION AND DATA ANALYSIS OF FOUR CHLOROBENZENES IN A LARGE LAKE SYSTEM (LAKE ONTARIO) WITH TOXFATE, A CONTAMINANT FATE MODEL

by
Efraim Halfon and Barry G. Oliver

13.1 INTRODUCTION

Lake Ontario, one of the Great Lakes of North America, receives large amounts of toxic contaminants from inflowing rivers and from the atmosphere (Allan et al., 1983). The problem of toxic contaminants is of particular concern since Lake Ontario is a source of freshwater to many towns along its shores. The lake is also a potentially rich source for commercial fishing. Unfortunately, the continuous inputs of various chemicals has lead to significant residue levels in fish (Whittle and Fitzsimons, 1983). From a practical or management point of view the prediction of concentrations in water, used as potable water supply, in fish, used for human consumption, and in bottom sediments, which can remain as a major source of internal loadings after external inputs are reduced, is of fundamental importance in a large lake that may respond very slowly (average water residence time is 7.8 years) to reduced contaminant loadings.

Lake Ontario is a large lake and the spatial scales involved - the lake is 300 km long with a volume of 10^{12} m^3 and wind driven currents transport a water mass ten times greater than the Niagara River flow (6000 m^3/s; Murthy et al., 1984) - make any modelling effort difficult. Physical transport in the lake plays an important role in determining the spatial distribution, fate and persistence of chemicals. The contaminants may undergo a number of chemical transformations and changes of phases. They may be biodegraded or photolyzed, they can move to the atmosphere by volatilization or be buried in the bottom sediments, they can leave the lake via the St. Lawrence River or bioaccumulate in the food chain. A mathematical model must take all these factors into consideration to produce an integrated and reliable prediction of a contaminant's fate.

TOXFATE, an organic TOXic contaminants FATE model, originally developed (Halfon, 1984a) in 1983 (version 1.0), has now an updated configuration (version 2.1), which includes the water circulation of Lake Ontario. The Niagara River receives large discharges from industries and leachings from dumps located along its shores and has been a source of contaminants to the lake since the beginning of the century (Durham and Oliver,1983). From the analysis of a radiodated sediment core and measurements of contaminant loadings from 1981 to 1983 (Oliver and Nicol, 1984), we have been able to estimate long term past loadings of several contaminants to Lake Ontario from the Niagara River. In this paper we compare the model with field data for four chlorobenzenes. The model includes degradation processes but biodegradation and photolysis terms are set to zero since these processes are *presumed* to be very low or negligible for the studied chlorobenzenes. The four chemicals represent a class of volatile contaminants that may remain in the lake temporarily or possibly be persistent: their final sink may be the atmosphere rather than the bottom sediments or the lake waters.

13.2 THEORETICAL CONSIDERATIONS

TOXFATE simulates the time-varying concentrations of a toxic contaminant in the water column. The equations can be parameterized to represent a variety of contaminants and the model is formalized as a system of ordinary differential equations. The state variables (Fig. 13.1) are organic contaminant concentrations in suspended clay and colloidal matter, suspended silt, suspended sand, water, plankton, and fish, bottom sediments and benthos compartments. **TOXFATE** also includes a resuspension submodel, which was not used in this exercise, since Charlton (1983) provided field measurements.

Most process relationships are nonlinear and reflect the continuous movement of the contaminants between sediments, water and biota. The transport processes of advection in water are handled by a box model (Fig. 13.2), the pollutant is assumed to be immediately and completely mixed within each compartment. Two classes of fish are included in the model, small fish of about 5 g in wet weight and large fish with a wet weight of 3.1 kg. These weight classes were chosen according to field data. The contaminant absorption from water (Eqs. 13.10-13.11) is conceptualized according to Neely's (1979) model, while uptake from food sources (Equation 13.12) follows a formulation introduced by Elliot (1975). TOXFATE includes a novel equation (13.13) to describe excretion of contaminants by fish. This formulation explains the observed increasing bioconcentration factors as the degree of chlorine substitution in the aromatic ring increases. For example in fish the bioconcentration factor for 1,2,4-trichlorobenzene is 5,800, for 1,2,3,4-tetrachlorobenzene 28,500, for pentachlorobenzene 260,000 and for hexachlorobenzene 2,400,000. Clearly these bioconcentration factors are *not* a linear function of log k_{ow}.

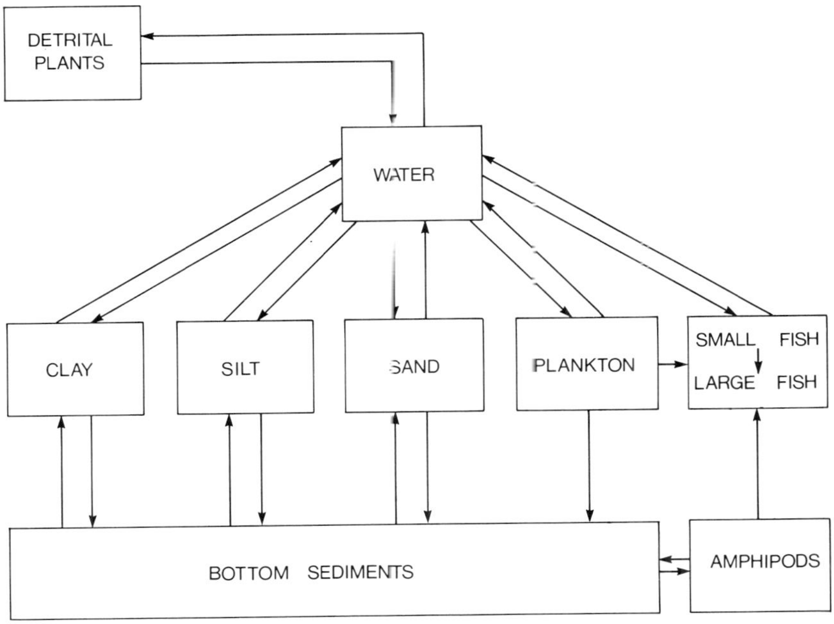

Fig. 13.1: Structure and state variables of the model TOXFATE. The arrows indicate the possible direction of flow of toxic contaminants between compartments. The state variables are the concentrations of the contaminant in each part of the ecosystem.

Volatilization

The volatilization rate is modelled using the well known two film representation of the water surface (Liss, 1973). Equation 13.1 shows the calculation of the volatilization parameter k_v [h^{-1}].

$$k_v = \frac{1}{K_{liq} + K_{gas}} \quad \frac{A}{V} \tag{13.1}$$

where K_{liq} [hour/m] is the reciprocal of the liquid phase mass transfer coefficient, K_{gas} [h/m] is the gas-phase mass transfer coefficient, and A [m^2] and V [m^3] are the areas and volumes of the lake, or lake compartments if the lake is divided into spatial compartments. The parameter K_{liq} is computed as:

$$K_{liq} = \frac{1}{k_{02} \sqrt{\frac{32}{MW}}} \qquad (13.2)$$

where k_{02} is the oxygen (molecular weight 32) exchange constant [m/hour] computed according to measured wind speeds (u) at 10 metres over the water surface (Banks, 1975),

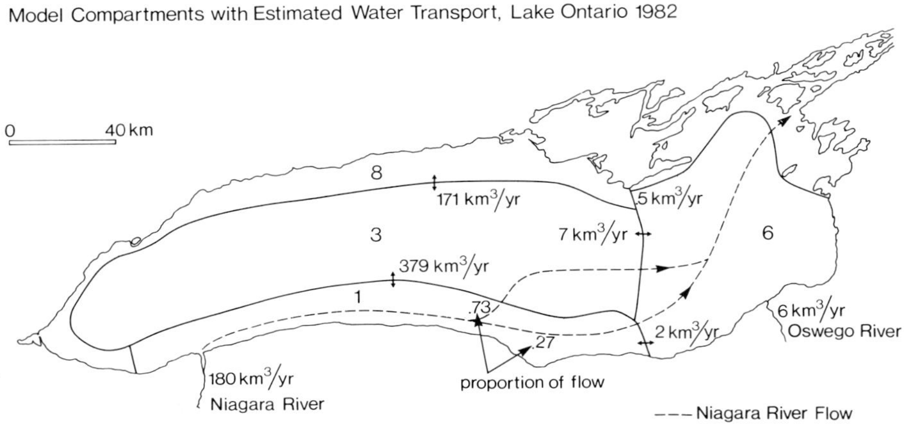

Fig. 13.2: Sections used to describe Lake Ontario environment and circulation. The flow among different sections of the lake, is several times the flow of the Niagara River. This lake flow is wind driven.

$$k_{O2} = 1.51 * 10^{-2}\ u, \text{ for } u \leq 5.5 \text{ m/s} \qquad (13.3)$$

$$k_{O2} = 1.15 * 10^{-3}\ u^2, \text{ for } u > 5.5 \text{ m/s} \qquad (13.4)$$

In this model the effects of intermittent turbulent and advective transport events are not included since, as Burns et al. (1981) noted, Whitman models usually differ very little from more complex (e.g. surface renewal) models (Danckwerts, 1970). In lake Ontario the average wind velocity is 6.8 metres per second at a height of 10 metres and therefore k_{O2} has a value of 0.053 [m/h].

The gas resistance, k_{O2}, is

$$K_{gas} = \frac{1}{\frac{WH}{RT}\sqrt{\frac{18}{MW}}} \quad (13.5)$$

where W [m/h] = 0.1857 + 11.36 u [m/sec] at 10 metres over the water surface (Liss, 1973), 18 is the molecular weight of water, R is the gas constant [m^3 atm/mol] and T is the water temperature [°K]. Should H be not known, it can be computed from vapour pressure V_p (in mm Hg) and the contaminant solubility S [mole/m^3].

$$H = \frac{V_p}{760\ S} \quad (13.6)$$

The factor 760 in Equation 13.6 converts the vapor pressure from mm Hg to atmospheres.

Sorption on suspended sediments

Suspended sediments in Lake Ontario consist of three fractions, namely clay and colloidal matter, silt, sand. To describe sorption we assume none-quilibrium between water and suspended sediments. In TOXFATE, adsorption and desorption from suspended sediments are quantified as

$$VS \frac{ds}{dt} = V\, k_1\, k_{oc}\, org\, w - k_2\, s\, V\, S \quad (13.7)$$

where s is the concentration of contaminants in suspended sediments [$mg_{contaminant}/kg_{suspended\ sediments}$], k_1 and k_2 are parameters whose respective value of $4.56 \cdot 10^{-7}$ [1000/h] and $1.14 \cdot 10^{-4}$ [hr^{-1}] were obtained by calibration, "org" is the fraction of organic carbon in suspended sediments, about 0.14 in Lake Ontario, and w is the concentration of the contaminant in water [$mg_{contaminant}/m^3_{water}$], V is the water volume [m^3] and S is the concentration of suspended sediments [kg/m^3].

Living organisms

Plankton (Phytoplankton and zooplankton combined)
Equation 13.7 also describes the uptake of toxic contaminants by

plankton (phytoplankton and zooplankton are combined in one compartment) from water, the only difference between equation 13.7 and 13.8 is that excretion is temperature dependent or:

$$V\,P\,\frac{dp}{dt} = V\,k_1\,k_{oc}\,\text{org}\,w - k_2\,f(T)\,p\,V\,P \tag{13.8}$$

p is the concentration of contaminants in plankton $[mg_{contaminant}/kg_{plankton}]$, org is the fraction of organic carbon in plankton, about 0.20; f(T) is 1.024^T; and all the other parameters are the same as in Equation 13.7. P is the concentration of plankton $[kg/m^3]$.

Fish (two weight classes)

Fish absorb organic contaminants both from water through the gills and from their food, i.e. plankton, benthos, or smaller fish. To follow the movement of organic contaminants into two fish size classes, small fish like smelt and large fish like trout, the following nonequilibrium mass balance model is included (Leidy and Ploskey, 1980)

$$V\,F\,\frac{df}{dt} = V\,k_3\,w + A\,F\,V - Ef\,F\,V \tag{13.9}$$

where k_3 $[m^3{}_{water}/(kg_{fish}\ hour)]$ represents the uptake directly from water; f is the concentration of the contaminant in fish $[mg_{contaminant}/kg(dry\ weight)_{fish}]$; A (Eq. 13.12) is the assimilation rate from food $[kg_{prey}/(kg_{predator}\ hour)]$, where predator is the fish itself and prey might be plankton, benthos, and for large fish, small fish. E $[h^{-1}]$ is the excretion rate computed with Equation 13.13 below. F is the concentration of fish $[kg/m^3]$. This model formulation assumes that the fish has not necessarily achieved equilibrium with the current toxicant concentration in the natural water from which it was taken.

More complex models have been presented in the literature (Jensen et al., 1982) to account for blood circulation and oxygen in the water. Given the lack of data or blood water partition coefficients and fat blood partition coefficient for most contaminants and fish, an unnecessarily complex model including such factors is deemed unreasonable until a definite need is shown. Intake of organic contaminants from water is parameterized as a function of k_{ow} (Neely, 1979):

$$k_3 = \frac{(0.07 \log k_{ow} - 0.02) * RV}{W * 1000} \tag{13.10}$$

where W, the individual fish weight [g], is 5 grams for the small fish compartment and 3180 grams for the large fish compartment and 1000 is a constant to convert grams to kilograms. Neely derived the function (0.07 log k_{ow} - 0.02) to account for the different assimilation efficiency of different contaminants. Note that the absorption efficiency is *a function of log* k_{ow}, *not* k_{ow} itself, therefore small adsorption efficiency differences between two chemicals, e.g. 1,2,4-TCB and HCB, may not result in the large differences observed in bioconcentration factors (5,800 for 1,2,4-TCB and 2,400,000 for HCB); this problem has been solved in TOXFATE with a new formulation of the excretion processes, Equation 13.13 below. RV is the filtration rate $[m^3_{water}/h]$ of fish,

$$RV = k_4 * T * W^{0.8} \qquad (13.11)$$

where T is the water temperature [°C], k_4 = 10.3 $[m^3_{water}/(kg^{-0.8}_{fish}\ °C\ h)]$ for the big fish and k_4 = 6.45 $[m^3_{water}/(kg^{-0.8}_{fish}\ °C\ h)]$ for the small fish compartments (modified from Neely, 1979).

The second major source of contaminants is food. In **TOXFATE**, small fish feed on plankton and benthos while the large fish feed on small fish and plankton. These feeding rates were computed using Elliot's (1975) model assuming 70% growth (Elliot provides model formulation for growth rates from 10% to 90%); the general formulation of Elliot's model is

$$A = k_5\ W^{k_6}\ e^{(k_7 * T)} \qquad (13.12)$$

where A is the feeding rate $[kg_{prey}/(kg_{predator}\ h)]$ of fish, the para- meters k_5 $[kg_{prey}/(kg_{predator}\ h)]$, k_6 [dimensionless] and k_7 $[°C^{-1}]$ have the relative values of [k_6: $3.58*10^{-5}$; $1.33*10^{-4}$; $3.16*10^{-3}$], [k_7: 0.769; 0.781; 0.764], [k_8: 0.335; 0.129; -0.76] for the three temperature ranges (0-6.7 °C, 6.7 - 14.9 °C, and higher than 14.9 °C.

To reproduce the very different bioconcentration factors for the four chlorobenzenes, relatively small for 1,2,4-TCB, 1,2,3,4-TeCB and QCB and very high for HCB, the excretion rate process in **TOXFATE** was quantified as

$$E = 2.28\ 10^3\ W^{-0.75} / k_{ow}\ [h^{-1}] \qquad (13.13)$$

where 2.28 10^3 $[g^{0.75}\ hr^{-1}]$ is an empirical scaling constant. The scaling constant is very large since k_{ow} can range from 10,000 to 1,000,000. The justification for Equation 13.13 is that contaminants with *a high* k_{ow} *are retained* in the fish much more than less hydrophobic compounds and bigger fish retain more contaminants than smaller fish since their metabolism is relatively slower. Before developing Equation 13.13 we tried many other formulations published in the literature but none worked since they are all proportional to log k_{ow} rather than to k_{ow} itself.

Sedimentation and resuspension

Kemp and Harper (1976) estimated that $4.56*10^9$ kg of fine grained material enter Lake Ontario from the Niagara River each year, most of this allochthonus suspended sediment deposits to the bottom of Lake Ontario. The average concentration of suspended sediments in Lake Ontario is about 3 mg/L (Nriagu et al., 1981) and most of these sediments are of autochthonus origin. Suspended sediments are therefore always created as metabolic products of primary production and destroyed by microbial degradation as they sediment to the bottom. These processes pose the most problems to the modeler since the present version of TOXFATE does not include a eutrophication model of Lake Ontario and therefore plankton productivity is not included as a process to create new suspended sediments to replenish those which sediment. In TOXFATE, this problem is solved by keeping a constant plankton standing stock in the lake ($kg_{plankton}/m^3{}_{water}$) but computing the losses of seston to the bottom sediments as if some plankton were continually created by primary and secondary production (as mentioned before phytoplankton and zooplankton are combined into one comartment). Losses by biodegradation of plankton in the water column are accounted for in Equation 13.8 (parameter k_2). Losses from the water column by sedimentation were computed by the method of Dietrich (1982) for sands and silts. The sedimentation velocity of clay, silt and sand was set at 0.5, 0.8 and 19.4 m/day respectively, and the sedimentation of zooplankton at 0.2 m/day.

Losses by sedimentation are important process but Charlton (1983) showed that resuspension is also an important process. In the model resuspension rates from the bottom sediments are set at 5 g $m^{-2}day^{-1}$ (Charlton, 1983); desorption from suspended sediments is computed in the model but passive diffusion of chemicals into the water column from "in place" bed sediments is not considered. In the past years other sampling programs took place to determine the fate of suspended sediments: Oliver and Charlton (1984) measured the proportion of Niagara River contaminant input that sediments in Lake Ontario using sediment traps over a six month period in 1982. This short term study yielded somewhat lower amounts of chlorobenzenes (CB's) sedimenting than Oliver's (1984) other study. The average suspended sediment concentration in the Niagara River over the trap deployment period was 5 mg/L, somewhat lower than values reported in other studies - 8.4 mg/L (Kuntz and Warry, 1983) and 22 mg/L (Kemp and Harper, 1976). Higher suspended sediment concentrations in the river increase the percentage of CB's sedimenting to the lake bottom. In any case, considering the limited number of measurements in both field studies, the agreement between them can be considered fair. For the modelling effort only long term mean resuspended sediment concentrations in the river (which hopefully have not changed dramatically in the 1900's) are being significant. Results of a sensitivity analysis of the model to suspended sediment concentrations and resuspension rates are presented in the discussion section.

The physical model

A finite section approach describes the lake, each section or compartment is completely mixed (Figs. 13.1 and 13.2). Within each volume the differential equations describing the distribution of the contaminant are then solved. The finite section approach is essentially a finite difference approximation to an ordinary differential equation (Thomann, 1972). The hydrodynamic description includes a eastward current on the south side of the lake. Simons and Murthy (1985) found that the total transport in the belt of eastward flow is ten times larger than the hydraulic transport associated with the Niagara River flow, thus suggesting a recirculation of 90% of the river inflow.

13.3 PAST LOADINGS ESTIMATES

The approximate loadings of chlorinated contaminants to Lake Ontario from the Niagara River for the period 1981 to 1983 can be estimated using the weekly sampling data of Oliver and Nicol (1984). During the course of that study, a sediment core was also collected, analyzed and dated using radiochemical methods (Durham and Oliver, 1983). The major chlorinated compounds found in both of these studies were chlorobenzenes, or CB's, so we chose the four chemicals, hexachlorobenzene (HCB), pentachlorobenzene (QCB), 1,2,3,4-tetrachlorobenzene (1,2,3,4-TeCB), and 1,2,4-trichlorobenzene (1,2,4-TCB) to test and validate the model. In addition to their presence in all samples, these compounds span a fairly wide range of physical chemical properties.

Present loadings (1981-1983) and historical loadings (1909-1981) were estimated from the sediment core analysis (Durham and Oliver, 1983) using the assumption that a constant percentage of incoming contaminants was sequestered to settling particulates, which became bottom sediments. For example, the 1,2,4-TCB concentration in the 7 to 8 cm interval of the sediment core (1966-67) is approximately 3 times higher than in the superficial sediment (0 to 1 cm). Thus, we estimated loadings for this compound in 1966/67 to be 7400 kg per year, three times the measured 1981 to 1983 loadings of approximately 2400 kg per year. These year to year loadings estimates drive the model (Table 13.1) to produce the simulations whose results are presented in Tables 13.3 and 13.4. The sum of these year to year loadings (Table 13.3) compare well with the overall loadings estimates made by Oliver (1984) and also reported in Table 13.3. The laboratory studies of Karickhoff (1981) have shown that at low concentrations, such as those observed in the Niagara River (ng/L range), the sorption of hydrophobic pollutants by sediments is linear, reversible and characterized by a partition coefficient.

Table 13.1
Estimated loadings (kg/year) to Lake Ontario from the Niagara River 1909-1983

Year/compound	1,2,4-TCB	1,2,3,4-TeCB	QCB	HCB
1909-1920	1500	150	60	46
1921-1931	2400	290	140	230
1932-1939	3300	490	160	150
1940-1946	2800	500	410	180
1947-1949	4400	880	450	180
1950-1953	3900	670	280	110
1954-1956	3900	1000	440	190
1957-1959	6100	1200	560	260
1960-1962	8900	2100	770	920
1963-1965	9000	3800	870	710
1966-1967	7400	3200	710	390
1968-1969	5700	1400	420	350
1970-1972	3900	1300	300	200
1973-1974	3700	940	200	160
1975-1976	3700	880	200	100
1977-1978	2400	610	150	100
1979	2200	760	240	140
1980	2400	760	240	140
1981	2400	760	240	160
1982-1983	2400	760	240	120

We have performed a similar detailed analysis of three other sediment cores from the Niagara, Mississauga and Rochester basins of Lake Ontario and these cores show virtually the same trends as the above core which was collected near the mouth of the Niagara River. Historically there have been no long-term changes in flow volume in the river (Inland Waters Directorate, 1984). Since Lake Erie is the major source of suspended sediments to the Niagara River seasonal changes in suspended sediments concentrations with weather conditions are expected. Nevertheless, long term mean suspended sediment concentrations probably will be fairly constant.

The chemical properties for 1,2,4-TCB, 1,2,3,4-TeCB, QCB and HCB are presented in Table 13.2. The k_{ow} were measured in the laboratory (Oliver and Niimi, 1983) and the k_{oc} were derived from field measurements in the Niagara River and Lake Ontario (Oliver, 1987). Note that in this environment measured k_{oc}'s are higher than k_{ow}'s.

Table 13.2
Physico-chemical properties of the four chlorobenzenes

	1,2,4-TCB	1,2,3,4-TeCB	QCB	HCB
Molecular Weight	181.45	215.9	250.3	289.81
Solubility [mg/L]	30	4.32	0.55	0.005
Solubility [mole/m^3]	$1.7*10^{-1}$	$2*10^{-2}$	$2.2*10^{-3}$	$3.5*10^{-5}$
log k_{ow}	4	4.5	4.9	5.5
log k_{oc}*	5.1	5.2	5.8	6.1
Vapour Pressure [TORR]	0.4200	0.0100	0.0017	0.00001
Henry's Law Constant [m^3-Atm/mole]	0.00334	0.00066	0.00102	0.00037

13.4 LAKE ONTARIO DATA

Table 13.3 shows two sets of results. The first set was collected by Oliver (1984) who estimated the overall inputs to the lake, the amounts present in the lake water and bottom sediments. He also computed the percentage losses of chemicals due to sedimentation, volatilization and transportation (St. Lawrence River). This set of data was obtained without using a mathematical model and full results are presented in Oliver (1984). The model TOXFATE produced the second set (Table 13.3). Note that year to year loadings (Table 13.1) drive the model simulations and that total loadings are shown in Table 13.3 only for comparison with Oliver's (1984) estimates.

Oliver estimated the percentages of CB's lost via the St. Lawrence River by analyzing two large volume water samples near the outflow of the lake. The CB's in suspended sediments were estimated by analyzing centrifuged material from three stations at two depths on one occasion. The chlorobenzenes concentrations in sediment trap material collected for six-month intervals between 1982 and 1985 in the three major lake basins were in close agreement to those concentrations in the above centrifuged material. The CB's in lake biota were taken from the literature (Oliver and Niimi, 1983; Fox et al., 1983). Bottom sediments were sampled more extensively - one core from each of the major sedimentation basins (Niagara, Mississauga, Rochester) was analyzed in detail together with 15 superficial sediment samples from each of these basins as well as three samples from the Kingston Basin to obtain the CB masses in lake bottom sediments (Oliver, 1984).

Table 13.3
Comparison of simulations and field measurements of removal processes

	Method	Input (kg)	Water (kg)	Bottom Sediment (kg)	Sedimentation (%)	Volatilization (%)	Transport (%)
1,2,4-TCB	Model	270,000**	1050	39,000	14	76	10
Oliver, 1984		310,000	700	11,000	4 (1)***	93	3
1,2,3,4-TeCB	Model	64,000**	360	13,000	21	69	10
Oliver, 1984		66,000	210	3,400*	5 (2)***	93	2
QCB	Model	22,000**	71	10,500	48	43	9
Oliver, 1984		23,000	90	4,100*	18 (4)***	79	3
HCB	Model	16,000**	31	9,900	61	31	8
Oliver, 1984		15,000	90	8,500*	57 (15)***	38	5

* In sedimentation basins.

** From core data using one centimetre slices and used as inputs to the model. Model inputs are different from Oliver's estimates since they are based on yearly estimates rather than on a global 75 years estimate. Yearly loading estimates (Table 13.1) are used to drive the model.

*** Second (in parenthesis) sedimentation estimate. From Reference 27.

Table 13.3 also shows Oliver's estimates of volatilization losses from the field data; these estimates can be obtained by difference in two ways. The sum of the short-term (May to November, 1982) sedimenting material (Oliver and Charlton, 1984) plus the amount of material lost via the St. Lawrence River can be subtracted from the total loading to give volatilization: 1,2,4-TCB (96%), 1,2,3,4-TeCB (96%), QCB (93%) and HCB (80%). Volatilization losses calculated more precisely from long term bottom sediment masses (Oliver, 1984) and from losses via the St. Lawrence River (Oliver, 1984); these losses, reported in Table 13.3, are somewhat different: 1,2,4-TCB (93%), 1,2,3,4-TeCB (93%), QCB (79%) and HCB (38%).

13.5 MODEL SIMULATION

TOXFATE can simulate the contaminants' fate and the percentage of the loadings lost to the atmosphere or present in the lake; a good agreement between predicted and observed processes, and between computed and observed mass balance, is confirmation of the proper numerical behaviour of the mode.

Table 13.4.
Concentration of toxic contaminants in Lake Ontario in 1983, predicted and (observed from ref. 24). The standard deviation indicates spatial variability (concentrations in ng/L for water and ng/g dry weight for other compartments). Total loadings are presented to show that contaminant concentrations in the lake depend more on the physico-chemical characteristics than on loadings. * No observations available .

Contaminant/Compartment	1,2,4-TCB	1,2,3,4-TeCB	QCB	HCB
Water	.66 ± .40 (.52 ± .18)	.22 ± .14 (.14 ± .07)	.045 ± .028 (.054 ± .026)	.020 ± .012 (.054 ± .032)
Suspended sediments	22 ± 13 (15 ± 10)	9.5 ± 5.6 (5.3 ± 4.4)	7.7 ± 4.5 (5.3 ± 3.5)	6.7 ± 3.8 (14 ± 8)
Bottom sediments	156 ± 94 (98 ± 50)	66 ± 40 (36 ± 19)	51 ± 31 (36 ± 21)	46 ± 28 (100 ± 79)
Macro-invertebrates	14 ± 9 (79 ± 71)	22 ± 13 (46 ± 38)	45 ± 27 (42 ± 22)	175 ± 105 (170 ± 140)
Plankton	25 ± 15 (7 ± 4)	11 ± 6 (9 ± 7)	11 ± 7 (10 ± 10)	10 ± 6 (11 ± 12)
Small fish (5 g)	.22 ± .14 *	.40 ± .23 *	1.1 ± 0.6 *	12 ± 7 *
Trout (3.1 kg)	6.2 ± 3.8 (3 ± 2)	9.7 ± 5.9 (4 ± 2)	9.7 ± 5.8 (14 ± 5)	65 ± 38 (130 ± 60)
Total loadings in 75 years	270,000	64,000	22,000	16,000

The simulations were run for the period of 1909 to 1983 and took six hours each on a CDC Cyber 171 computer with a time step of one hour. Tables 13.3 and 13.4 show a comparison of computed concentrations and removal processes with data collected by Oliver (1984).

Concentrations and masses of contaminants in Lake Ontario

Table 13.4 shows computed and literature (Oliver, 1984) values for concentrations of several lake compartments. Simulated water concentrations in 1983 agree well with measured values. Suspended sediment concentrations vary widely in the lake with coefficients of variation being about 50% of the mean value. The agreement between calculated and observed concentrations is good for 1,2,3,4-TeCB and QCB while average computed 1,2,4-TCB concentrations are 30% higher and HCB concentrations are about 50% lower. Nevertheless, given the variability in both data and simulations the two sets cannot be considered different. Concentrations in bottom sediments similarly agree in measured and observed data sets. For all contaminants the agreement for fish concentrations is within a factor of one. The only possible exception is the average low concentration computed for HCB in trout (65 ± 38) versus observed concentrations (130 ± 60). Nevertheless, when standard deviations are considered the discrepancy is considerably smaller.

A mass balance of contaminants in the lake is important to assess the fate of the chemicals once they enter the lake. Table 13.3 shows the masses of the four chlorobenzenes in two lake compartments, water and bottom sediments. The agreement between Oliver's (1984) estimates and the model's are quite good for water. Given the very low percentage of loadings that remains in the lake - 0.28% and 0.40 % in the water compartment for 1,2,4-TCB and for HCB, respectively - the agreement is quite good. Similarly good comparisons are obtained for the other contaminants. Notwithstanding the very large inputs to the lake (15000 to 310000 kg over the examined 75 year period), the water fraction in 1983 contains only a minute proportion, 30 to 1000 kg for each of the chemicals over the whole lake, resulting in very low water concentrations in the order of 0.02 to 0.7 ng/L.

Mathematical models can determine which water fraction, suspended sediments or water, should be sampled to find a given contaminant. An analysis of Tables 13.3 and 13.4 shows that **TOXFATE** computes that 9.0% of 1,2,4-TCB can be found in suspended sediments and 90% in water. Our calculations using Oliver's data show that 9.2% of the mass of the contaminant is in suspended sediments and 89.8% in water, a close agreement. For the other three chlorobenzenes the computed percentages in suspended sediments are as follows: 1,2,3,4- TeCB 11.1% (10.6% observed); QCB 33.8% (21.7% observed); HCB 50.5% (42.3% observed).

TOXFATE simulates higher concentrations in the bottom sediments than estimated by Oliver (1984) and therefore higher quantities of the four chlorobenzenes at the bottom. These results are crucial for the determination of the importance of volatilization as a removal process. In fact if more contaminants are present in the bottom sediments, less are lost to the atmosphere. Simulations and data agree well for HCB but they disagree for the less chlorinated benzenes. For example the model estimates that 38,000 kg of 1,2,4-TCB are in the bottom sediments of the whole lake in 1983 versus 11,000 kg estimated by Oliver (1984) in the sedimentary basins. Therefore the model implies that less HCB volatilizes to the atmosphere than Oliver's calculations suggest.

A source of uncertainty is the estimate of suspended sediments entering the lake. In the present exercise suspended loadings were assumed constant over the whole 75 years, but this assumption might have not been valid. For assessing the relative influence of estimated suspended loadings concentrations and bottom sediments resuspension rates, a sensitivity analysis was performed on TOXFATE using the chemical hexachlorobenzene. Results show that the estimated volatilization rate is lower with lower resuspension rates but it is higher with lower suspended loading concentrations, i.e. more HCB is in the water fraction and therefore might be removed by volatilization. Similarly, low suspended loading rates reduce the amounts of HCB sedimenting to the bottom while low resuspension rates increase the amounts in the bottom sediments. Outflow to the St. Lawrence River is lower with low resuspension rates and higher with low suspended loadings.

Contaminants' fate and the lake's self-cleaning ability

The fact that only a low percentage of the four contaminants is in the lake confirms Oliver's (1984) estimate that volatilization is the most important route of removal for 1,2,4-TCB and 1,2,3,4-TeCB. The model suggests that volatilization is as important as sedimentation as a removal process for the other two chlorobenzenes QCB and HCB, while Oliver (1984) suggests that volatilization is the most important removal process also for QCB and HCB.

Volatilization rates vary according to the chemical, more for 1,2,4-TCB and less for HCB. For example the average yearly loadings since 1977 have been about 120 kg for HCB and 2300 kg for 1,2,4-TCB (Table 13.1) but concentrations in the bottom sediments top layer are very similar for the two compounds, about 0.1 μg/g, indicating that 1,2,4-TCB is much less persistent than HCB. As for fluxes, TOXFATE predicts that about 31% of HCB loadings were lost from the lake by volatilization processes, while Oliver (1984) estimated 38% during the period May-November 1982. Similarly for the other contaminants, the model produces lower volatilization estimates than Oliver's.

One last question that we investigated is the time frame for contaminants to be permanently buried in the bottom sediments if loadings were completely stopped. Two chemicals HCB and 1,2,4-TCB are chosen as representative of relatively persistent and more volatile compounds. As mentioned above, 1,2,4-TCB is very volatile and therefore only about 4 to 14% of the loadings are in the bottom sediments; if loadings were completely stopped the model predicts that the four chlorobenzenes could be buried under 8 cm of clean sediments in about 38 year.

13.6 DISCUSSION

A simulation model, TOXFATE, quantifies the behaviour of four chlorobenzenes, 1,2,4-TCB, 1,2,3,4-TeCB, QCB and HCB, in Lake Ontario. The purpose of this study is to verify the correct formulation of the model using four chemicals which belong to the same family but which have different physico-chemical characteristics. The computed and observed concentrations in the different components of the ecosystem agree within 50% for averages and are completely indistinguishable if standard deviations are considered (Table 13.4). The ranking by importance of sedimentation, volatilization and transport rates from the St. Lawrence River is the same as field estimates (Table 13.3) even if the relative percentages may not agree exactly.

As pointed out by Halfon (1984b) large lake toxic contaminant models are sensitive to one parameter, mainly the resuspension rate from the sediments; 'fiddling' or calibrating the model with that parameter allows an almost perfect fit to the data, within a few percentages for the mass of the four contaminants. Therefore, the question is whether we should only use measured parameters, such as those available from Charlton (1983). Here we used the measured value even if a calibrated model would have given a better fit. Comparison with other contaminant data in the future will permit the resolution of this problem. The inescapable conclusion, however, is that resuspension rates should be measured accurately wherever possible since they have fundamental role in controlling contaminants' fate in Lake Ontario (see also Halfon, 1984b).

The main problem of the prediction of toxic contaminants' fate, once a model has been satisfactorily validated for a few compounds, is the lack of loading data; this problem is particularly evident in a large lake system. Presently, the Niagara River is regularly monitored for compounds in the water and in suspended sediments (Kuntz and Warry, 1983), however; such monitoring has taken place only since the late 1970's, and earlier loadings data can only be inferred from bottom sediment data. For new or recently released compounds, or for compounds not yet identified in the lake, these data are not available and may not be available in the foreseeable future; for contaminants which are already entering the lake, the bottom sediments are a good source of past loadings data provided that a comprehensive data base with concentrations and dating is available, as shown by Oliver (1984).

13.7 ACKNOWLEDGMENTS

Many individuals at the National Water Research Institute provided their knowledge and scientific insight to make this model a realistic representation of Lake Ontario. In particular, we would like to thank the late T.J. Simons, R.J. Allan, J. Carey, M. Charlton, F.C. Elder, F. Rosa, M.E. Fox, K. Kaiser, B. Krishnappan, J. Magurie, A. Mudroch N.A. Rukavina, And W.M. Strachan.

Six anonymous reviewers provided useful comments to improve the paper readability and understanding.

REFERENCES

Allan, R.J., A. Mudroch and M. Munawar, eds. 1983. The Niagara River - Lake Ontario pollution problem. J. Great Lake Res. (Special Issue on the Niagara River), 9: 109-340.

Banks, R.B., 1975. Some features of wind action on shallow lakes. J.Environ. Eng. Div., Proc. ASCE 101(EE5): 813-827.

Burns, L., Cline, D.M. and Lassiter, R.R., 1981. Exposure Analysis Modelling System (EXAMS): User manual and system documentation. U.S.EPA, Athens, Georgia, 440 pp.

Charlton, M.N., 1983. Downflux of sediment, organic matter, and phosphorus in the Niagara River area of Lake Ontario. J. Great Lakes Res., 9: 210-211.

Danckwerts, P.V., 1970. Gas-liquid reactions. McGraw-Hill Book Co., New York, 276 pp.

Dietrich, W.E., 1982. Settling velocity of material particles. Wat. Res., 18: 1615-1626.

Durham, R.W. and B.G. Oliver, 1983. History of Lake Ontario contamination from the Niagara River by sediment radiodating and chlorinated hydrocarbon analysis. J. Great Lakes Res., 9: 160-168.

Elliott, J.M., 1975. The growth rate of brown trout *(Salmon trutta L.)* fed on reduced ratios. J. Animal Ecol., 44: 823-842.

Fox, M.E., J.H. Carey and B.G. Oliver, 1983. Compartmental distribution of organochlorine contaminants in the Niagara River and the western basin of Lake Ontario. J. Great Lakes Res., 9: 287:294.

Inland Waters Directorate, Environment Canada. Surface Water Data Ontario. Issues 1926 to 1983.

Halfon, E., 1984a. Modelling the fate of toxic contaminants in the Niagara River and Lake Ontario. Part I and Part II, Environment Canada, NWRI report No. 84-39.

Halfon, E., 1984b. Error analysis and simulation of mirex behaviour in Lake Ontario, Ecological Modelling, 22: 213-252.

Jensen, A.L., S.A. Spigarelli and M.M. Thommes, 1982. PCB uptake by five species of fish in Lake Michigan, Green Bay of Lake Michigan and Caiuga Lake, New York. Can. J. Fish. Aquat. Sci., 39: 700-709.

Karickhoff, S.W., 1981. Semi-empirical estimation of sorption of hydrophobic pollutants on natural sediments and soils. Chemosphere, 10: 833-846.

Kemp, A.L.W. and N.S. Harper, 1976. Sedimentation rates and a sediment budget for Lake Ontario. J. Great Lakes Res., 2: 324-340.

Kuntz, P.W and N.D. Warry, 1983. Chlorinated organic contaminants in water and suspended sediments of the lower Niagara River. J. Great Lakes Res., 9: 241-248.

Leidy, G.R. and G.R. Ploskey, 1980. Simulation modelling of zooplankton and benthos in reservoirs: documentation and development of model constructs. US Army Engineer Waterways Experiment Station Environmental Laboratory, Vicksburg, Miss., Technical Report E-80-4, 300 pp.

Liss, P.S., 1973. Processes of gas exchange across an air-water interface. Deep-Ses Res., 20: 221-238.

Murthy, R.J. D.C.L. Lam, T.J. Simons, J.A. Bull and W.M. Schertzer, 1984. Dynamics of the Niagara River Plume in Lake Ontario. Environment Canada, NWRI report # 84-7.

Neely, W.B., 1979. Estimating rate constants for the uptake and clearance of chemicals by fish. Environ. Sci. Technol., 13: 1506-1510.

Nriagu, J.O. H.K.T. Wong and R.D. Coker, 1981. Particulate and dissolved trace metals in Lake Ontario. Wat. Res., 15: 91-96.

Oliver, B.G., 1984. Distribution and pathways of some chlorinated benzenes in the Niagara River and Lake Ontario. Water Poll. Res. J. Canada, 19: 47-58.

Oliver, B.G., 1987. Partitioning relationships for chlorinated organics between water and particulates in the St. Clair, Detroit and Niagara Rivers. In QSAR in Environmental Toxicology, Vol. II ed. K.L.E. Kaiser, D. Reidel Publ., Holland 251-260.

Oliver, B.G. and A.J. Niimi, 1983. Bioconcentration of chlorobenzenes from water by rainbow trout: correlations with partition coefficients and environmental residues. Environ. Sci. Technol., 17: 287-291.

Oliver, B.G. and N.M. Charlton, 1984. Chlorinated organic contaminants on settling particulates in the Niagara River vicinity of Lake Ontario. Environ. Sci. Technol., 18: 903-908.

Oliver, B.G. and K.D. Nicol, 1984. Chlorinated contaminants in the Niagara River, 1981-1983. Sci. Tot. Environ., 39: 57-70.

Simons, T.J. and C.R. Murthy, 1985. Winter Circulation in Lake Ontario. J. Great Lakes Res., 11: 423-433.

Thomann, R.V., 1972. Systems analysis and water quality measurement. Environmental Research and Applications, Inc., New York, 286 pp.

Whittle, D.M. and J.D. Fitzsimons, 1983. The influence of the Niagara River and contaminant burdens of Lake Ontario biota. J. Great Lakes Res., 9: 295-302.

CHAPTER 14

PESTICIDE TRANSFORMATIONS AND MOVEMENT IN SOILS

by
Alessandro Marani & Gordon Chesters

14.1 INTRODUCTION

The recent discovery in several countries of pesticides in groundwater has aroused concern about the toxicological significance of these findings, particularly where groundwater is a primary drinking water source. At the national level, it is important that areas highly susceptible to leaching of pesticides be located. To achieve a predictive capability it is critical to understand the modes and rates of decomposition of pesticides in both the root zone and vadose zone (the unsaturated soil between the root zone and water table) of soils and to determine factors and conditions which promote migration of pesticides through soils. To establish some boundaries for the discussion a few pesticides actually found in groundwater have been selected. At this time, one of the pesticides which has been most studied is aldicarb [2- methyl- 2- (methylthio) propionaldehyde *O*- (methylcarbamyl) oxime], formulated under the trade name Temik. Other compounds of concern are carbofuran (2,3-dihydro-2,2-dimethyl-7-benzofuranyl *N*-methylcarbamate) with the trade name Furadan; 1,2-dibromo-3-chloropropane (DBCP); ethylene dibromide (EDB); alachlor [2-chloro-2',6'-diethyl-*N*-(methoxymethyl) acetanilide] marketed as Lasso.

To move through soil to groundwater and be a health concern, a pesticide must be water-soluble, not strongly adsorbed to soil particles, and have sufficient persistence to remain as the parent compound or a closely related toxic degradation product, e.g., a simple oxidation or hydrolysis product. When organochlorine pesticides, such as DDT, methoxychlor, heptachlor and its epoxide, were widely used there was little or no evidence that these materials reached groundwater. This could have been because monitoring for pesticides in groundwater was uncommon or analytical methods were insufficiently sensitive, but most likely because these compounds were very insoluble in water, strongly adsorbed to soil particles and had very little mobility in soils.

The discussion is divided into two parts relative to pesticide residue occurrence in groundwater: pesticide transformations in soils and pesticide movement through soils.

14.2 PESTICIDE TRANSFORMATIONS IN SOILS

A pesticide in soils can volatilize to the atmosphere; be strongly adsorbed to soil particles; be taken up by plants where it may or may not be degraded, and whence it might be returned to the soil in crop residue; be degraded by chemical, photochemical and/or microbiological processes; be leached to lower depths in the profile and/or to groundwater; be transported by erosion/runoff mechanisms and redeposited on land or reach surface waterbodies (usually river systems and eventually lakes). The fate of most pesticides combines a number of these processes, during which the chemical structure may or may not be changed. Degradation products may be more or less toxic than the parent compound (Chesters & Konrad, 1971; Browman & Chesters, 1977).

Adsorption may profoundly affect degradation processes. Adsorption involves a chemical or physical adherence of compounds at the surface of soil particles and reduces pesticide leachability, volatility and bioavailability. The degree of adsorption varies with the structure of the pesticide, and the pH, temperature and clay and organic matter contents (i.e., the colloidal or fine particle fractions) of the soil. Because pesticides are usually added to soils in very small amounts, the adsorption of some pesticides, e.g., diquat and paraquat is almost quantitative (diquat is 1,1'-ethylene-2,2'-dipyridinium dibromide and paraquat is 1,1'-dimethyl-4,4'-dipyridinium dichloride). At the pH-values encountered in soils, these compounds carry a positive charge which accounts for their strong adsorption on clay and organic colloids). Adsorption is *de facto* an indirect method of dissipation of pesticidal effects from the environment because adsorption reduces bioavailability. Adsorption usually reduces the rate of microbiological degradation of a pesticide (Armstrong & Konrad, 1974; Saltzman & Mingelgrin, 1984), but can catalyze chemical degradation (Steen, Paris & Baughman, 1980; Scow, 1982).

Aldicarb intrusion into groundwater in the potato-growing areas of Suffolk County, NY led to its withdrawal from use on Long Island. In the Central Sand Plain of Wisconsin, where potatoes are also grown under irrigation, a rule was invoked to create a moratorium on aldicarb use within a 1.6-km (1 mile) radius of any well producing water containing $> 10 \mu g$/liter of aldicarb and its oxidation products - aldicarb sulfoxide and aldicarb sulfone (Ehart, Chesters & Sherman, 1986). Trace amounts of aldicarb residues also have been identified in groundwater in potato-growing regions of California, Connecticut, Delaware, Maine, Maryland, New Jersey and Rhode

Island, in the citrus-growing region of Florida, under cotton in Texas, and peanuts in Virginia (Harkin et al., 1986; Zaki, Moran & Harris, 1982; Guerrera, 1981; Chesters et al., 1982; Institute of Food and Agricultural Sciences, 1983; Miller et al., 1985). It is perhaps the most pervasive pesticide in groundwater in the United States and its registration is presently under Special Review by the U.S. Environmental Protection Agency. However, it is very valuable because it provides excellent systemic protection against phytopredatory insects and phytopathological nematodes throughout the growing season. It is claimed to improve tuber quality and increase yields by 15 to 25% over those obtained using a soil fumigant for nematode control and multiple applications of organophosphorus insecticides for potato beetle and leaf hopper control.

Aldicarb has a molecular weight of 190, aqueous solubility of 6 g/liter, melting point of 99°C, vapour pressure of 8 * 10^{-5} torr, K_{ow} of 5 (Rao & Davidson, 1980) (K_{ow} is the distribution coefficient of an organic chemical between *n*-octanol and water. K_{ow} for aldicarb is 5; the value for DDT is about 10^6), and K_d frequently < 1 (Hornsby, Rao & Wheeler, 1983) (K_d is the partition coefficient of the pesticide between soil and water and usually is measured using the Freundlich equation K_d = $x/mC^{1/n}$. For 26 soils 1/n averaged 0.87 and if the value is not available it is taken to be 1 for convenience. K_d then equals x/mC where x/m is µg pesticide/g of soil and C is µg pesticide/ml of soil solution), K_{oc} range of 10 to 50 ml/g of organic carbon (oc) for Florida soils (Hornsby, Rao and Wheeler, 1983; Lyman, 1982) (K_{oc} may be thought of as the ratio of chemical adsorbed/unit weight of organic carbon (oc) in the soil to the concentration of chemical in solution at equilibrium. Values are different for each soil type. K_{oc} = µg adsorbed/g oc divided by µg/ml solution. Values of K_{oc} can be as high as 10^7 ml/g organic carbon); for the Central Sands of Wisconsin the calculated K_{oc} is 3.1 ml/g (Harkin et al., 1986). Thus, aldicarb has a higher aqueous solubility than most pesticides in use today and its low K_{ow}, K_d and K_{oc} suggest that it would be weakly adsorbed, particularly in the sandy or structurally creviced soils encountered in most groundwater contamination situations. Because of these properties aldicarb should have high bioavailability to plants, insects, nematodes and microorganisms and be degraded rapidly by microorganisms in surface soils.

Important chemical and physical properties of the selected pesticides as they relate to transformation and movement are given in Table 14.1 (data compiled for many compounds by Cohen et al., 1984; additional information taken from Hornsby, Rao & Wheeler,1983, Rao & Davidson,1980; Verschueren, 1983; Wilson et al., 1981; Kenaga & Goring, 1980; Pionke & DeAngelis, 1980).

Table 14.1
Properties of pesticides controlling their environmental fate

Properties	Group 1. Aldicarb and its oxidation products		
	Aldicarb	Its sulfoxide	Its sulfone
Solubility (mg/l)	6,000	43,000	7,800
Vapour Pressure, torr	$8 * 10^{-5}$	NA	NA
Henry's Law Constant ($atm\text{-}m^3/mol$)	$2.5 * 10^{-6}$	NA	NA
K_{ow}	5	NA	NA
K_d	<4, 0.03-0.7	0.34, 3.3	0.02-0.20
K_{oc}	29 ± 11	42, 405	11 ± 6
Hydrolysis, $t_{1/2}$ in weeks	20-650	52-114	15-130
Soil/Field Dissipation, $t_{1/2}$ in weeks	~10	NA	NA
Retardation Factor	1.0-4.0	NA	1-1.9
Amount in Groundwater ($\mu g/l$)	total residue range is 1-50, parent compound rarely if ever found		
States where found	AZ, CA, FL, ME, MD, NC, NJ, NY, OR, TX, VA, WA, WI		
Chemical Class	Carbamate	Carbamate	Carbamate

NA is not available

Properties	Group 2. Other Mobile Compounds			
	Carbofuran	DBCP	EDB	Alachlor
Solubility (mg/l)	320-700	700-1,000	4,300	242
Vapour Pressure, torr	$8.3 * 10^{-6}$	0.8	11	$2.2 * 10^{-5}$
Henry's Law Constant ($atm\text{-}m^3/mol$)	$3.4 * 10^{-9}$	$4 * 10^{-4}$	$6.3 * 10^{-4}$	$3.2 * 10^{-8}$
K_{ow}	207	NA	NA	434
K_d	<2, 0.25-8.7	0.13-2.1	0.25-0.80	0.6-8.1
K_{oc}	29 ± 9, 30-62	55-150	36-160	120 ± 49, 190
Hydrolysis, $t_{1/2}$ in weeks	<1-52	2,000-7,300	300	NA
Soil/Field Dissipation, $t_{1/2}$ in weeks	1-37	10	1.5-18	1-10
Retardation Factor	NA	2.1 ± 0.5	NA	NA
Amount in Groundwater ($\mu g/l$)	1-5	0.02-20	0.05-20	0.04
States where found	NY, WI	AZ, CA, HI, MD, SC	CA, FL, GA, SC	NE
Chemical Class	Carbamate	Halogenated alkane	Brominated alkene	Acetanilide

Table 14.1 Continued

Properties	Group 3. Relatively Immobile Compounds		
	Atrazine	DDT	Parathion
Solubility (mg/l)	33	$1.7 * 10^{-3}$	24
vapour Pressure, torr	$3 * 10^{-7}$	$1.9 * 10^{-7}$	$3 * 10^{-3}$
Henry's Law Constant ($atm\text{-}m^3/mol$)	$2.6 * 10^{-9}$	$5.2 * 10^{-5}$	$4.8 * 10^{-5}$
K_{ow}	174-212	$3.7*10^5$ - $1.5*10^6$	$6.5 * 10^3$
K_d	0.4-8.0	$4.2*10^4 \pm 4.0*10^3$	22 ± 14
K_{oc}	51, 214, 163 ± 80	$2.4*10^5 \pm 1.6*10^5$	$1.1*10^4 \pm 8.0*10^3$
Hydrolysis, $t_{1/2}$ in weeks	10-106	NA	NA
Soil/Field Dissipation, $t_{1/2}$ in weeks	<4-57	237 ± 233	2 ± 1 - 5 ± 4
Retardation Factor	NA	NA	NA
Amount in Groundwater (μg/l)	0.3-3.0	Not found	Not found
States where found	IA, NE, WI		
Chemical Class	Triazine	Organochlorine	Organophosphorus

NA is not available

The compounds have been divided into three groupings,

Group 1 aldicarb and its oxidation products;

Group 2 other mobile pesticides found in groundwater in relatively high concentrations;

Group 3 pesticides found only in low concentration or not at all in groundwater. Atrazine, DDT and parathion are not discussed but are included for comparative purposes.

The initial conversion of aldicarb in soil is by oxidation. The principal products are the sulfoxide and sulfone. The sulfoxide is slightly more toxic, the sulfone appreciably less toxic than aldicarb. Toxicity associated with aldicarb residues is acute/subacute neurotoxicity. The acute toxicity (LD_{50}) of aldicarb in rats is about 1 mg/kg, the sulfoxide 0.9 mg/kg and the sulfone about 25 mg/kg (Union Carbide Agricultural Products Company). Other conversion products of aldicarb and its oxides found in Florida are fragments formed by hydrolysis: the sulfoxide oxime and nitrile and the sulfone oxime (Ou et al., 1985a). The oxime is formed under aerobic and anaerobic conditions, the nitrile only under strictly anaerobic conditions. The oxime is found in surface and subsurface, the nitrile only in subsurface soils. Using ^{14}C-labeled aldicarb, rates of mineralization of the pesticide were measured from rates of $^{14}CO_2$ evolution. After 63 days of incubation, $^{14}CO_2$ generated ranged from 12 to 90%. In general - though not exclusively - CO_2 was generated more rapidly in surface than subsurface soils. In the vadose zone, anaerobic degradation - although slower than aerobic degradation - was substantial; in 63 days CO_2 production ranged from 3.5 to 19%. In contrast, a recent report has shown more rapid rates of degradation under anaerobic conditions (Smelt et al., 1978). Several studies have measured "first-order" rate constants for the removal of aldicarb from soils ranging from 0.08 to 0.80/day (Bromilow et al., 1980; Leistra, Smelt & Lexmond,1976), while it was found that after 70 days of aerobic incubation, "total toxic residues" of aldicarb were 8 to 74% of the initial concentration (Richey, Bartley & Sheets,1977). Following 63 days of incubation of aldicarb sulfone in Florida soils, the ranges of CO_2 produced (1 to 50%) were similar under aerobic and anaerobic conditions. The rate of CO_2 formation was much higher in the surface than subsurface soil (Ou et al.,1985b).

Thus, aldicarb can undergo oxidation to the sulfoxide and sulfone, hydrolysis, reduction and eventually complete mineralization. Organisms responsible for these changes have not been well characterized, particularly at aldicarb concentrations found in field soils. Organisms have been isolated from the unsaturated zone/groundwater interface which metabolize aldicarb as a sole source of carbon (Dzantor & Harkin, 1985). Three isolates were *Bacillus* spp., three were *Flavobacterium* spp., two were *Corynebacterium* spp. and one was a *Cytophaga* sp. Metabolism occurred on yeast extract agar and nutrient agar with aldicarb or its sulfoxide or sulfone at concentrations of 1 ppm. The size and properties of the isolated organisms suggest that they are indigenous to the vadose zone, although arguments could be made that the *Bacillus* spp. are present as endospores leached from the surface soil. Whether the organisms are indigenous to the subsurface is irrelevant if they adapt to these conditions and remain metabolically active. Because the ^{14}C label was on the S-methyl group of the aldicarb used, $^{14}CO_2$ could arise only from complete mineralization of aldicarb or its oxides. To better define rates of aldicarb degradation by groundwater organisms, additional experiments are needed using labels at different locations on the molecule.

Thus, the scheme of aldicarb decomposition is complex and site-specific; rates are controlled by soil pH (alkalinity), redox status, soil temperature and moisture and organic matter content, soil microfloral composition and other factors.

Carbofuran is also highly water-soluble and has a very low vapour pressure and Henry's Law Constant. Volatilization losses are negligible as determined using simulated paddy water (Deuel, 1979). A complex pathway of carbofuran degradation was revealed (Rajagopal et al., 1984) involving 3-hydroxy- and 3-keto-carbofuran and at least six phenols before mineralization to CO_2. The half-life (time required for 50% of the pesticide to degrade, $t_{1/2}$) was 35 days in sandy soils (Harris, 1969; Onsager & Rusk,

1969) and about 80 days in a muck soil (Harris, 1969). In acidic or neutral soils $t_{1/2}$ was 350 days, but in alkaline soils (pH about 8), degradation was 7 to 10 times faster by hydrolysis at the carbamate linkage to form carbofuran phenol (Getzin, 1976). This information suggests that soils of high colloidal (mineral or organic matter) content adsorb carbofuran and increase its persistence. However, alkaline conditions promote hydrolysis to the phenol. Controversy exists concerning the relative rates of chemical and microbial degradation of carbofuran. Carbofuran was only slightly more persistent in sterile than nonsterile soils indicating that chemical degradation is very important (Miles, Tu & Harris, 1981) The 3-hydroxycarbofuran degradation product is highly unstable even in nonsterile soils and is transformed quantitatively within 3 days to 3-ketocarbofuran. The latter reached maximum concentrations in 1 day and was removed completely in 4 days in a loam and 7 days in a muck soil. Mineralization of carbofuran follows approximately first-order kinetics (Ou et al., 1982) with a half-life in the field of 1 to 37 weeks (Table 14.1), i.e., very rapid to moderately slow rates. The low values of K_d and K_{oc} and the moderately low K_{ow} value for carbofuran (Table 14.1) indicate that its transformations in soils should be relatively rapid particularly under alkaline conditions.

Little information is available on the transformation of DBCP and EDB in soils. DBCP is very stable to hydrolysis (Table 14.1) but will hydrolyze slowly with elimination of HBr and HCl to allylic halide intermediates; the estimated $t_{1/2}$ is 140 years at pH 7 and 15°C, and 38 years at pH 7 and 25°C (Burlinson, Lee & Rosenblatt, 1982). In a soil-water culture, DBCP was biodegraded through reductive dehalogenation within weeks (Castro & Belser, 1968); up to 80% was degraded in 3 weeks at a soil:solution ratio of 0.5 when cultures were fortified with glycerol, biotin and thiamine. EDB was also shown to debrominate almost quantitatively to form ethylene in 8 weeks (Castro & Belser, 1968). Recent investigations of EDB at different acidities and temperatures provided indications that the $t_{1/2}$ of EDB at ambient groundwater temperatures is >6 years and the most likely degradation product is vinyl bromide (Cohen et al., 1984).

Alachlor degradation rates have been determined for several soil types under laboratory and field conditions. Alachlor has a solubility of 240 mg/liter, low vapour pressure and very low Henry's Law Constant. Its K_{ow}-, K_d- and K_{oc}-values are intermediate between those of aldicarb and the persistent organochlorine insecticides (Table 14.1). In laboratory experiments with a sandy loam soil, $t_{1/2}$ for alachlor ranged from 6 (25°C and 15% moisture) to 40 days (5°C and 12% moisture) depending on temperature and moisture levels (Walker & Brown, 1985). In field experiments where alachlor was applied at a rate of 1 kg a.i./ha (a.i. is an abbreviation for active ingredient) in sandy loam, 50% degradation occurred in 11 to 24 days and

80% in 21 to 48 days. The ranges in degradation rates resulted from three experiments prepared at different times. Half-lives of alachlor in a laboratory experiment at a moisture content equivalent to 50% field moisture capacity (FMC) where 17, 25 and 37 days in sandy loam and 6, 13 and 15 days in clay loam at 10, 20 and 30°C, respectively (Zimbdahl & Clark, 1982)(field moisture capacity is the water remaining in a soil after it has been saturated and allowed to drain until no further drainage occurs, usually 24 to 48 hours). At 20°C in sandy loam, $t_{1/2}$ was 19, 25 and 43 days at 80, 50 and 20% of FMC; in clay loam the $t_{1/2}$-values were 11, 13 and 25 days. In a field experiment, $t_{1/2}$ agreed closely with the laboratory experiments showing values of 11 and 9 days in the sandy loam and clay loam. In a model ecosystem alachlor was completely degraded in 35 days (Yu et al., 1975) and in the field 90% dissipation occurred in 40-70 days (Stewart et al., 1975). The rates of these transformations are quite similar with some variations at different soil temperatures, moisture and colloidal particle contents.

14.3 MOVEMENT OF PESTICIDES

Pesticide movement to groundwater is determined by the interactions (e.g., adsorption, leaching, degradation, volatilization) they undergo in soil. The processes are affected by the physical/chemical properties of both pesticide and soil (Table 14.1) and by climatic variables.

The Freundlich equation is used to relate adsorbed- and solution-phase concentrations of pesticides at equilibrium. The soil to solution partition coefficient, K_d is controlled *inter alia* by organic carbon and clay contents, cation exchange capacity and pH of the soil. Studies have consistently shown that organic carbon content is the soil characteristic best correlated with K_d for organic pesticides (Rao & Davidson, 1980; Kenaga & Goring, 1980). The adsorption partition coefficient based on soil organic carbon content (K_{oc}) for a given pesticide is relatively constant and independent of soil type - $K_{OC} = 100\ (K_d/\%\ oc)$. Thus, K_{oc} is important for assessing transformations and movement of pesticides and is used frequently in mathematical models describing overland flow of materials in solution and adsorbed to soil particles and downward movement through the unsaturated zone (Carsel et al., 1984; Enfield et al., 1982).

Aqueous solubilities of pesticides limit the concentration of the chemical in the soil-water phase and pesticide solubility is correlated inversely with adsorptivity (Kenaga & Goring, 1980). Pesticides of high water solubility are subject to greater downward movement and can be transported to greater depths in the vadose zone. Another pesticide property related to adsorption and solubility is K_{ow} (Rao & Davidson, 1980; Kenaga & Goring, 1980). Pesticides with low K_{ow} have high solubilities and low adsorption coefficients. When K_{oc} of a soil is not known it can be estimated using either water solubility (W_s) and/or K_{ow} data (Rao & Davidson, 1980; Kenaga & Goring, 1980).

$$\log K_{oc} = -0.55 \log Ws + 3.64 \quad (Ws \text{ in mg/liter})$$
$$\log K_{oc} = 0.544 \log K_{ow} + 1.377$$

The adsorption and mobility of such relatively soluble pesticides as aldicarb, carbofuran, DBCP, EDB and alachlor are now considered in more detail because of the high potential for these pesticides to move through soil to groundwater.

Aldicarb adsorption by soil organic matter is more important for restraining movement than adsorption by mineral constituents. Adsorption at organic surfaces is physical and occurs mainly through hydrophobic interactions and increases with increasing soil organic matter content (Felsot & Dahm, 1979; Supak, Swaboda & Dixon, 1978). In sandy soils in Florida, small amounts of aldicarb are adsorbed by top soils but not by subsurface soils (Hornsby, Rao & Wheeler, 1983). Mean adsorption of aldicarb in 51 Belgian soils was 22% for surface horizons and 5% for subsurface horizons (Moreale & Bladel, 1981a). Differences in adsorption emphasize the importance of comparing root and vadose zones separately when assessing downward migration of pesticides (Hornsby, Rao & Wheeler, 1983; Bromilow, 1980; Leistra, Smelt & Lexmond, 1976).

Aldicarb and its oxides are only weakly adsorbed by soils (8, 15). The high mobility of aldicarb and its metabolites has been noted in leaching experiments using soil columns (Leistra, Smelt & Lexmond, 1976; Smelt et al., 1983; Smelt et al., 1981; Bromilow, 1973), lysimeters (Snegaroff & Jamet, 1976), an irrigated field (Hough, Thomason & Farmer, 1975), and arable fields (Bilkert & Rao, 1985; Bromilow & Leistra, 1980; Smelt et al., 1981). Some leaching of aldicarb residues was found in sandy soils but not in clay loam or muck soils (Awad, Kilgore & Winterlin, 1984; Coppedge, Bull & Ridgway, 1977). Downward movement of aldicarb has been observed in sands, sandy loams, and silty clay loams of the south-eastern United States but deepest penetration invariably occurred in sandy soils (Bilkert & Rao, 1985). Aldicarb sulfoxide and sulfone were retained mainly in the top 30 cm in two loams and a humic sand (Smelt et al., 1981; Leistra & Smelt, 1981) under field conditions. The behavior of aldicarb and its oxides during summer in these soils was evaluated by a computer model; the concentration patterns agreed closely with field measurements (Leistra & Smelt, 1981). In loam, most residue remained in the 10- to 20-cm layer. In the humic sand most was in the upper 30 cm though small concentrations penetrated to 60 cm. The oxides were retained in the upper 20 cm of grassed loam and humic sand 165 days after application (30 cm total rainfall); similar data were obtained for a fallow loam soil (Smelt, Schut & Leistra, 1983). However, 2 to 15% of the dose from fallow humic sand was leached to a depth of 60 cm. Most residues remaining in the humic sand in the fall were leached beyond depths of 80 cm by winter precipitation (36 cm) but only 0.5 to 2.2% of the dose leached to depths of 100 cm in the loams.

The amount of water (precipitation and irrigation) influences the extent of aldicarb movement in soils (Leistra, Dekker & van der Burg, 1984; Kejonckheere et al., 1983; Leistra, Smelt & Lexmond, 1976; Leistra & Dekker, 1976). Mobility of aldicarb in soil columns at moderate (12 to 13 mm/day) or high water flow rates (23 to 28 mm/day) was similar. After 5 days at high water flux, 20 and 72% of applied aldicarb was leached from a humic sand and a silt loam (Leistra, Smelt & Lexmond, 1976). The total residue of sulfoxide and sulfone leached from columns of two loams after 5 days of moderate water flow was 44 and 66%. The computer model (Leistra & Dekker, 1976) was used to simulate the application of aldicarb (2.5 kg a.i./ha) at the beginning of April to a sandy loam used for growing potatoes. During the first month, with little rainfall (15 mm in 30 days) on initially moist soil, sulfoxide penetrated to a limited extent due to convection, dispersion and diffusion. As the crop extracted water, very little downward movement occurred during the second month. With more rainfall in the first month (35 mm in 30 days), downward movement to a depth of 25 cm was considerable. After a month with high rainfall (60 mm in 30 days) or sprinkler irrigation, sulfoxide peaks occurred at 15 to 25 cm depth.

These investigations of aldicarb movement under different conditions in various soil types indicate that the potential for aldicarb or its oxides to leach from the root zone/topsoil is high in coarse-textured soils but low in soils of high organic matter content. Leaching of aldicarb residues is conditioned by amount of water available for net infiltration [i.e., (precipitation + irrigation) - (evapotranspiration + runoff + soil storage)]. Sandy soils have low water-holding capacities and allow greater infiltration of water than fine-textured soils. Thus, groundwater beneath sandy soils is vulnerable to aldicarb residue contamination. Small amounts of aldicarb sulfoxide and sulfone leach through sandy soils to shallow groundwater in Central Wisconsin where aldicarb is applied to irrigated potatoes (Harkin et al., 1986). Leaching occurs mainly during late summer and fall of the year of application. Aldicarb residues apparently do not penetrate to groundwater by bleeding through the vadose zone at a constant rate, but rather by pulses of varying concentrations.

Aldicarb and its oxides volatilize to a small degree: volatilization of residues from thin layers of two aldicarb-treated clay soils during drying cycles of up to 28 weeks amounted to <0.2% of the dose (Supak, Swaboda & Dixon, 1978). Some losses may occur for surface-applied aldicarb, but soil incorporation drastically reduces volatilization. Furthermore, aldicarb is rapidly converted to aldicarb sulfoxide and sulfone which are less volatile than the parent compound.

Carbofuran is a benzofuranyl carbamate nematicide/insecticide with high water solubility (320 to 700 mg/liter), very low vapour pressure and Henry's Law Constant, low K_d and K_{oc} and rapid dissipation rate from soils (Table 14.1). K_d values ranging from 0.25 to 8.7 were reported for seven soils with organic carbon contents ranging from 0.4 to 17% (Felsot & Wilson, 1980) giving K_{oc}-values of 30 to 63.

Adsorption and desorption isotherms of labeled carbofuran in nine types of soil satisfied the Freundlich equation; adsorption increased with organic carbon content (Jamet & Pidallu, 1975). Movement of carbofuran on soil thin-layer plates indicates that it is very mobile in seven soils and that adsorption correlates positively with organic carbon content and cation exchange capacity (Felsot & Wilson, 1980). Adsorption of carbofuran by sandy loam and loam soils was linear, but stronger on the sandy loam (Moreale & van Bladel, 1981b). Significant adsorption of carbofuran by clay-sized particles decreased carbofuran uptake by plants (Abdellatif, Hermanson & Reynolds, 1967).

Soil-bound residues have been encountered after application of carbofuran (Fuhremann & Lichtenstein, 1980; Venkateswarlu & Sethunathan, 1979). The bound residues from ring-labeled ^{14}C-carbofuran was 10% in loam but <1% in a sandy soil after 13 days. The nonextractability of some carbofuran residues (e.g., carbofuran phenol) suggests that they are bound chemically rather than by physical adsorption through van der Waals forces, hydrogen bonding, or cation exchange mechanisms.

Carbofuran was more readily desorbed from soils than other insecticides; order of desorption was carbofuran > carbaryl > lindane > diazinon > parathion (Sharon et al, 1980). Moreover, carbofuran completely leached from the soil in the first 2*200 ml-washings with water, indicating high mobility. Under percolating conditions, 50% of soil-applied carbofuran leached into the aquatic component of an agromicrocosm after 3 weeks incubation (Koeppe & Lichtenstein, 1982). Under nonpercolating conditions, 80% of ^{14}C was associated with the terrestrial component of the microcosm. Carbofuran leached to 45 cm depth in small field plots receiving 50 mm of rainfall during storms (Read & Gaul, 1983). Under normal weather conditions most carbofuran residues remained near or above the point of application. Lateral or vertical movement of carbofuran was not significant in an organic soil under different water regimes (Chapman et al., 1984).

Carbofuran can move upwards from the site of application (Read & Gaul, 1983). It moved upwards with soil moisture from the subsoil in small greenhouse plots. Residues reaching the soil surface from 5 to 10 cm depth increased over 8 to 12 months and gradually decreased thereafter.

Carbofuran has been found at levels between 1 and 50 μg/liter in groundwater in Wisconsin and New York, in areas of sandy soils and shallow water tables (Data summarized from U.S.-EPA pesticide registration files, unpublished). However, the residues of carbofuran in private drinking wells in the Central Sands of Wisconsin could not be substantiated when fresh samples were carefully analyzed (Harkin et al., 1984).

DBCP has a water solubility range of 700 to 1,200 mg/liter, a vapour pressure of 0.8 torr and a Henry's Law Constant of $4*10^{-4}$ atm-m^3/mol; it has a low K_d and a medium K_{oc} (Table 14.1). Various reports attest to its mobility in soil (Biggar, Nielson & Tillotson, 1984; Wilson et al., 1981; McHenry & Naylor, 1979; Johnson & Lear, 1968). Movement is greater in coarse-textured soils and soils of low organic matter content. Although adsorption may inhibit its movement, evidence exists that the degree of adsorption is low and leaching below the root zone occurs in many soils (Biggar, Nielson & Tillotson, 1984). Some movement of DBCP may occur through the vapour phase but most movement in soil occurs in the liquid phase suggesting that water flux determines the extent of leaching. Once DBCP is in the subsurface, attenuation by adsorption and biodegradation becomes less effective because of low organic matter content and low microbial activity. This may explain why DBCP is found in some California aquifers several years after its use was banned (Cohen, Eiden & Lorber, 1986).

One set of California soil core samples showed that mg/liter amounts of DBCP were leached as much as 15 m through the unsaturated zone (Nelson et al., 1981), but no DBCP was detected in other cores sampled down to 10 m five years after its last application (Zalkin, Wilkerson & Oshima, 1983). All the DBCP had possibly passed below the 10 m depth.

EDB has very high aqueous solubility, vapour pressure and Henry's Law constant. It rapidly dissipates from soil and has low K_d and medium K_{oc} (Table 14.1). EDB has little affinity for soil particles and residues from agricultural fumigation probably leach to groundwater. California soil core data show that EDB leached about 10 m at the μg/liter level (Zalkin, Wilkerson & Oshima, 1983); this was a driving factor behind U.S.-EPA's emergency suspension of soil fumigation uses of EDB (U.S.-EPA, 1983). In Georgia, EDB leached through a confining layer to an artesian aquifer (U.S.-EPA, 1983).

Sampling throughout profiles of a coarse-textured soil and a silty clay indicated that bulk flow of water can be associated with EDB movement through soils (Duncan & Oshima, 1986). EDB concentrations in the top 45 cm correlated with organic carbon content. EDB was not found in the deeper soil layers of the coarse-textured soil which theoretically had a higher percolation rate than the silty clay at a mean moisture content of 16%. At 23% moisture, a band of EDB was found between 230 and 300 cm in the silty clay. The amounts of EDB in the band did not correlate to any measured variables, including organic carbon content. Penetration of EDB was more rapid in loamy sand than silt loam; no movement was evident in clay, possibly due to adsorption to clay and organic matter (Townsend, Dirks & Marks, 1980). Movement in the coarse-textured soils is affected by such edaphic factors as moisture content, temperature and bulk density.

Alachlor is an acetanilide herbicide with a water solubility of 24 mg/l; low vapour pressure and Henry's Law Constant; K_d ranging from 0.6 to 8.1; and K_{oc} of 70 to 190 (Table 14.1). The primary factors affecting its persistence in soils are adsorption and biodegradation. Adsorption increases with increasing organic matter and clay content of soils (Spillner et al., 1983; Strek & Weber, 1981; Weed Sci. Soc. Am., 1979; Eshel, 1969). Alachlor adsorbed on soil colloids is partially inactivated in soils high in organic matter and clay, rendering the chemical less available for plant uptake, biodegradation and volatilization (Putnam & Rice, 1979; Eshel, 1969). Laboratory and field evaluations reveal its mobility/leachability potential. Applied at 1 and 2 kg/ha, alachlor leached 5 to 8 and 2.5 to 5 cm in a sandy soil and clay, respectively, following a 5-cm rainfall (Eshel, 1969). In a fine sandy loam cornfield, alachlor sprayed at 2.3 kg/ha moved but remained in the top 8 cm (Wu, 1980). Comparatively, atrazine sprayed at 1.7 kg/ha leached into the vadose zone. Low leachability of alachlor in field soil is attributed to its strong affinity for soil organic matter. The order of leaching of alachlor in soil columns was sand > sandy loam > silty clay loam (Spillner et al., 1983). Leaching of alachlor in these soils was less than a metolachlor related acetanilide herbicide. Factors affecting the diffusion of these chemicals are temperature, bulk density, and moisture content of the soils.

Evidence of alachlor contamination of groundwater is limited even in areas of extensive agricultural use (Libra et al., 1984; Spalding, Junk & Richard, 1980; Junk, Spalding & Richard, 1980). Low concentrations were detected in 2 of 14 wells sampled (represenative of over 1,000 wells located in the study area) in the corn-growing areas of Nebraska (Spalding, Junk & Richard, 1980). These corn fields have sandy soils and shallow, unconfined aquifiers. Although alachlor has had a U.S.-EPA Special Review, additional evidence of groundwater contamination is needed to reach a decisive conclusion on continued use of the compound.

14.4 CONCLUSION

Factors controlling the transformations and movement of pesticides in soils are similar. Research needs of highest priority for providing a solution to problems of pesticide intrusion into groundwater are: 1. prediction on a site-specific and compound-specific basis of those portions of the landscape most susceptible to pesticide leaching; 2. improvement in the assessment of rates of chemical and microbial transformation of pesticides in the vadose zone and groundwater; 3. determination of whether better quantitative structure-activity relations can be described which interlink pesticide structure, ease of pesticide movement and rates of degradation.

The foregoing discussion emphasizes that the degradation and transport of pesticides in soils depend on the chemical structure of each compound and a set of site-specific soil characteristics. It is consequently very difficult to set generalized policy for establishing moratoria or bans on the use of specific pesticides. To establish policy at the national level, governments ensure that models used to regulate pesticides in relation to groundwater pollution are not so sophisticated that the data needed to apply them cannot be gathered, nor so simplistic that all problems are trivialized or that all chemicals are deemed a major threat to the nations's wellbeing. This problem is critical since it must balance human and environmental health questions against assurances that the nation has a plentiful supply and diversity of high-quality foods at reasonable prices.

Specificity and toxicity aside, there are several paradoxical physical-chemical principles that the synthesizers of pesticides must take into account in the design of an "ideal" compound. Soluble compounds have the drawback that they leach rapidly through the soil profile and might eventually contaminate groundwater. However, such compounds are the most biodegradable, particularly by microorganisms. Because of their high bioavailability they can be used in smallest amounts to perform their pesticidal functions. Conversely, very insoluble, highly hydrophobic compounds exemplified by DDT are not mobile in soils but have the disadvantage of being highly persistent. Furthermore, these compounds are bioaccumulated to substantial amounts in the food chain. Apart from its pesticidal effects, the "ideal" pesticide from an environmental point of view should have properties which strike a balance between the highly mobile, aldicarb and the immobile, persistent organochlorine insecticides such as DDT. Perhaps this balance is found with some of the commonly used organophosphorus compounds.

Our deepest gratitude is expressed to Geronimo V. Simsiman for his inestimable help in retrieving documents, evaluating and reviewing data and suggesting manuscript changes. John M. Harkin has provided unstintingly of his time, experience and editorial wisdom. They are both of the Water Resources Center, University of Wisconsin-Madison. Andrea Rinaldo of the University of Trento has generously provided editorial comment and assistance in the preparation of the Italian summary.

REFERENCES

Abdellatif, M.A., H.P. Hermanson & H.T. Reynolds, 1967. J. Econ. Entomol. 60, 1445.
Armstrong, D.E. & J.G. Konrad, 1974. in Pesticides in Soil and Water. W.D. Guenzi Ed. Soil Science Society of America, Inc., Madison, WI. pp. 123-132.
Awad, T.M., W.W. Kilgore & W. Winterlin, 1984. Bull. Contam. Toxicol. 32, 377.

Biggar, J.W., D.R. Nielson & W.R. Tillotson, 1984. Environ. Geol. 5, 127.
Bilkert, J.N. & P.S.C. Rao, 1985. J. Environ. Sci. Health B 20, 1.
Bromilow, R.H., et al., 1980. Pesticide Science 11, 371.
Bromilow, R.H., 1973. Ann. Appl. Biol. 75, 473.
Bromilow, R.H. & M. Leistra, 1980. Pesticide Science 11, 389.
Browman, M.G. & G. Chesters, 1977. in Fate of Pollutants in the Air and Water Environments. I.H. Suffet, Ed. John Wiley & Sons, New York pp. 49-105.
Burlinson, N.E., L.A. Lee & D.H. Rosenblatt, 1982. Environ. Sci. Technol. 16, 627.

Carsel, R.F., et al., 1984. Users Manual for the Pesticide Root Zone Model (PRZM). EPA-600/3-84-109. U.S. Environmental Protection Agency, Washington, D.C. 216 p.
Castro, C.E. & N.O. Belser, 1968. Environ. Sci. Technol. 2, 779.
Chapman, R.A. et al., 1984. J. Environ. Sci. Health B 19, 259.
Chesters, G. et al., 1982. Aldicarb in Groundwater. Water Resources Center, University of Wisconsin-Madison. 38 p.
Chesters, G. & J.G. Konrad,1971. BioScience 21, 565.
Cohen, S.Z., C. Eiden & M.N. Lorber, 1986. in Evaluation of Pesticides in Ground Water. W.Y. Garner, R.C. Honeycutt, H. Nigg, Eds. American Chemical Society Books, Symposium Series, Washington, D.C. pp. 170-196.
Cohen, S.Z. et al., 1984. in Treatment and Disposal of Pesticide Wastes. R.F. Krueger & J.N. Seiber, Eds. American Chemical Society Books, Symposium Series 259, Washington D.C. pp. 297-325.
Coppedge, J.R., D.L. Bull & R.L. Ridgway, 1977. Arch. Environ. Contam. Toxicol. 5, 129.

Deuel, L.E. Jr., 1979. J. Environ. Qual. 8, 23.
Duncan, D. & R.J. Oshima, 1986. in Evaluation of Pesticides in Groundwater. W.Y. Garner, R.C. Honeycutt, H. Nigg, Eds. American Chemical Society Books, Symposium Series, Washington, D.C. pp. 282-293.
Dzantor, E.K. & J.M. Harkin, 1985. Wisconsin Acad. Rev. 32, 40.

Ehart, O.R., G. Chesters & K.J. Sherman, 1986. in Evaluation of Pesticides in Groundwater. W.Y. Garner, R.C. Honeycutt, H.C. Nigg, Eds. American Chemical Society Books, Symposium Series, Washington, D.C. pp. 488-498.
Enfield, C.F., et al., 1982. Ground Water 20, 711.
Eshel, Y., 1969. Weed Sci. 17, 441.

Felsot, A. & J. Wilson, 1980. Bull. Environ. Contam. Toxicol. 24, 778.
Felsot, A. & P.A. Dahm, 1979. J. Agric. Food Chem. 27, 557
Fuhremann, T. & E.P. Lichtenstein, 1980. J. Agric. Food Chem. 28, 446.

Getzin, L.W., 1976. Environ. Entomol. 2, 461.

Guerrera, A.A., 1981. J. Am. Water Works Assoc. 73, 190.

Harkin, J.M. et al., 1986. in Evaluation of Pesticides in Groundwater. W.Y. Garner, R.C. Honeycutt, H.C. Nigg, Eds. American Chemical Society Books, Symposium Series, Washington, D.C. pp. 219-255.

Harkin, J.M. et al., 1984. Pesticides in Groundwater Beneath the Central Sand Plain in Wisconsin. Tech. Rep. WIS WRC 84-01. Water Resources Center, University of Wisconsin-Madison. 46 p.

Harris, C.R., 1969. J. Econ. Entomol. 62, 1437.

Hornsby, A.G., P.S.C. RAo & W.B. Wheeler, 1983. in Characterization and Monitoring of the Vadose (Unsaturated) Zone. D.M. Nielsen, Ed. National Water Well Assoc., Worthington, OH. pp. 959-978.

Hough, A., I.J. Thomason & W.J. Farmer, 1975. J. Nematol. 7, 214.

Institute of Food and Agricultural Sciences, 1983. Aldicarb Research Task Force Report, University of Florida, Gainesville, FL. 58 p.

Jamet, P. & M.A. Pidallu, 1975. Phytiatr.-Phytopharm. 24, 279.

Johnson, D.E. & B. Lear, 1968. Soil Sci. 105, 31.

Junk, G.A., R.F. Spalding & J.J. Richard, 1980. J. Environ. Qual. 9, 479.

Kejonckheere, W., et al., 1983. Pesticide Science 14, 99.

Kenaga, E.E. & C.A.I. Goring, 1980. in Aquatic Toxicology ASTM STP 707. J.G. Eaton et al., Eds. Amer. Soc. for Testing and Materials, Washington, D.C. pp. 78-115.

Koeppe, M.K. & E.P. Lichtenstein, 1982. J. Agric. Food Chem. 30, 116.

Leistra, M., W.A. Dekker & A.M.M. van der Burg, 1984. Arch. Environ. Contam. Toxicol. 13, 327.

Leistra, M. & J.H. Smelt, 1981. in Studies in Environmental Science, Vol. 17, Quality of Groundwater. W. van Duijvenbooden, P. Glasbergen, H. Van Lelyveld, Eds. Elsevier, Amsterdam. pp. 941-952.

Leistra, M. & W.A. Dekker, 1976. Water Air Soil Pollut. 5, 491.

Leistra, M., J., H. Smelt & T.M. Lexmond, 1976. Pesticide Science 7, 471.

Libra, R.D. et al., 1984. I. Groundwater Quality and Hydrogeology of Devonian-Carbonate Aquifers in Floyd and Mitchell Counties, Iowa. Iowa Geological Survey, Open File Report 84-2.

Lyman, W.J., 1982. in Handbook of Chemical Property Estimation Methods. W.J. Lyman et al., Eds. McGraw-Hill Book Company, New York. 33 p.

McHenry, M.V. & P. Naylor, 1979. J. Nematol. 11, 260.

Miles, J.R.W., C.M. Tu & C.R. Harris, 1981. J. Environ. Sci. Health B 16, 409.

Miller, W.L. et al., 1985. The Florida Temik Study. Groundwater Monitoring. Final Report CR 811606 to U.S. Environmental Protection Agency. University of Florida, Gainesville, FL. 401 p.

Moreale, A. & R. van Bladel, 1981a. Rev. Agr. (Brussels) 34, 939.

Moreale, A. & R. van Bladel, 1981b. Meded Fac. Landbouwwet. Rijksuniv. Gent 46, 281.

Nelson, S. et al., 1981. Sci. Total Environ. 21, 35.

Onsager, J.A. & H.W. Rusk, 1969. J. Econ. Entomol. 62, 1060.

Ou, L.-T. et al., 1985a. J. Agric. Food Chem. 33, 72.

Ou, L.-T. et al., 1985b. J. Agric. Food Chem. 33, 545.

Ou, L.-T. et al., 1982. J. Environ. Qual. 11, 293.

Pionke, H.B. & R.J. DeAngelis, 1980. in CREAMS: A Field Scale Model for Chemicals, Runoff and Erosion from Agricultural Management Systems. W.G. Knisel, Ed. USDA Conservation Research Report 26. pp. 607-643.

Putnam, A.R. & R.P. Rice, Jr., 1979. Weed Sci. 27, 870.

Rajagopal, B.S., et al., 1984. Residue Revs. 93, 1.

Rao, P.S.C. & J.M. Davidson, 1980. in Environmental Impact of Nonpoint Source Pollution. M.R. Overcash and J.M. Davidson, Eds. Ann Arbor Science Publishers, Inc., Ann Arbor, MI. pp. 23-67.
Read, D.C. & S.O. Gaul, 1983. Agric. Ecosystems Environ. 9, 57.
Richey, F.A., Jr., W.J. Bartley & K.P. Sheets, 1977. J. Agric. Food Chem. 25, 47.

Saltzman, S. & U. Mingelgrin, 1984. in Pollutants in Porous Media. B. Yaron, G. Dagan, J. Goldshmid, Eds. Springer-Verlag, New York. pp. 153-161.
Scow, K.M., 1982. in Handbook of Chemical Property Estimation Methods. W.J. Lyman, W.F. Reehl, D.H. Rosenblatt, Eds. McGraw-Hill Book Company, New York. Chap. 9, 85 p.
Sharon, M.S., et al., 1980. Water Res. 14, 1095.
Smelt, J.H., C.J. Schut & M. Leistra, 1983. J. Environ. Sci. Health B 18, 645.
Smelt, J.H., et al., 1983. Pesticide Science 14, 173.
Smelt, J.H., et al., 1981. Netherlands J. Plant Path. 87, 177.
Smelt, J.H. et al., 1978. Pesticide Science 9, 293.
Snegaroff, J. & P. Jamet, 1976. Phytiatr.-Phytopharm. 25, 223.
Spalding, R.F., E.A. Junk & J.J. Richard, 1980. Pesticide Monitoring J. 14, 70.
Spillner, C.J. et al., 1983. in Fate of Chemicals in the Environment. R.L. Swann and A. Eschenroeder, Eds. American Chemical Society Books, Symposium Series 225, Washington, D.C. pp. 231-247.
Steen, W.C., D.F. Paris & G.L. Baughman, 1980. in Contaminants and Sediments, Volume I. R.A. Baker, Ed. Ann Arbor Science Publishers Inc., Ann Arbor, MI. pp. 477-482.
Stewart, B.A., et al., 1975. Control of Water Pollution from Cropland: Vol.I. EPA-600/2-75-026a; USDA ARS-H-5-1. U.S. Environmental Protection Agency and Agricultural Research Service, U.S. Department of Agriculture Washington, D.C.
Strek, H.J. & J.B. Weber, 1981. Proc. South. Weed Sci. Soc. 34, 33.
Supak, J.R., A.R. Swaboda & J.B. Dixon, 1978. Soil Sci. Soc. Am. J. 42, 244.

Townsend, J.L., V.A. Dirks & C.F. Marks, 1980. Can. J. Soil Sci. 60, 177.

Union Carbide Agricultural Products Company. Temik Aldicarb Pesticide: A Scientific Assessment. Research Trinagle Park, NC. 71 p.
U.S.-EPA, 1983. Ethylene Dibromide (EDB) - Position Document 4. U.S. Environmental Protection Agency, Office of Pesticide Programs, Washington, D.C.

Venkateswarlu, K. & N. Sethunathan, 1979. J. Environ. Qual. 8, 365.
Verschueren, K., 1983. Handbook of Environmental Data on Organic Chemicals. Van Nostrand Rheinhold Co., New York. 1310 p.

Walker, A. & P.A. Brown, 1985. Bull. Environ. Contam. Toxicol. 34, 143.
Weed Sci. Soc. Am., Herbicide Handbook, 4th ed., 1979. Weed Science Society of America, Champaign, IL. 479 p.
Wilson, J.T. et al., 1981. J. Environ. Qual. 10, 501.
Wu, T.L., 1980. J. Environ. Qual. 9, 459.

Yu, C., et al., 1975. J. Agric. Food Chem. 23, 877.

Zaki, M.H., D. Moran & D. Harris, 1982. Am. J. Publ. Health 72, 1391.
Zalkin, F., M. Wilkerson & R.J. Oshima, 1983. Pesticide Movement to Groundwater, Vol.II. California Dept. of Food and Agr. Sacramento.
Zimbdahl, R.L. & S.K. Clark, 1982. Weed Sci. 30, 545.

CHAPTER 15

A SOFTWARE 'ECOTOX' FOR ESTIMATION OF ECOTOXICOLOGICAL PARAMETERS

by
Bent Halling-Sørensen

15.1 WHY USE A SOFTWARE PROGRAM TO ESTIMATE PARAMETERS IN ECOTOXICOLOGY?

A software program has been developed on the basis of the physico-chemical and biological equations described in Chapter 4, Chapter 5, and in Appendices 1 and 4.

The idea behind this software is to provide a tool that readily estimates values on all, or some, of the parameters described in Chapters 4 and 5, and in Appendices 1 and 4.

Before using this software it is important to realise that the values found by means of it are only guide values. The estimated values will always be different from those found by laboratory experiments. This will be illustrated later in this chapter.

Working with ecotoxicological questions however, it is often better to have guide values for the properties of a considered component, than to have no values whatsoever. Today there are complete data on the physical, chemical and biological properties, for only a few chemicals.

15.2 HOW DOES THE SOFTWARE FUNCTION?

The software ecotox is developed on the basis of the SuperCalc4 spreadsheet, and "Skærm design" a menu program. As to the operative system the user may apply PC-DOS or MS, (referring to TF SuperCalc4, Willis and Pacewark).

Starting with ecotox may be a little complicated if the user does not already know some of the principles of the program. Firstly Ecotox provides the premises for estimating a specific parameter in the program. Second, the software enters a dialogue with the user, in which the user is informed about the different ways for calculating a specific parameter (see below). After this choice is made, the software often asks for unknown constants, or asks the user to "move the cursor" into a special table (tables from Appendix

2) in the program to estimate component specific constants, to be used later in the calculation of the parameters.

This need for specific constants implies the user knows the structural form of the component under consideration.

All tables (see Appendix 2) are based upon values for atomic and structural parts of a molecule.

Example: calculating Ω for the compound 1,1,1 trichloroethane (one of EPA's 12 Benchmark Chemicals).

Fig. 15.1

Group	Ω contribution	Number of	Considered component
Carbon	-0.5592	2	-1.1184
Hydrogen	0.2875	3	0.8625
Hydroxyl group	0.3382		0
Oxygen	0.0165		0
Nitrogen	-0.2625		0
Sulphur	0.014		0
Fluorine	0.3042		0
Chlorine	0.2983	3	0.8949
Bromine	0.299		0
Iodine	0.2994		0
Double bond (C=C)	0.564		0
Double bond (C=O)	0.564		0
Double bond (C=S)	0.564		0
Double bond (S=O)	0.564		0
Triple bond (C≡O)	1.1415		0
Triple bond (C≡N)	1.159		0
Triple bond (N≡)	1.159		0
6-membered ring (aromatic or naphthenic)	0.5666		0
Sum Ω			0.639

Before using the software it is of great importance to scrutinize all equation choices carefully. Normally the software offers several ways of reaching the same results.

In some circumstances it would give greater confidence in a value, to use several equations for the solution, and use the average of the various estimated values of the parameter.

It is not easy to indicate which equations are best, but the conclusion is: Reach as far as possible into the program with the knowledge available for the component under consideration, and use as many routes as possible to find the same parameter. The more the results, the greater confidence of the user in a specific value.

For all considered components there is no need to go through the software. If you do not know some of the values needed to estimate some of the more sophisticated chemical properties, then use parts of the program. Nearly all of the physical and biological equations can be estimated on the basis of simple parameters such as the boiling point, the melting point and the molecular weight of the considered component. For compounds in solution molarity will also be needed.

The software has a summation sheet where all calculated data for a considered component will be automatically transferred. After finishing work the user will only have to print out this summation sheet.

15.3 ABILITY OF THE PROGRAM

For a considered component the software is able to estimate values for the following physico-chemical and biological parameters:

- the boiling point, T(b)
- the melting point, T(m)
- the critical temperature, T(c)
- the partial pressure, p
- the pressure, P
- the critical pressure, P(c)
- the critical volume, V(c)
- the molal volume, V(m)
- the solubility in water, S
- the oil/water partition coefficient, K_{ow}
- the adsorption isotherm, K_{ac}
- the air/soil exchange coefficient, K_a
- the air/water exchange coefficient, KA and KL
- the evaporation rate, Q
- the hydrolysis rate coefficient
- the photochemical reaction rate coefficient
- the bioaccumulation, BCF
- the bioconcentration, CF
- the biomagnification, EM
- the biodegradation
- the MAC value
- the LD_{50} value
- the LC_{50} value
- the NC_{50} value
- the threshold concentration, C_{min}

An important question to answer before starting to work with the software is: at the least, what does the user need to know of a considered component to be able to carry out an estimation procedure?

Basically, only the structure, the molecular weight, and, for solutions, the molarity, are needed. But for the more sophisticated chemical parameters as, for instance, the adsorption isotherms, one also needs to know the concentration of the pollutant in both the sorbed phase, C(s), and in the solution, usually water, C(w).

For estimation of the air/soil exchange coefficient, K_a, one needs to know, furthermore, the compound concentration, C_s, of the considered component in the air layer adjacent to the surface, and Ca, the compound's concentration far from the surface. Additionally the user will need the flux rate, J, of the compound perpendicular to the surface, in g/m^2*s, the average velocity of air (m/s) parallel to the distance, L, in meters.

For estimation og the air/water exchange coefficient the user needs to know the U10 value, defined as the actual wind velocity measured at a height of 10 m, in the units m/s.

15.4 SOME OF THE ESTIMATION ROUTES THROUGH THE SOFTWARE ECOTOX

To illustrate the different routes through the software some remarks on the different parameters are outlined below.

<u>The boiling point, T(b)</u>
The boiling point, T(b) for the considered component is either known from the literature, or can be estimated by equation (4.7).

<u>The melting point, T(m)</u>
Will automatically be estimated by the software from the boiling point.

<u>The critical temperature, T(c)</u>
The critical temperature has several estimation methods by use of equation (4.1) or Ω, see equation (4.2).

<u>The partial pressure, p, and the vapour pressure, P</u>
If one knows, or can estimate, Henry's Constant and the molarity (in the case where the considered component is in solution), it is easy to estimate the partial pressure or the vapour pressure.

<u>The solubility, S</u>
To estimate the solubility one needs to estimate:

1. the parachor, PCH
2. the critical pressure, P(c)
3. the molar refraction, RD
4. the critical volume, V(c), and
5. the molal volume, V(m).

The parachor, PCH, and the molar refraction, RD, are found from additive values in tables based upon the structure of the considered component.

The critical pressure, P(c), can be found by 3 different equations (4.13, 4.14 and 4.15). All constants needed for insertion in the 3 equations are either already known by the software, or can be found in tables pointed out in the software.

For the critical volume, V(c), 3 different equations are available (4.16, 4.17 and 4.18).

The molal volume, V(m), can easily be found knowing the critical volume, V(c), and/or the critical pressure, P(c).

The solubility, S, will be estimated for the considered component on the basis of V(m), RD or the molecular weight, M.
Remember to choose the right equation for the regarded compound class.

<u>The oil/water partition coefficient, K_{ow}</u>
Use the same principle as for estimating solubility, S.

<u>The adsorption isotherm, K_{ac}</u>
The adsorption isotherm, K_{ac}, can be estimated by different equations. For some equations one will need specific values for the concentration of pollutant in both the sorbed phase, C_s, and in the solution, C_w, which is usually water. K_{ac} can also easily be estimated from K_{ow}, the oil/water partition coefficient, by different equations for various classes of organic compounds.
K_{ac} can also be estimated from the mole fraction solubility, SOL, when the solubility, S, in water is known.

<u>The air/soil exchange coefficient, K_a</u>
K_a, in m/s, can be estimated by 3 different equations. Before estimating the air/soil exchange coefficient, K_a, one needs to know various specific values for the considered component.

<u>The air/water exchange coefficient, KA and KL</u>
The overall mass transfer coefficients, KL and KA, can be estimated knowing values for the air-phase friction velocity. All other required values are already known by the software.

The evaporation rate, Q

For standard conditions, i.e. at a wind velocity of 4 m/s, a spill size of 75 m^2, and a temperature of 293°K, no other data are needed for estimating Q. Only remember to insert the vapor pressure in mm Hg in the equation.

The hydrolysis rate coefficient

All required constants can be found in tables indicated by the software based upon the structure of the considered component.

The photochemical reaction rate coefficient

The absorption of light in solutions due to photochemical reactions takes place in a narrow wavelength region. This absorption is, a, in the equation, and is an area specific value, which has to be known. This value has to be added to the Lambert-Beers equation.

The bioaccumulation, bioconcentration and biomagnification

The software offers quite different routes for estimating these parameters. This may be done either using the correlation between the concentration in the organism, C_o, and the concentration in the medium, C, under different circumstances for the considered component, or by using the great number of equations made on basis of relations to the solubility and the partition coefficient, K_{ac}. Take care of the different units involved in the solubility and K_{ac} in the different equations.

There is a fundamental relationship between size (surface) and the biochemical activity in the organisms. The surface determines the contact with the environment quantitatively, and thereby the possibility to take up food and excrete waste substances. For aquatic organisms the concentration factor CF, which indicates the ratio: concentration in the organism to the concentration in the medium, follows the same lines.

Therefore, it is possible to correlate the length (surface) of the animal with the CF-factor. (S.E. Jørgensen, 1986).

In other words it is possible to find the concentration factor for other species provided it is known for the considered compound for one - but preferably several species.

The biodegradation

The software offers a number of equations, dependent on the molecular weight and the number of aliphatic carbon atoms. Be very careful when estimating values for biodegradation. The applied equations are summarized in Appendix 6.

The maximum allowable concentration (MAC)
MAC is a value in accordance with environmental legislation. Though often dependent on time, the software offers the user other equations dependent on the melting point, boiling point, the molecular weight, the LD_{50}-value and the LC_{50}-value. Remember to use the correction table for log MAC for different types of compounds. MAC is often estimated with different units, so be careful when estimations are compared.

The LD_{50} (lethal dose fifty) value
The software offers the user a table with LD_{50}-values (mg per kg body) for many components. The values have been found for different animal species.
To estimate the LD_{50}-value for metals the software offers a number of equations to estimate the value for white mice, dependent on the standard potential (SP) of the metal, or on the solubility of sulfides (log S).

The LC_{50}- and NC_{50}-values
The software offers the user a table with LC_{50}-values (in µg/l) for many compounds. These values are found for different animal species.
To estimate the LC_{50}-value, or the NC_{50}-value, the software offers a number of equations for estimating these, dependent on the molecular weight, the boiling point, the RD value, and the melting point.

The threshold concentration, C_{min}
Estimation of C_{min} can be made, dependent on the molecular weight and the boiling point for particular classes of organic compounds.

15.5 CALCULATION EXAMPLES

To illustrate the estimation methods in Chapters 4 and 5, the application of the software to three Benchmark Chemicals will be shown: 1,1,1 trichloroethane, p-cresol and DDT.
For each of these Benchmark Chemicals most of the possible routes are followed and the resulting estimations compared with literature values.

15.5.1 1,1,1 trichloroethane

Structure:

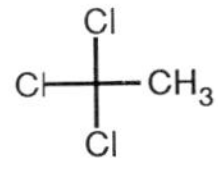

Firstly, the boiling point is found. This can be estimated by use of equation (4.7) or taken from the literature.
Using equation (4.7) demands that 4 constants, specific to 1,1,1 trichloroethane, are known, i.e. Ω, ß A, and ¥.

Fig. 15.2: Contributions to Ω

Group	Ω contribution	Number of	Considered component
Carbon	-0.5592	2	-1.1184
Hydrogen	0.2875	3	0.8625
Hydroxyl group	0.3382		0
Oxygen	0.0165		0
Nitrogen	-0.2625		0
Sulphur	0.014		0
Fluorine	0.3042		0
Chlorine	0.2983	3	0.8949
Bromine	0.299		0
Iodine	0.2994		0
Double bond (C=C)	0.564		0
Double bond (C=O)	0.564		0
Double bond (C=S)	0.564		0
Double bond (S=O)	0.564		0
Triple bond (C≡O)	1.1415		0
Triple bond (C≡N)	1.159		0
Triple bond (N≡)	1.159		0
6-membered ring (aromatic or naphthenic)	0.5666		0
Sum Ω			0.639

2 carbon atoms =
2 * (-0.5592) = -1.1184

3 hydrogen atoms =
3 * 0.2875 = 0.8625

3 chlorine atoms =
3 * 0.2983 = 0.8949

Sum Ω 0.6390

Ω can also be estimated by (4.3). ΔT is needed and is found from contributions to the structure.

Fig. 15.3: Contributions to ΔT

Group	ΔT contribution	Number of	Considered component
$-CH_3$, $-CH_2-$, $-CH_2-$ (in ring)	0.016	1	0.016
| - CH (in ring) |	0.013		0
| - C - (in ring) |	0.003	1	0.003
H	0		0

$-CH_3$	0.016
-C-	0.003
3 * -Cl	0.039
Sum Ω	0.058

Fig. 15.3 (continued)

Group	ΔT contribution	Number of	Considered component
=C-H, $=CH_2$ (with a bond drawn above C)	0.015		0
=C-, =C= (in ring) (with a bond drawn above C)			
≡C-, ≡CH			
Ring atom	-0.005		0
-F	0.015		0
-Cl	0.013	3	0.039
-Br	0.01		0
-I			
-O- (in ring)	0.02		0
-OH (alcohols)	0.07		0
-OH (phenols)	0.029		0
CO (in ring)	0.046		0
-CHO	0.046		0
-COO-	0.039		0
-COOH	0.07		0
$-NH_2$	0.027		0
NH (in ring)	0.027		0
N	0.012		0
-CN	0.053		0
-CN	0.053		0
-SH, -S-	0.012		0
Sum ΔT			0.058

A ΔT value of 0.058 gives by equation (4.3) an $\Omega = 0.632$.

Fig. 15.4: Contributions to ¥

Group	¥ contribution	Number of	Considered component
Carbon	1.02	2	2.04
Hydrogen	1.95	3	5.85
Oxygen	3		0
Hydrogen group (includes H and O)	23.1		0
Nitrogen	2.45		0
Sulphur	6.7		0
Fluorine	3.63		0
Chlorine	7.45	3	22.35
Bromine	9.38		0
Iodine	10.95		0
C=C	2.5		0
C=O	2.5		0
C=S	2.5		0
S=O	2.5		0
C≡C	5.59		0

2 carbon atoms = 2 * 1.02	2.04
3 hydrogen atoms = 3 * 1.95	5.85
3 chlorine atoms = 3 * 7.45	22.35
sum ¥	30.24

Fig. 15.4: Contributions to ¥ (continued)

Group	¥ contribution	Number of	Considered component
C≡N	29.1		0
N≡N	2.68		0
5- or 6-membered ring	-0.14		0
Sum ¥			30.24

Fig. 15.5: Contributions to ß

Group	ß contribution	Number of	Considered component
Hydrogen	0		0
Carbon	0.23		0
Fluorine	0.23		0
Chlorine	0.32	3	0.96
Bromine	0.54		0
Iodine	0.87		0
Oxygen	0.18		0
Nitrogen	0.14		0
Sulphur	0.27		0
C=C	-0.06		0
C=O	-0.12		0
C≡N	0		0
-O-H (alcohols)	-0.11		0
-OH (aromatic alcohols)	-0.21		0
OH (acids)	0		0
NH	-0.02		0
C≡C	-0.14		0
C≡S	-0.05		0
Each ring atom.	-0.043		0
Condensed ring	-0.1		0
$-CH_3$, $-CH_2-$	0.227	1	0.227
$-CH_2-$	0.184		0
$-\overset{\vert}{\underset{\vert}{C}}H$	0.21	1	0.21
When in ring	0.192		0
$=\overset{\vert}{C}H$, $=CH_2$	0.198		0
$=\overset{\vert}{C}H$ in ring	0.154		0
$-\overset{\vert}{\underset{\vert}{C}}-$	0.21		0
When in ring	0.154		0
$=\overset{\vert}{C}-$, $=C=$	0.198		0
When in ring	0.154		0
≡C-, ≡CH	0.153		0
-F	0.224		0
-Cl	0.32		0
-Br	0.5		0

3 * -Cl	0.960
$-CH_3$	0.227
-CH-	0.210
Sum ß	1.397

Fig. 15.5: continued

Group	ß contribution	Number of	Considered component
-I	0.83		0
-O-	0.16		0
When in ring	0.12		0
-OH (alcohols)	0.06		0
-OH (phenols)	-0.002		0
CO	0.29		0
CO in ring	0.2		0
-CHO	0.33		0
-COO-	0.47		0
-COOH	0.4		0
$-NH_2$	0.095		0
NH	0.135		0
When in ring	0.09		0
N-	0.17		0
When in ring	0.13		0
-CN	0.36		0
-SH, -S-	0.27		0
-S- in ring	0.24		0
=O (exceot for com. above)	0.12		0
$-NO_2$	0.42		0
=S	0.24		0
Sum ß			1.397

Fig. 15.6: Contributions to A

Family	Value of A	Considered component
Alcohols	0.65	
Phenols	0.665	
Nitriles	0.68	
Carbozylic acids	0.685	
All other compounds	0.8	0.8
		0.8

The most appropiate value for A (1,1,1 trichloroethane) is A = 0.80 ("all other compounds").

Now, having all the required constants, equation (4.7) gives a boiling point of 399,4°K. The literature value is 347.3°K.

<u>The melting point, T(m)</u>

Equation (4.10) gives a Tm-value of 233.2°K (Tb = 399.4°K) and Tm-value of 202.8°K (Tb = 347.3°K is the literature value).

<u>The critical temperature, T(c)</u>

Equation (4.2) gives a Tc-value of 625.0°K (543.5°K at a boiling point of

347.3°K).

Using equation (4.5), the gas constant has the value of 82.05 cm*atm/°K*g*mol.

Estimation of the two constants, a and b, is difficult.

1. Construct the longest normal chain present in the hydrocarbon.
2. Produce only single methyl substitutions directly on the normal chain starting from the left end of the normal chain and moving always in a clockwise direction.
3. If necessary, continue methyl-group substitutions on already substituted methyl groups by maintaining the previously established clockwise direction until the end.

Equation (4.5) is only valid for aliphatic hydrocarbons, and yields 576°K.

<u>Henry's constant, H</u>

A useful method to estimate H is equation (4.9). H has the unit atm*m^3/mol. Equation (4.9) gives a H-value of $3.9*10^{-2}$ at 298°K. ($9.4*10^{-2}$ at 298°K, using Tb = 347.3°K and Tm = 202.8°K).

<u>The vapour pressure, P</u>

The vapour pressure P in atm can be estimated by equation (4.11). At 298°K, T > Tm for 1,1,1 trichloroethane and therefore the last term in equation (4.11) is ignored. At 298°K, equation (4.11) gives P = 0.021 atm. (0.167 atm when Tb = 347.3°K). Literature value: 0.163 atm (Benchmark Chemicals).

<u>The critical pressure, Pc</u>

The critical pressure can be estimated by the equations (4.13), (4.14) and (4.15).

Using equation (4.13) Pc = 44.7 atm.

Using equation (4.14) Pc = 36.7 atm. at Tc = 541.8°K.

Using equation (4.15), the parachor, PCH, and the constants, C and D, have to be found.

Fig. 15.7: **Contributions to PCH**

Group	PCH contribution	Number of	Considered component
$-(CH_2)n$: n<12	40	1	40
$-(CH_2)n$: n>12 (1)	40.3		0
C	9		0
H	15.5		0
H in OH	10		0
H in HN	12.5		0
O	19.8		0
O:O in esters	54.8		0
N	17.5		0
S	49.1		0
P	40.5		0
F	26.1		0
Cl	55.2	3	165.6
Br	68		0
I	90.3		0
Se	63		0
Si	31		0
Al	55		0
Sn	64.5		0
As	54		0
Ethylenic bond:			
Terminal	19.1		0
2,3 position	17.1		0
3,4 position	40.6		0
Carbonyl bond in ketones:			
total nr of carbonatoms= 3	22.3		0
4	20		0
5	18.5		0
6	17.3		0
7	17.3		0
8	15.1		0
9	14.1		0
10	13		0
11	12.6		0
Single bond	0		0
Semipolar bond	0		0
Singlet linkage	-9.5		0
Hydrogen bridge (2)	-14.4		0
Chain branching, per branch.	-3.7	3	-11.1
Sec.-Sec. adjacency	-1.6		0
Sec.-Ter. adjacency	-2		0
Ter.-Ter. adjacency	-4.5		0
Alkyl groups: (3)			
1-Methylethyl	133.3		0
1-Methylpropyl	171.9		0
1-Methylbutyl	211.7		0
2-Methylpropyl	173.3		0
1-Ethylpropyl	209.5		0
1,1-Dimethylethyl	170.4		0
1,1-Dimethylpropyl	207.5		0
1,2-Dimethylpropyl	207.9		0
1,1,2-Trimethylpropyl	243.5		0

$-CH_2$ < 12	40
-Cl * 3	165.6
chain branching, 3 branches	-11.1
Sum PCH	194.5

Fig. 15.7: Continued

Group	PCH contribution	Number of	Considered component
Position differences in benzene:			
Ortho-meta	2.6		0
Meta-para	0.35		0
Ortho-para	2.9		0
Ring closure:			
3-membered ring	12.5		0
4-membered ring	6		0
5-membered ring	3		0
6-membered ring	0.8		0
7-membered ring	4		0
Sum PCH			194.5

(1) Somewhat greater beyond 20 $-CH_2-$
(2) Includes any ortho effect
(3) Complete group, including branching.

Contributions to C and D.
Using the values for aliphatic hydrocarbons:

C = 3.4271 D = -0.7829

Equation (4.15) yields Pc = 43.2 atm.
The mean value for Pc = 41.5 atm. Equations (4.13), (4.14) and (4.15).

<u>The critical volume, Vc</u>
By means of equations (4.16), (4.17) and (4.18) the critical volume is estimated (cm^3/g^*mol):
Equation (4.16) Vc = 236.2 cm^3/g^*mol (Tb = 399.4°K)
246.4 cm^3/g^*mol (Tb = 347.4°K)
Equation (4.17) Vc = 232.5 cm^3/g^*mol
Equation (4.18) Vc = 252.5 cm^3/g^*mol

Mean value for Vc = 243.8 cm^3/g^*mol

<u>The molal volume, Vm</u>
Using equation (4.21) the molal volume Vm = 91.5 cm^3/mol, at 347.5°K.

<u>The oil/water partition coefficient, K_{ow}</u>
Log K_{ow} is found by means of equation (4.22) and the table of structure contributions.

Fig. 15.8: Contribution to Log K_{ow} (oil/water partition coefficiert)

	B	C	D	E	F
$-CH_3$	0.142	-0.020	-0.016	-0.023	-0.015
-C-	-0.075	-0.076	0.091	0.043	0.034
3 * -Cl	0.495	0.261	-0.072	-0.036	-0.063
Sum	0.562	0.165	0.003	-0.016	-0.044

Thus, from equation (4.22), log Kow = 2.16.

The partition coefficient, log Kow, for 1,1,1-trichloroethane may also be estimated on the basis of Vm, molecular weight and RD. The regression equations for chlorinated hydrocarbons are used.

log Kow = 0.33 + 0.014 M r = 0.73 log Kow = 2.20

log Kow = 0.67 + 0.029 Vm r = 0.83 log Kow = 3.19

Fig. 15.9: Contributions to RD (molar refraction)

Group	RD contribution	Number of	Considered component
C	2.418	2	4.836
H	1.1	3	3.3
$-CH_2$	4.618		0
O (hydroxyl)	1.525		0
O (ether)	1.643		0
O (carboxyl) (1)	2.211		0
Cl	5.967	3	17.901
Br	8.865		0
I	13.9		0
N (primary amine)	2.322		0
N (secondary amine)	2.502		0
N (tertiary amine)	2.84		0
N (nitrile)	5.516		0
3-membered ring	0		0
4-membered ring	0		0
5-membered ring	0		0
6-membered ring	0		0
7-membered ring	0		0
Double bond	1.733		0
Semipolar double bond	1.733		0
Triple bond	2.398		0
Sum RD			26.037

-C * 2	4.836
-H * 3	3.300
-Cl * 3	17.901
Sum RD	26.037

log K_{ow} = -0.15 + 0.088 RD r = 0.83 log K_{ow} = 2.14

The 4 estimations of log K_{ow} give mean log Kow = 2.42

Literature value of log K_{ow} = 2.47.

The water solubility, S

When considering the water solubility, S, equations found in Table 11B, 16 and 17 App. 2 can be used. Vm, RD, the molar weight and log K_{ow} are used for the regression equations.

The regression equations involving Vm, RD and M are developed for hydrocarbons.

log S in mmol/l	
log S = 0.61 - 0.011 M	S = 18,5 mg/l
log S = 1.52 - 0.016 Vm	S = 180.0 mg/l
log S = 0.60 - 0.03 RD	S = 88.1 mg/l

The literature values: 1334 mg/l (Benchmark Chemicals)
3450 mg/l (Environmental rapport no. 100)

Comparison of the estimated values with the literature values for the solubility of 1,1,1 trichloroethane shows that the estimated values are much smaller than the actual literature values. The probable reason for this is that the equations used are valid for hydrocarbons.

By use of the equations based on log K_{ow}, better results are obtained.

$\log K_{ow} = 7.10 - 1.05 \log S$ (S in μmol/l)

S = 4830 mg/l ($\log K_{ow} = 2.42$) S = 3420 mg/l ($\log K_{ow} = 2.47$)

For halogenated hydrocarbons (Table 17, Appendix 2)

$\log S = -0.962 \log K_{ow} + 6.50$ (S in μmol/l)

S = 1770 mg/l ($\log K_{ow} = 2.47$)

$\log S = -1.37 \log K_{ow} + 7.26$ (S in μmol/l)

S = 1003 mg/l ($\log K_{ow} = 2.47$).

The mean solubility S = 2755 mg/l. This value is much closer to the literature value, and demonstrates how careful the selection of regression parameters has to be.

The adsorption isotherm, log K_{ac}

This is found by the following equation parameters devised for hydrocarbons (Table 13, Appendix 2).

$\log K_{ac} = 0.72 \log K_{ow} + 0.49$

$\log K_{ac} = 2.23$ ($\log K_{ow} = 2.42$)

The literature value for $\log K_{ac} = 2.25$

The molecular diffusivity of the chemical in air and liquid

Da is the molecular diffusivity in air (m^2/s).

Dl is the molecular diffusivity in liquid (m^2/s).

Da can be estimated with equations (4.27) and (4.28). Using equation (4.27) Ω has to be found. Ω corresponds to the value of 1.30 * T_r and $T_r = T/T_c$. Using equation (4.27) Da = $5.83*10^{-6}$ m^2/s

Using equation (4.28) V_b and V_{ba} has to be estimated by means of contribution tables.

Fig. 15.10: Contributions to Vb

Compound	Contribution m</g mole Schroeder	Number of	Considered component
Carbon	7.0	2.0	14
Hydrogen	7.0	3.0	21
Oxygen	7.0		0
Double bound	-		0
Methyl esters end ethers	-		0
Ethyl esters and ethers	-		0
Higher esters and ethers	-		0
Acids	-		0
Joined to S, P, N	-		0
Nitrogen	7.0		0
Primary amines	-		0
Secondary amines	-		0
Flourine	10.5		0
Chlorine	24.5	3.0	73.5
Bromine	31.5		0
Iodine	38.5		0
Sulphur	21.0		0
Ring	-7.0		0
3-membered	-		0
4-membered	-		0
5-membered	-		0
6-membered	-		0
Naphtalene	-		0
Anthracene	-		0
Double bond between carbon atoms	7.0		0
Triple bond between carbon atoms	14.0		0
Sum V_b			108.5

-C * 2	14.0
-H * 3	21.0
-Cl * 3	73.5
Sum V_b	108.5

V_{ba} = 28

V_{ba} is the molal volume at the normal boiling point for air. From equation (4.28), Da = 7.45 $*10^{-6}$ m^2/s

Dl can be estimated by equation (4.39), where μ is the viscosity of water at the estimation temperature, in centipoise.

At 298.15°K Dl = 15.72 $*10^{-10}$ m^2/s

The literature value for Dl = $9.32 \ast 10^{-10}$ m^2/s

The evaporation rate, Q

The evaporation rate, Q, is estimated from equation (4.40) in $g/s^\ast m^2$.
For 1,1,1 trichloroethane Q = 1.21 $g/s^\ast m^2$.

Estimation of BCF, CF and EM

A number of equations are outlined (Table 15.1) on the basis of log K_{ow}, log K_{ac} and the solubility S. When choosing a useful equation, notice that the different equations have limited indicator ranges, and are species specific.

The biological concentration factor BCF (indicator K_{ac}):

$\log BCF = 2.024 + 1.225 \log K_{ac}$ $r = 0.91$ $\log BCF = 4.76$

$\log BCF = 1.579 + 1.191 \log K_{ac}$ $r = 0.87$ $\log BCF = 4.24$

Thus the mean value for log BCF = 4.50 (fish species).

The concentration factor CF (indicator K_{ow}):

The indicator range allows the use of the following regression equations:

$\log CF = -0.973 + 0.767 \log K_{ow}$ $r = 0.76$ (fish species)
$\log CF = 0.88$

$\log CF = -1.495 + 0.935 \log K_{ow}$ $r = 0.87$ (fish species)
$\log CF = 0.77$

Thus the mean value for log CF (fish species) = 0.83

$\log CF = -0.70 + 0.85 \log K_{ow}$ $r = 0.95$ (fathead minnow)
$\log CF = 1.36$

$\log CF = 0.124 + 0.542 \log K_{ow}$ $r = 0.90$ (fathead minnow)
$\log CF = 1.44$

Thus the mean value for log CF (fathead minnow) = 1.40

The ecological magnification factor EM, (indicator S in µmol/l):

S = 2755 mg/l S = 20650 µmol/l log S = 4.31

$\log EM = -0.4732 \log S + 4.4806$ (mosquito fish, whole)
$\log EM = 2.44$

$\log EM = -0.3891 \log S + 3.995$ (mosquito fish, whole)
$\log EM = 2.32$

The mean value for log EM (mosquito fish, whole) = 2.38

Estimation of MAC, LC_{50}, LD_{50} and C_{min}

Estimations of log MAC can be made by different regression equations which are valid for all compounds, remembering to use the correction table in appendix.

log MAC = -1.2 - 0.012 t-melt + log M
(correction +0.5) log MAC = 2.27

log MAC = 0.40 - 0.01 M + log M
(correction +0.5) log MAC = 1.69

log MAC = -0.4 - 0.006 t-boil + log M
(correction +0.5) log MAC = 1.91

Mean value for log MAC = 1.91.

Estimation of LC_{50} in mmol/l can be made by different regression equations valid for all compounds:

log LC_{50} = -0.02 - 0.009 t-boil	log LC_{50} = -0.688
log LC_{50} = 0.08 - 0.011 M	log LC_{50} = -1.387
log LC_{50} = -1.6 - 0.010 t-melt	log LC_{50} = -0.895

Estimation of LC_{50} in mmol/l can also be made by two different regression equations made for chlorohydrocarbons:

log LC_{50} = 0.20 - 0.012 M	log LC_{50} = -1.400
log LC_{50} = -0.1 - 0.011 t-boil	log LC_{50} = -0.917

All temperture used in the equations for LC_{50} are in degrees Celcius.
Mean value for log LC_{50} = -1.058 = LC_{50} = 11.7 mg/l

Two equations for hydrocarbons can be used to estimate C_{min}:

log C_{min} = -0.95 - 0.010 t-boil	log C_{min} = -1.69
log C_{min} = -0.51 - 0.015 M	log C_{min} = -2.51

Mean value for log C_{min} = -2.1

On basis of the calculated value for C_{min}, the LD_{50}-value is estimated by the outlined regression equation:

log C_{min} = 0.68 log LD_{50} -3.4 (LD_{50} in mmol/kg) log Cmin = -2.1 log LD_{50} = 1.91 LD_{50} = 10.9 g/kg

Table 15.1
1,1,1 trichloroethane. Comparison of estimated values with literature values, using both known boiling points and estimated boiling points

	Estimated value given by the boiling point from the literature	Deviation %	Estimated value given by the boiling point from Equation (4.7)	Deviation %	Literature value
Tb (4.7)			399.4°K	14.8	347.3°K
Tm (4.10)	202.8°K	0.9	233.2°K	13.9	204.7°K
Tc (4.1)	521°K	2,3	599,1°K	12.4	
(4.2)	543.5°K	1.9	625,0°K	17,3	
(4,3)	549,5°K	3,1	631,9°K	18,4	
(4,4)	553,0°K	3,7	635,4°K	19,2	533,2°K
(4,5)			576.0°K	8.0	
Tc mean value	541.8°K	1.6	613.5°K	15.1	
H (4.9) atm m^3/mol	$9.4 \cdot 10^{-2}$		$3.9 \cdot 10^{-2}$		1.650 Pa/m^3mol
P (4.11) atm (298°K)	0.167	2.5	0.021	87.1	0.163
Pc (4.13)	44.7	10.5			
(4.14)	36.7	26.6	44.1	11.9	50.0
(4.15)	43.2	13.7			
Pc mean value	41.5	16.9			
Vc (4.16)	246.4	1.0	236.2	3.2	
(4.17)	232.5	4.7			244.0
(4.18)	252.5	3.5			
Vc mean value	243.8	0.1			
Vm (4.21)	91.5	20.1	88.3	22.8	114.5
log K_{ow}	2.42	10.9			2.47
Solubility S mg/l	2755	20.1			3450
log K_{ac}	2.23	4.5			2.25
Da m^2/s (4.27)	$5.83 \cdot 10^{-6}$	30.0			$8.32 \cdot 10^{-6}$
(4.28)	$7.45 \cdot 10^{-6}$	10.5			

Table 15.1. Continued

	Estimated value given by the boiling point from the literature	Deviation %	Estimated value given by the boiling point from Equation (4.7)	Deviation %	Literature value
DI m^2/s	$15.72 \cdot 10^{-10}$	71.6			$9.32 \cdot 10^{-10}$
Q $g/s\ m^2$	1.21				
log BCF	4.50 (fish species)				
log CF	0.83 (fish species) 1.40 (fathead minnow)				
log EM	2.38 (fathead minnow)				
MAC mg/m^3	10850				440
LD_{50} g/kg	10.9				10.3-12.3 (rat) single oral dose
LC_{50} mg/l	11.7				31.2 (crustaceans)
C_{min} mg/l	0.008				

15.5.2. p-Cresol

Structure: 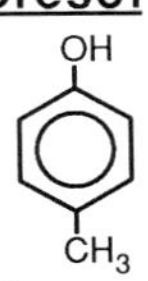

Firstly, the boiling point is found. This can be estimated using equation (4.7), or is available from the literature. Using equation (4.7) demands 4 constants, specific for p-Cresol, are known, i.e. Ω, ß, A and ¥.

Fig. 15.11: Contributions to Ω

Group	Ω contribution	Number of	Considered component
Carbon	-0.5592	7	-3.9144
Hydrogen	0.2875	7	2.0125
Hydroxyl group	0.3382	1	0.3382
Oxygen	0.0165		0
Nitrogen	-0.2625		0
Sulphur	0.014		0

7 carbon atoms =
7 * (-0.5592) = -3.9144

7 hydrogen atoms =
7 * 0.2875 = 2.0125

Fig. 15.11: Continued

Group	Ω contribution	Number of	Considered component
Fluorine	0.3042		0
Chlorine	0.2983		0
Bromine	0.299		0
Iodine	0.2994		0
Double bond (C=C)	0.564	3	1.692
Double bond (C=O)	0.564		0
Double bond (C=S)	0.564		0
Double bond (S=O)	0.564		0
Triple bond (C≡O)	1.1415		0
Triple bond (C≡N)	1.159		0
Triple bond (N≡)	1.159		0
6-membered ring (aromatic or naphthenic)	0.5666	1	0.5666
Sum Ω			0.6949

-OH	0.3382
C=C bonds 3 * 0.5640	1.6920
6-membered ring	0.5666
Sum Ω	0.6949

Ω can also be estimated by equation (4.3). For this ΔT is needed and is found from contributions to the structure.

Fig. 15.12: Contributions to ΔT

Group	ΔT contribution	Number of	Considered component
$-CH_3$, $-CH_2-$, $-CH_2-$ (in ring)	0.016	1	0.016
>CH– (in ring)	0.013		0
>C< (in ring)	0.003		0
H	0		0
=C(–)-H, $=CH_2$, =C(–)-H (in ring)	0.015	6	0.09
=C(–)-, =C= (in ring), ≡C-, ≡CH	0.003		0
Ring atom	-0.005	1	-0.005
-F	0.015		0
-Cl	0.013		0
-Br	0.01		0
-I			

$-CH_3$	0.016
-CH * 6	0.090
Ring atom	-0.005
-OH	0.029
Sum ΔT	0.130

Fig. 15.12: continued

Group	ΔT contribution	Number of	Considered component
-O- (in ring)	0.02		0
-OH (alcohols)	0.07		0
-OH (phenols)	0.029	1	0.029
CO (in ring)	0.046		0
-CHO	0.046		0
-COO-	0.039		0
-COOH	0.07		0
$-NH_2$	0.027		0
NH (in ring)	0.027		0
N	0.012		0
-CN	0.053		0
-CN	0.053		0
-SH, -S-	0.012		0
Sum ΔT			0.13

By equation (4.3) a ΔT-value of 0.130 gives an Ω = 0.704.

Fig. 15.13: Contributions to ¥

Group	¥ contribution	Number of	Considered component
Carbon	1.02	7	7.14
Hydrogen	1.95	7	13.65
Oxygen	3		0
Hydrogen group (includes H and O)	23.1	1	23.1
Nitrogen	2.45		0
Sulphur	6.7		0
Fluorine	3.63		0
Chlorine	7.45		0
Bromine	9.38		0
Iodine	10.95		0
C=C	2.5	3	7.5
C=O	2.5		0
C=S	2.5		0
S=O	2.5		0
C≡C	5.59		0
C≡N	29.1		0
N≡N	2.68		0
5- or 6-membered ring	-0.14	1	-0.14
Sum ¥			51.25

7 carbon atoms = 7 * 1.02	7.14
7 hydrogen atoms = 7 * 1.95	13.65
-OH	23.10
3 * C=C	7.50
Ring	-0.14
Sum ¥	51.25

Fig. 15.14: Contributions to ß

Group	ß contribution	Number of	Considered component
Hydrogen	0		0
Carbon	0.23	7	1.61
Fluorine	0.23		0
Chlorine	0.32		0
Bromine	0.54		0
Iodine	0.87		0
Oxygen	0.18	1	0.18
Nitrogen	0.14		0
Sulphur	0.27		0
C=C	-0.06	3	-0.18
C=O	-0.12		0
C≡N	0		0
-O-H (alcohols)	-0.11		0
-OH (aromatic alcohols)	-0.21	1	-0.21
OH (acids)	0		0
NH	-0.02		0
C≡C	-0.14		0
C≡S	-0.05		0
Each ring atom	-0.043	6	-0.258
Condensed ring	-0.1		0
$-CH_3$, $-CH_2-$	0.227		0
$-CH_2-$	0.184		0
-CH	0.21		0
When in ring	0.192		0
=CH, $=CH_2$	0.198		0
=CH in ring	0.154		0
-C-	0.21		0
When in ring	0.154		0
=C-, =C=	0.198		0
When in ring	0.154		0
≡C-, ≡CH	0.153		0
-F	0.224		0
-Cl	0.32		0
-Br	0.5		0
-I	0.83		0
-O-	0.16		0
When in ring	0.12		0
-OH (alcohols)	0.06		0
-OH (phenols)	-0.002		0
CO	0.29		0
CO in ring	0.2		0
-CHO	0.33		0
-COO-	0.47		0
-COOH	0.4		0

7 * -C-	1.61
-O-	0.18
3 * C=C	-0.18
-OH aromatic	-0.21
6 * (ring atom)	-0.26
Sum ß	1.14

Fig. 15.14: continued

Group	ß contribution	Number of	Considered component
$-NH_2$	0.095		0
NH	0.135		0
When in ring	0.09		0
N-	0.17		0
When in ring	0.13		0
-CN	0.36		0
-SH, -S-	0.27		0
-S- in ring	0.24		0
=O (exceot for com. above)	0.12		0
$-NO_2$	0.42		0
=S	0.24		0
Sum ß			1.142

Fig. 15.15: Contributions to A

Family	Value of A	Considered component
Alcohols	0.65	
Phenols	0.665	0.665
Nitriles	0.68	
Carbozylix acids	0.685	
All other compounds	0.8	
		0.665

The most appropriate value for A (p-Cresol) is A = 0.665 ("phenols").

Now, having all the necessary constants, equation (4.7) gives a boiling point of 480.5°K. The literature value is 475.0°K.

The melting point, Tm

Equation (4.10) gives a Tm-value of 280.7°K (Tb = 480.5°K) and Tm-value of 277.3°K (Tb = 475.0; literature value).
The literature value for Tm = 308°K.

The critical temperature, Tc

Equation (4.1) yields 720.8°K. Equation (4.2) gives a Tc-value of 691.5°K (683.6°K at a boiling point of 475.0°K).
Mean value for Tc = 689.8°K (Tb = 475°K).

Henry's Constant, H

A useful method for estimating H is equation (4.9) H has the units of atm*m^3/mol. Equation (4.9) gives a H-value of $9.69*10^3$ at 298°K. ($1.07*10^{-2}$

at 298°K using Tb = 475.0°K and Tm = 277.3°K).

The vapour pressure, P

The vapour pressure P in atm can be estimated by equation (4.11). At 298°K T > Tm for p-Cresol and therefore the last term in equation (4.11) is considered.

At 298°K equation (4.11) gives $P = 4.98*10^{-4}$ atm ($6.46*10^{-4}$ atm when Tb = 475.0°K). Literature value: $1.16*10^{-4}$ atm (Benchmark Chemicals).

The critical pressure, Pc

The critical pressure can be estimated by equations (4.13), (4.14) and (4.15).

Using (4.13) Pc = 50.1 atm

Using (4.14) Pc = 49.9 atm   at Tc = 689.8°K.

Using (4.15), the parachor, PCH, and the constants, C and D, have to be found.

PCH = 259.4

Contributions to C and D, using the values for phenols: C=3.8584, D= -0.9215.

Equation (4.15) yields Pc = 42.7 atm.

The mean value for Pc = 47.6 atm. Equations (4.13), (4.14) and (4.15).

The critical volume, Vc

By mean of equations (4.16), (4.17) and (4.18) the critical volume is estimated ($cm^3/g*mol$):

Equation (4.16)   Vc = 370.6 $cm^3/g*mol$ (Tb = 475.0°K)

Equation (4.17)   Vc = 334.4 $cm^3/g*mol$

Equation (4.18)   Vc = 336.9 $cm^3/g*mol$

Mean value for Vc = 347.3 $cm^3/g*mol$

The molal volume, Vm

Using equation (4.21) the molal volume Vm = 129.2 cm^3/mol.

The oil/water partition coefficient, K_{ow}

Log K_{ow} is found by means of equation (4.22) and the table of structure contributions.

Fig. 15.16: Contributions to the estimation of log K_{ow}

	B	C	D	E	F
$-C_6H_5$	0.467	-0.007	0.012	0.007	-0.017
$-CH_3$	0.142	-0.020	-0.016	-0.023	-0.015
-OH	0.202	0.324	-0.012	-0.015	0.003
Sum	0.811	0.297	-0.016	-0.031	-0.029

From equation (4.22), log Kow = 2.24. The literature value for log Kow = 2.30.

Fig. 15.17: Contributions to RD (molar refraction).

Group	RD contribution	Number of	Considered component
C	2.418	7	16.926
H	1.1	8	8.8
$-CH_2$	4.618		0
O (hydroxyl)	1.525	1	1.525
O (ether)	1.643		0
O (carboxyl) (1)	2.211		0
Cl	5.967		0
Br	8.865		0
I	13.9		0
N (primary amine)	2.322		0
N (secondary amine)	2.502		0
N (tertiary amine)	2.84		0
N (nitrile)	5.516		0
3-membered ring	0		0
4-membered ring	0		0
5-membered ring	0		0
6-membered ring	0		0
7-membered ring	0		0
Double bond	1.733	3	5.199
Semipolar double bond	1.733		0
Triple bond	2.398		0
Sum RD			32.450

-C * 7	16.926
-H * 8	8.800
O (hydroxy)	1.525
Double bond:	
3 * C=C	5.199
Sum RD	32.450

<u>The water solubility, S</u>

When considering the water solubility S, the equations found in table 17, Appendix 2 ("mixed classes; aromatic and chlorinated hydrocarbons") are used. By use of the equations based on log Kow:

log S = -1.37 log Kow + 7.26 (S in µmol/l)

S = 1675 mg/l (log Kow = 2.24)

"All chemicals represented" (Table 17, Appendix 2):

log 1/S = 1.214 log Kow - 0.850 (S in mol/l)

S = 1460 mg/l (log Kow = 2.24).

"Aromatics" (Table 17, Appendix 2)

$\log 1/S = 0.996 \log K_{ow} - 0.339$ (S in mol/l)

$S = 1387$ mg/l ($\log K_{ow} = 2.24$).

The mean solubility S = 1510 mg/l. The literature value is 1840 mg/l (Benchmark Chemicals).

<u>The adsorption isotherm, log K_{ac}</u>

This is found by the following equation parameter built for "predominantly methylated and halogenated benzens" (Table 13, Appendix 2).

$\log K_{ac} = 0.72 \log K_{ow} + 0.49$

$\log K_{ac} = 2.10$ ($\log K_{ow} = 2.24$).

<u>The molecular diffusivity of the chemical in air and liquid</u>

Da is the molecular diffusivity in air (m^2/s). Dl is the molecular diffusivity in liquid (m^2/s).

Da can be estimated from equation (4.27) and (4.28).

When using equation (4.27) Ω has to be found. Ω corresponds to the value of 1.30 * Tr, where Tr = T/Tc. From equation (4.27), Da = 4.44 $*10^{-6}$ m^2/s. The litertaure value is 7.77 $*10^{-6}$ m^2/s.

When using equation (4.28), Vb and Vba have to be estimated by means of contribution tables.

Fig. 15.18: Contributions to Vb

Compound	Contribution m</g mole Schroeder	Number of	Considered component
Carbon	7.0	7.0	49
Hydrogen	7.0	8.0	56
Oxygen	7.0	1.0	7
Double bound	-		0
Methyl esters end ethers	-		0
Ethyl esters and ethers	-		0
Higher esters and ethers	-		0
Acids	-		0
Joined to S, P, N	-		0
Nitrogen	7.0		0
Primary amines	-		0
Secondary amines	-		0
Flourine	10.5		0
Chlorine	24.5		0
Bromine	31.5		0

-C * 7	49.0
-H * 8	56.0
-O-	7.0
Ring	-7.0
C=C * 3	21.0
Sum Vb	126.0

Fig. 15.18: Continued

Compound	Contribution m</g mole Schroeder	Number of	Considered component
Iodine	38.5		0
Sulphur	21.0		0
Ring	-7.0	1.0	-7
3-membered	-		0
4-membered	-		0
5-membered	-		0
6-membered	-		0
Naphtalene	-		0
Anthracene	-		0
Double bond between carbon atoms	7.0	3.0	21
Triple bond between carbon atoms	14.0		0
Sum Vb			126

Vba = 28

Vba is the molal volume at the normal boiling point for air. From equation (4.28), Da = 7.15 $*10^{-6}$ m^2/s.

Dl can be estimated by equation (4.39), where μ is the viscosity of water at the estimation temperture, in centipoise.

At 298.15°K   Dl = 15.0 $*10^{-10}$ m^2/s

The evaporation rate, Q

The evaporation rate, Q, is estimated from equation (4.40) in $g/s*m^2$. For p-Cresol, Q = 0.003 $g/s*m^2$.

Estimation of BCF, CF and EM

A number of equations are outlined n Table 5.- on the basis of log K_{ow}, log K_{ac} and the solubility, S. When choosing a useful equation, notice that the different equations have limited indicator ranges, and are species specific.

The biological concentration factor, BCF (indicator K_{ac}):

log BCF = 2.024 + 1.225 log K_{ac} r = 0.97   (fish species)

log BCF = 4.60

log BCF = 1.579 + 1.191 log K_{ac} r = 0.87   (fish species)

log BCF = 4.08

Mean value for log BCF (fish species) = 4.34.

The concentration factor, CF (indicator Kow):

The indicator range allows use of following regression equations:

$\log CF = 0.124 + 0.542 \log K_{ow}$ r = 0.90 (Fathead minnow)
$\log CF = 1.34$

$\log CF = -0.70 + 0.85 \log K_{ow}$ r = 0.95 (Fathead minnow)
$\log CF = 1.20$

Mean value for log CF (Fathead minnow) = 1.27.

The ecological magnification factor, EM (indicator S in µmol/l).
S = 1597 mg/l S = 13940 µmol/l log S = 4.14

$\log EM = -0.4732 \log S + 4.4806$ (Mosquito fish, whole)
$\log EM = 2.52$

$\log EM = -0.3891 \log S + 3.995$ (Mosquito fish, whole)
$\log EM = 2.38$

Mean value for log EM (Mosquito fish, whole) = 2.45.

Estimation of MAC

Log MAC can be estimated from different regression equations which are valid for all compounds, and the values are in mmol/m^3. Remember to use the correction table.

log MAC = -1.2 - 0.012 t-melt + log M
(correction +0.5) log MAC = 0.91

log MAC = 0.40 - 0.01 M + log M
(correction +0.5) log MAC = 1.85

log MAC = -0.4 - 0.006 t-boil + log M
(correction +0.5) log MAC = 0.92

Mean value for log MAC = 1.23.

Estimation of LC_{50}

The estimation of LC_{50} in mmol/l is by different regression equation valid for all compounds.

$\log LC_{50} = -0.02 - 0.009$ t-boil $\log LC_{50} = -1.838$
$\log LC_{50} = 0.08 - 0.011$ M $\log LC_{50} = -1.110$
$\log LC_{50} = -1.6 - 0.010$ t-melt $\log LC_{50} = -1.950$

Mean value for log LC_{50} = -1.633 = LC_{50} = 2.52 mg/l

<u>Estimation of LD_{50}</u>

The estimation of LD_{50} in mmol/kg involves the regression equation based on the molecular weight valid for phenols.

$$\log LD_{50} = 0.18 + 0.0036\ M \qquad r = 0.55$$
$$\log LD_{50} = 0.57 \qquad LD_{50} = 0.7\ g/kg.$$

Table 15.2
p-Cresol. Comparison of estimated values with literature values, using both known boiling points and estimated boiling points

	Estimated value given by the boiling point from the literature	Deviation %	Estimated value given by the boiling point from Equation (4.7)	Deviation %	Literature value
Tb (4.7)			480.5°K	1.2	475.0°K
Tm (4.10)	277.3°K	9.7	280.7°K	8.7	308.0°K
Tc (4.1)	712.5°K	1.9	720.3°K	3.1	
(4.2)	683.6°K	2.2	691.5°K	1.1	
(4,3)	669.0°K	4.3	677.7°K	3.0	699.0°K
(4,4)	694.0°K	0.7	702.5°K	0.5	
Tc mean value	689.8°K	1.3	698.1°K	0.1	
H (4.9) atm m^3/mol	$1.07 \cdot 10^{-2}$		$9.69 \cdot 10^{-3}$		0.690 Pa·m^3/mol
P (4.11) atm (298°K)	$6.46 \cdot 10^{-4}$		$4.98 \cdot 10^{-4}$		$1.16 \cdot 10^{-4}$
Pc (4.13)	50.1	1.5			
(4.14)	49.9	1.8	50.8	0	50.8
(4.15)	42.7	16.0			
Pc mean value	47.6	6.3			
Vc (4.16)	370.6				
(4.17)	334.4				
(4.18)	336.9				
Vc mean value	347.3				
Vm (4.21)	129.2	2.9	136.7	8.8	125.6
log K_{ow}	2.24	12.9			2.30
Solubility S mg/l	1510	17.9			1840

Table 15.2. **Continued**

	Estimated value given by the boiling point from the literature	Deviation %	Estimated value given by the boiling point from Equation (4.7)	Deviation %	Literature value
log K_{ac}	2.10				
Da m^2/s (4.27) (4.28)	$4.44*10^{-6}$ $7.15*10^{-6}$	42.9 8.0			$7.77*10^{-6}$
Dl m^2/s	$15.0*10^{-10}$	72.1			$8.71*10^{-10}$
Q g/s m^2	0.003				
log BCF	4.34 (fish species)				
log CF	1.27 (fathead minnow)				
log EM	2.45 (mosquito fish)				
MAC mg/m^3	1.8				22.0
LD_{50} g/kg	0.7				1.1 (rabbit lethal dose)
LC_{50} mg/l	2.5				2.1 crucian carp 24h
C_{min} mg/l					

15.5.3. DDT (1,1,1 trichloro-2,2-bis(p-chlorophenyl)ethane)

Structure: 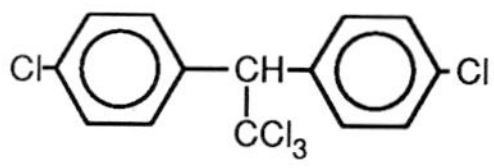

Firstly, the boiling point is found. This can be estimated by use of equation (4.7), or taken from the literature. Using equation (4.7) demands that 4 constants, specific for DDT are known, that is Ω, ß, A and ¥.

Fig. 15.19: Contributions to Ω

Group	Ω contribution	Number of	Considered component
Carbon	-0.5592	14	-7.8288
Hydrogen	0.2875	9	2.5875
Hydroxyl group	0.3382		0
Oxygen	0.0165		0
Nitrogen	-0.2625		0
Sulphur	0.014		0
Fluorine	0.3042		0
Chlorine	0.2983	5	1.4915
Bromine	0.299		0
Iodine	0.2994		0
Double bond (C=C)	0.564	6	3.384
Double bond (C=O)	0.564		0
Double bond (C=S)	0.564		0
Double bond (S=O)	0.564		0
Triple bond (C≡O)	1.1415		0
Triple bond (C≡N)	1.159		0
Triple bond (N≡)	1.159		0
6-membered ring (aromatic or naphthenic)	0.5666	2	1.1332
Sum Ω			0.7674

14 carbon atoms = 14 * (-0.5592) =	-7.8288
9 hydrogen atoms = 9 * 0.2875 =	2.5875
5 chlorine atoms= 5 * 0.2983	1.4915
Double bonds: 6 * C=C	3.3840
6-membered ring 2 * 0 5666	1.1332
Sum Ω	0.7674

Ω can also be estimated by equation (4.3). For this ΔT is needed and is found from contributions to the structure.

Fig. 15.20: Contributions to ΔT

Group	ΔT contribution	Number of	Considered component
$-CH_3$, $-CH_2-$ $-CH_2-$ (in ring)	0.016		0
| - CH (in ring) |	0.013	1	0.013
| - C - (in ring) |	0.003	1	0.003
H	0		0
| =C-H, $=CH_2$	0.015		0
| =C-H (in ring) | =C-, =C= (in ring) ≡C-, ≡CH	0.003	12	0.036
Ring atom	-0.005	2	-0.01

-CH	0.013
-C-	0.003
=CH- * 12	0.036
Ring atom	-0.010
-Cl * 5	0.065
Sum ΔT	0.107

Fig. 15.20 continued

Group	ΔT contribution	Number of	Considered component
-F	0.015		0
-Cl	0.013	5	0.065
-Br	0.01		0
-I			
-O- (in ring)	0.02		0
-OH (alcohols)	0.07		0
-OH (phenols)	0.029		0
CO (in ring)	0.046		0
-CHO	0.046		0
-COO-	0.039		0
-COOH	0.07		0
$-NH_2$	0.027		0
NH (in ring)	0.027		0
N	0.012		0
-CN	0.053		0
-CN	0.053		0
-SH, -S-	0.012		0
Sum ΔT			0.107

By equation (4.3) a ΔT-value of 0.107 gives an Ω = 0.681.

Fig. 15.21: Contributions to ¥

Group	¥ contribution	Number of	Considered component
Carbon	1.02	14	14.28
Hydrogen	1.95	9	17.55
Oxygen	3		0
Hydrogen group (includes H and O)	23.1		0
Nitrogen	2.45		0
Sulphur	6.7		0
Fluorine	3.63		0
Chlorine	7.45	5	37.25
Bromine	9.38		0
Iodine	10.95		0
C=C	2.5	6	15
C=O	2.5		0
C=S	2.5		0
S=O	2.5		0
C≡C	5.59		0
C≡N	29.1		0
N≡N	2.68		0
5- or 6-membered ring	-0.14	2	-0.28
Sum ¥			83.80

14 carbon atoms = 14 * 1.02	14.28
9 hydrogen atoms = 9 * 1.95	17.55
5 chlorine atoms= 5 * 7.45	37.25
Double bonds: 6 * C=C	15.00
Ring *2	-0.28
Sum ¥	83.80

Fig. 15.22: Contributions to ß

Group	ß contribution	Number of	Considered component
Hydrogen	0		0
Carbon	0.23	14	3.22
Fluorine	0.23		0
Chlorine	0.32	5	1.6
Bromine	0.54		0
Iodine	0.87		0
Oxygen	0.18		0
Nitrogen	0.14		0
Sulphur	0.27		0
C=C	-0.06	6	-0.36
C=O	-0.12		0
C≡N	0		0
-O-H (alcohols)	-0.11		0
-OH (aromatic alcohols)	-0.21		0
OH (acids)	0		0
NH	-0.02		0
C≡C	-0.14		0
C≡S	-0.05		0
Each ring atom	-0.043	12	-0.516
Condensed ring	-0.1		0
$-CH_3$, $-CH_2-$	0.227		0
$-CH_2-$	0.184		0
-CH< (with vertical bonds)	0.21		0
When in ring	0.192		0
=CH-, $=CH_2$	0.198		0
=CH- in ring	0.154		0
>C<	0.21		0
When in ring	0.154		0
=C<, =C=	0.198		0
When in ring	0.154		0
≡C-, ≡CH	0.153		0
-F	0.224		0
-Cl	0.32		0
-Br	0.5		0
-I	0.83		0
-O-	0.16		0
When in ring	0.12		0
-OH (alcohols)	0.06		0
-OH (phenols)	-0.002		0
CO	0.29		0
CO in ring	0.2		0
-CHO	0.33		0
-COO-	0.47		0
-COOH	0.4		0

-C- * 14	3.22
5 * -Cl	1.60
6 * C=C	-0.36
12 * (ring atom)	-0.52
Sum ß	3.94

Fig. 15.22: continued

Group	ß contribution	Number of	Considered component
$-NH_2$	0.095		0
NH	0.135		0
When in ring	0.09		0
N-	0.17		0
When in ring	0.13		0
-CN	0.36		0
-SH, -S-	0.27		0
-S- in ring	0.24		0
=O (exceot for com. above)	0.12		0
$-NO_2$	0.42		0
=S	0.24		0
Sum ß			3.944

Fig. 15.23: Contributions to A

Family	Value of A	Considered component
Alcohols	0.65	
Phenols	0.665	
Nitriles	0.68	
Carbozylix acids	0.685	
All other compounds	0.8	0.8
		0.8

The most appropriate value for A (DDT) is A = 0.80 ("all other compounds").

Now, having all the necessary constants equation (4.7) gives a boiling point of 604.8°K. The literature value is 533.0°K.

The melting point, Tm

Equation (4.10) gives a Tm-value of 353.1°K (Tb = 604.8°K) and Tm-value of 311.2°K (Tb = 533.0°K; literature value).

The critical temperature, Tc

Equation (4.2) gives a Tc-value of 788.5°K (694.9°K at a boiling point of 533.0°K).

Henry's Constant, H

A useful method for estimating H is equation (4.9). H has the units of $atm*m^3/mol$. Equation (4.9) gives a H-value of $1.17*10^{-3}$ at 298°K. ($4.0*10^{-3}$ at 298°K using Tb = 533.0°K and Tm = 311.2°K).

The vapour pressure, P

The vapour pressure, P, in atm can be estimated by equation (4.11). At 298°K T < Tm for DDT and, therefore, the last term in equation (4.11) is considered. At 298°K equation (4.11) gives $P = 2.62*10^{-7}$ atm ($2.80*10^{-5}$ atm when Tb = 533.0°K). Literature value: $2.60*10^{-10}$ atm (Benchmark Chemicals).

The critical pressure, Pc

The critical pressure can be estimated by equations (4.13), (4.14) and (4.15).
Using (4.13) Pc = 19.4 atm
Using (4.14) Pc = 16.6 atm at Tc = 766.5°K.

Using (4.15), the parachor, PCH, and the constants C and D, have to be found.

Fig. 15.24: Contributions to PCH

Group		PCH contribution	Number of	Considered component
$-(CH_2)n$: n<12		40	2	80
$-(CH_2)n$: n>12 (1)		40.3		0
C		9	12	108
H		15.5	9	139.5
H in OH		10		0
H in HN		12.5		0
O		19.8		0
O:O in esters		54.8		0
N		17.5		0
S		49.1		0
P		40.5		0
F		26.1		0
Cl		55.2	5	276
Br		68		0
I		90.3		0
Se		63		0
Si		31		0
Al		55		0
Sn		64.5		0
As		54		0
Ethylenic bond:				
Terminal		19.1		0
2,3 position		17.1		0
3,4 position		40.6		0
Carbonyl bond in ketones:				
total nr of carbonatoms=	3	22.3		0
	4	20		0
	5	18.5		0
	6	17.3		0
	7	17.3		0
	8	15.1		0
	9	14.1		0
	10	13		0
	11	12.6		0
Single bond		0		0

$-CH_2$ * 2	80
-C * 12	108.0
-H * 9	139.5
-Cl * 5	276.0
3 branhes	-11.1
6-membered ring	
2 * 0.8	1.6
Sum PCH	594.0

Fig. 15.24: Continued

Group	PCH contribution	Number of	Considered component
Semipolar bond	0		0
Singlet linkage	-9.5		0
Hydrogen bridge (2)	-14.4		0
Chain branching, per branch.	-3.7	3	-11.1
Sec.-Sec. adjacency	-1.6		0
Sec.-Ter. adjacency	-2		0
Ter.-Ter. adjacency	-4.5		0
Alkyl groups: (3)			
1-Methylethyl	133.3		0
1-Methylpropyl	171.9		0
1-Methylbutyl	211.7		0
2-Methylpropyl	173.3		0
1-Ethylpropyl	209.5		0
1,1-Dimethylethyl	170.4		0
1,1-Dimethylpropyl	207.5		0
1,2-Dimethylpropyl	207.9		0
1,1,2-Trimethylpropyl	243.5		0
Position differences in benzene:			
Ortho-meta	2.6		0
Meta-para	0.35		0
Ortho-para	2.9		0
Ring closure:			
3-membered ring	12.5		0
4-membered ring	6		0
5-membered ring	3		0
6-membered ring	0.8	2	1.6
7-membered ring	4		0
Sum PCH			594

(1) Somewhat greater beyond 20 $-CH_2-$
(2) Includes any ortho effect
(3) Complete group, including branching.

Contributions to C and D. C = 3.8584, D = -0.9215.

Equation (4.15) yields Pc = 20.1 atm.
The mean value for Pc = 18.7 atm. Equations (4.13), (4.14) and (4.15).

The critical volume, Vc

By mean of equations (4.16), (4.17) and (4.18) the critical volume is estimated ($cm^3/g*mol$):

Equation (4.16) Vc = 839.7 $cm^3/g*mol$ (Tb = 475.0°K)
Equation (4.17) Vc = 815.4 $cm^3/g*mol$
Equation (4.18) Vc = 771.4 $cm^3/g*mol$

Mean value for Vc = 808.8 $cm^3/g*mol$

The molal volume, Vm

Using equation (4.21) the molal volume Vm = 321.2 cm^3/mol.

The oil/water partition coefficient, K_{ow}

Log K_{ow} is found by means of equation (4.22) and the table of structure contributions.

Fig. 15.25: Contributions to the estimation of log K_{ow}

	B	C	D	E	F
=CH *12	1.056	0.024	-0.084	0.012	-0.042
$-CCl_3$	0.410	0.015	-0.009	-0.017	-0.017
-CH-	0.003	-0.058	0.053	0.018	0.015
-Cl * 2	0.330	0.174	-0.048	-0.024	-0.042
Sum	1.799	0.155	-0.088	-0.011	-0.080

From equation (4.22) log Kow = 7.06. The literature value for log K_{ow} is 6.19 (Benchmark Chemicals).

The oil/water partition coefficient, K_{ow} is also found by the following regression equation which is valid for chlorinated hydrocarbons

$\log K_{ow} = -0.15 + 0.088\ RD$ RD = 73.5

$\log K_{ow} = 6.32$

This yields a mean value for log Kow = 6.69.

The adsorption isotherm, log Kac

This is found by the following equation parameters built for hydrocarbons (Table 13, Appendix 2)

$\log K_{ac} = 0.72 \log K_{ow} + 0.49$

$\log K_{ac} = 5.31$ ($\log K_{ow} = 6.69$)

The literature value for log Kac is 5.39.

The molecular diffusivity of the chemical in air and liquid

Da is the molecular diffusivity in air (m^2/s). Dl is the molecular diffusivity in liquid (m^2/s).

Da can be estimated from equation (4.27) and (4.28).

Using equation (4.27) Ω has to be found. Ω corresponds to the value of 1.30 * Tr, where Tr = T/Tc.

Equation (4.27), Da = 2.43 $*10^{-6}$ m^2/s.
Equation (4.28), Da = 4.63 $*10^{-6}$ m^2/s.

Dl can be estimated by equation (4.39), where μ is the viscosity of water at the estimation temperture, in centipoise.
At 298.15°K   Dl = 11.44 $*10^{-10}$ m^2/s.

<u>The evaporation rate, Q</u>
The evaporation rate, Q, is estimated from equation (4.40) in $g/s*m^2$. For DDT, Q = $9.9*10^{-6}$ $g/s*m^2$.

<u>Estimation of BCF, CF and EM</u>
A number of equations are outlined in Table 5.1 on the basis of log K_{ow}, log K_{ac} and the solubility, S. When choosing a useful equation, notice that the different equations have limited indicator ranges, and are species specific.

The biological concentration factor, BCF (indicator K_{ac}):

log BCF = 2.024 + 1.225 log K_{ac} r = 0.91   (fish species)
log BCF = 8.53

log BCF = 1.579 + 1.191 log K_{ac} r = 0.87   (fish species)
log BCF = 7.90

Mean value for log BCF (fish species) = 8.22.

The concentration factor, CF (indicator K_{ow}):
The indicator range allows use of the following regression equation:

log CF = -0.973 + 0.767 log K_{ow} r = 0.76   log CF = 0.88
log CF = 4.16   (fish species)

The ecological magnification factor, EM (indicator S in μmol/l).
S = $9.75*10^{-2}$ mg/l
log EM = -0.4732 log S + 4.4806   (Fathead minnow, whole)
log EM = 4.74

log EM = -0.3891 log S + 3.995   (Fathead minnow, whole)
log EM = 4.21

Mean value for log EM (Fathead minnow) = 4.48.

Estimation of MAC, LC_{50}, LD_{50} and C_{min}

Log MAC can be estimated from different regression equations which are valid for all compounds.

$\log MAC = 0.87 \log LC_{50} - 0.36$ LC_{50} in $mmol/m^3$

$LC_{50} = 0.67$ mg/l MAC = 0.66 mg/m^3

Estimation of log LC_{50}

The estimation of LC_{50} in mmol/l is by different regression equation valid for all compounds:

$\log LC_{50} = -0.02 - 0.009$ t-boil og $LC_{50} = -2.36$
$\log LC_{50} = 0.08 - 0.011$ M og $LC_{50} = -3.82$
$\log LC_{50} = -1.6 - 0.010$ t-melt og $LC_{50} = -1.98$

Mean value for log $LC_{50} = -2.72$ = $LC_{50} = 0.67$ mg/l

Estimation of LD_{50} in mg/kg by one regression equation valid for all compounds.

$\log LD_{50} = \log MAC + 2.29 / 0.88$
log MAC = -2.72 which yields $\log LD_{50} = -0.49$
$LD_{50} = 114$ mg/kg.

Two equations for hydrocarbons are used to estimate C_{min}.

$\log C_{min} = -0.95 - 0.010$ t-boil $\log C_{min} = -3.55$
$\log C_{min} = -0.51 - 0.015$ M $\log C_{min} = -5.83$

Mean value for $\log C_{min} = -4.69$
$C_{min} = 7.24 * 10^{-3}$ mg/l.

Table 15.3
DDT. Comparison of estimated values with literature values, using both known boiling points and estimated boiling points

	Estimated value given by the boiling point from the literature	Deviation %	Estimated value given by the boiling point from Equation (4.7)	Deviation %	Literature value
Tb (4.7)			604.8°K	13.5	533.0°K
Tm (4.10)	311.2°K	18.5	353.1°K	7.5	381.7°K
Tc (4.1)	799.5°K		907.2°K		
(4.2)	694.9°K		788.5°K		
(4.3)	783.0°K		888.1°K		
(4.4)	788.5°K		894.7°K		
Tc mean value	766.5°K		869.6°K		
H (4.9) atm m^3/mol	$4.0*10^{-3}$		$1.17*10^{-3}$		2.84 Pa*m^3/mol
P (4.11) atm (298°K)	$2.80*10^{-5}$		$2.62*10^{-7}$		$2.60*10^{-10}$
Pc (4.13)	19.4				
(4.14)	16.6		19.4		
(4.15)	20.1				
Pc mean value	18.7				
Vc (4.16)	839.7		808.4		
(4.17)	815.4				
(4.18)	771.4				
Vc mean value	808.8				
Vm (4.21)	321.2	3.7	320.2	4.0	333.5
log K_{ow}	6.69	29.7			6.19
Solubility S mg/l	$9.7*10^{-2}$				$3.29*10^{-3}$
log K_{ac}	5.31				5.39
Da m^2/s (4.27)	$2.43*10^{-6}$	89.0			$4.68*10^{-6}$
(4.28)	$4.63*10^{-6}$	1.1			
Dl m^2/s	$11.44*10^{-10}$				$4.85*10^{-10}$
Q g/s m^2	$9.9*10^{-6}$				
log BCF	8.22 (fish species)				
log CF	4.16 (fish species)				

Table 15.3. Continued

	Estimated value given by the boiling point from the literature	Deviation %	Estimated value given by the boiling point from Equation (4.7)	Deviation %	Literature value
log EM	4.48 (fathead minnow, whole)				
MAC mg/m^3	0.66				1.0
LD_{50} mg/kg	114.0				113.0 mg/kg
LC_{50} mg/l	0.67				
C_{min} mg/l	$7.24 \cdot 10^{-3}$				

15.6 DISCUSSION OF THE RESULTS

Tables 15.1, 15.2 and 15.3 show the corresponding literature and estimated values, where the literature values are known, for the three Benchmark Chemicals. *Generally the physical estimations, found by the software, have a deviation of less than 10% from the literature values,* when the boiling point is known from the literature. Only few literature values are found for DDT, therefore the deviations are not always indicated.

The differences between the estimated values and literature values are acceptable, and the estimations are sufficiently accurate to solve ecotoxicological problems.

The calculations in section 15.5 show that better results are for most compounds obtained when the boiling point is already known from the literature. Equation (4.7) estimated a higher boiling point, a feature which will be reflected in many equations involving use of the boiling point.

The atomic and structural contributions required to estimate constants are sometimes missing from tables. The value for an aromatic ring, for instance is often mising. In this case 6 times the contribution for a carbon atom has been used.

Different calculations have shown that such approximations do not disturb the results unduly.

It is difficult to choose between the large number of available equations for calculating log Kow, log Kac, BCF, CF, EM, MAC, LD_{50} and LC_{50}. A wrong choice may give quite different results for the same parameter. Therefore, use only equations specific for the class to which the considered component belongs.

Where possible, use 3-4 appropriate equations and do not hesitate to use the mean value of the estimations. It is often found that the mean value gives an estimation quite close to the literature value. It is almost predictable that, the more parameters for a compound known already from the literature, the more accurately the missing parameter will be estimated. On the other hand, when 3 or 4 constants have to be estimated on basis of structure, the results will be less accurate. As an example, see the estimated values for 1,1,1 trichloroethane, both when the boiling point is unknown and when calculated on the basis of values in the literature.

Table 15.4

	Estimated values when the value for the boiling point is available in the literature	Deviation %	Estimated values when only the structure is known	Deviation %
Boiling point (4.7)	(347.3)°K		399.4°K	14.8
Melting point	202.8°K	0.9	233.2°K	13.9
The vapour pressure	0.167 atm	2.5	0.021 atm	87.1

The differences between results obtained from the boiling point estimated by (4.7) and the results obtained from literature values of the boiling point will generally be marked.

15.7 CONCLUSION

The system is recommended for estimating guide values for a compound when experimental data are lacking, particularly when the precautions outlined are followed.

The calculations for the 3 Benchmark Chemicals show that more experience is needed in selecting the right equations for calculating the biological parameters. Equations for a wider spectrum of compound classes ought to be developed. On the other hand, the results obtained for the physical and chemical parameters are very acceptable.

APPENDIX 1

PROPERTIES AND CHARACTERISTICS OF BENCHMARK CHEMICALS

Properties of the Benchmark Chemicals

Chemical	Environment	Rate constant
Carbon tetrachloride	Freshwater + sediment	0.05 day^{-1}
1,1,1-Trichloroethane	Freshwater + sediment	0.043 day^{-1}
Chlorobenzene	River water	0.0045 day^{-1} (22°C)
	River water + sediment	0.0092 day^{-1} (22°C)
	River water	0.065-0.31 day^{-1} (24°C)
	Esturaine water	0.038-0,22 day^{-1} (27°C)
	Marine water	0.0096 day^{-1} (24°C)
1,2,4-Trichlorobenzene	River water	0.029 day^{-1} (12°C)
	Estuarine water	0.026 day^{-1} (28°C)
	Marine water	0.012 day^{-1} (24°C)
	Soil	0.0013 day^{-1} (20°C)
2-Chlorobiphenyl	Lake water	0.17 day^{-1} (20°C)
2,4,2',4'-Tetrachlorobiphenyl	River water	0.0 (20°C)
Quinoline	Enrichment culture	7.4*10^{-5} ml cell^{-1} day^{-1}
p-Cresol	Enrichment culture	1.24*10^{-5} m cell^{-1} day^{-1}
	River water	1.67 day^{-1} (19°C)
	River water + sediment	2.37 day^{-1} (19°C)
	Estuarine watert	0.51-4.73 day^{-1} (19°C)
	Estuarine water - sediment	4.46-12.62 day^{-1} (19°C)
	Marine water	1.66-4.85 day^{-1} (19°C)
	Marine water + sediment	0.98-5.88 day^{-1} (19°C)
m-Cresol	River water	0.0096-28.1 day^{-1} (7-32°C)
	Estuarine water	0.0096-0.74 day^{-1} (7-32°C)
	Marine water	0.0096-0.041 day^{-1} (7-32°C)
2,4-Dichlorophenoxyacetic acid (1-29°C)	Natural waters	3.6-28.8*10^{-6} ml cell^{-1} day^{-1}
	River water	<0.14-0.07 day^{-1} (25°C)
	Lake water	0.058 ± 0.006 day^{-1} (29°C)
	Soil	0.08-0.46 day^{-1} (25°C)
Chlorpyrifos	Soil	0.008-0.025 day^{-1} (25°C)
	Soil	0.014 day^{-1} (28°C)
DDT	Soil	1.3-1.8*10^{-4} day^{-1}
	Soil	0.021-0.025 day^{-1}
Di-(2-ethylhexyl)phthalate	River water	0.023 day^{-1}
	Pond water + sediment	0.091 day^{-1}
	Lake water	0.028 ± 0.013 day^{-1}

(continued)

Chemical	Molal volume ($m^3/mol*10^{-3}$)	Molecular diffusivity Water ($m^2/s*10^{10}$)	Air ($m^2/s*10^6$)	Experimental melting point (K)
1,1,1-Trichloroethane	0.1145	9.21	8.32	204.7
Chlorobenzene	0.1169	9.09	8.10	228.
1,2,4-Trichlorobenzene	0.1587	7.57	6.76	290.
2-Chlorobiphenyl	0.2055	6.48	5.94	307.
2,4,2',4'-Tetrachlorobiphenyl	0.2682	5.52	5.24	314.
Quinoline	0.1411	8.12	7.23	253.7
p-Cresol	0.1256	8.71	7.77	307.9
m-Cresol	0.2052	6.49	6.03	
2,4-Dichlorophenoxyacetic acid				411.
Chlorpyrifos	0.3546	4.67	4.55	315.
DDT	0.3335	4.85	4.68	381.7
Di-(2-ethylhexyl)phthalate				325.7
Trichlorofluromethane	0.0973	10.20	9.02	162.
Trichlorobenzene				

Experimental boiling point (K)	Mol weight (g/mol)	Water solubility mg/l	Water solubility mol/m^3	H (Pa/m^3/mol)	K_w (m/s)	Flux N mol/m^2 year*10^3
347.3	133.4	1334	10.0	1,650	$9.8*10^{-6}$	3.1
404.9						
486.7	181.4	48.8	0.269	195	$8.9*10^{-6}$	2.8
547.	188.7	5.0	0.264	42	$6.3*10^{-6}$	2.0
-	291.9	0.068	$2.33*10^{-4}$	49	$6.6*10^{-6}$	2.1
510.9	129.2	6110	47.3	0.026	$1.8*10^{-8}$	0.0033
475.1	108.1	1840	17.0	0.690	$2.71*10^{-7}$	0.086
-	221.0	690	3.12	$2.56*10^{-5}$	$1.03*10^{-11}$	$3.3*10^{-6}$
-	350.6	0.87	$2.48*10^{-3}$	1.00	$3.90*10^{-7}$	0.12
-	354.5	$3.29*10^{-3}$	$9.36*10^{-6}$	2.84	$1.01*10^{-6}$	0.32
-	390.6	2.49	$6.37*10^{-3}$	$4.13*10^{-3}$	$1.67*10^{-9}$	$5.3*10^{-4}$
296.9	137.4	1240	9.02	11,800	$9.9*10^{-6}$	3.1
	112.6	503	4.47	352		

(continued)

Chemical	Resistance (%)		Measured properties		
	Water	Air	log K_{ow}	-log SOL (mole fraction)	log K_{oc}
1,1,1-Trichloroethane	98.5	1.5	2.47	3.74	2.25
Chlorobenzene			2.84	4.11	
1,2,4-Trichlorobenzene	88.8	11.2	3.97	5.53	
2-Chlorobiphenyl	62.9	37.1	4.72	6.32	
2,4,2',4'-Tetrachlorobiphenyl	66.5	33.5	6.44	8.38	
Quinoline	0.1	99.9	2.04	3.07	
p-Cresol	2.7	97.3	2.30	3.52	
2,4-Dichlorophenoxyacetic acid	0.01	99.99			
Chlorpyrifos	3.9	96.1	4.6	6.97	4.24
DDT	10.3	89.7	6.19	9.79	5.39
Di-(2-ethylhexyl)phthalate	0.01	99.9	5.3	7.22	
Trichlorofluromethane	99.8	0.2	2.53	3.84	

(continued)

Experimental vapor pressure, P, 25°C (atm.)	log K_{OW}	Experimental solubility (S) at 25°C (mol/l)
0.163	2.48	0.0100
$1.55*10^{-2}$	2.84	$4.47*10^{-3}$
$5.2*10^{-4}$	4.10	$2.69*10^{-4}$
$1.1*10^{-5}$	4.54	$2.6*10^{-5}$
$1.13*10^{-7}$	6.31	$2.3*10^{-7}$
$1.2*10^{-5}$	2.04	0.0473
$1.16*10^{-4}$	1.95	0.0170
$7.9*10^{-10}$	2.81	$3.1*10^{-3}$
$2.46*10^{-8}$	4.90	$2.5*10^{-6}$
$2.6*10^{-10}$	6.36	$9.28*10^{-9}$
$2.6*10^{-10}$	5.3	$6.38*10^{-6}$
1.05	2.53	$9.02*10^{-3}$

(continued)

Chemical	Hydrolytic rate constant 1) Experimental (days^{-1})	Estimated (days^{-1})	Quantum yield	Hydrolysis at pH 7.0 (day^{-1})
1,1,1-Trichloroethane	$4*10^{-3}$ (pH 7.0)	Resistant		
Chlorobenzene	-			
1,2,4-Trichlorobenzene	-			
2-Chlorobiphenyl	-			
2,4,2',4'-Tetrachlorobiphenyl	-			
Quinoline	-		$1.03*10^{-5}$	
p-Cresol	-		0.079	
2,4-Dichlorophenoxyacetic acid	-		0.044	
Chlorpyrifos	$3*10^{-2}$ (pH 8.0)	$2.5*10^{-4}$	0.0048	0.023
DDT	$2.5*10^{-4}$ (pH 7.0)	Resistant		$1.57*10^{-4}$ (pH 5.0)
Di-(2-ethylhexyl)phthalate	$9.5*10^{-7}$ (pH 7.0)	$1.46*10^{-6}$		
Trichlorofluromethane	$9.5*10^{-7}$ (based on CCl_4)			

1) All values estimated at 25°C except Di-(2-ethylhexyl)phthalate which is 30°C.

Sources: Irish (1963), Windholz (1976), Reid et al. (1977), MacKay et al. (1980), Verschneeren (1977), Callahan and Slimak (1979), Habasch and Leo (1979), Kenaqa and Goring (1980), Smity et al. (1978), Dilling (1977), Chion et al. (1982), and Benergve (1980).

APPENDIX 2

TABLES FOR ESTIMATION OF PHYSICAL-CHEMICAL PARAMETERS

Table 1
Atomic and structural contribution to Ω in equation 4.2

Carbon	-0.5592
Hydrogen	0.2875
Hydroxyl group	0.3382
Oxygen	0.0165
Nitrogen	-0.2625
Sulphur	0.0140
Fluorine	0.3042
Chlorine	0.2983
Bromine	0.2990
Iodine	0.2994
Double bond (C=C)	0.5640
Double bond (C=O)	0.5640
Double bond (C=S)	0.5640
Double bond (S=O)	0.5640
Triple bond (C≡C)	1.1415
Triple bond (C≡N)	1.1590
Triple bond (N≡)	1.1590
6-membered ring (aromatic or naphthenic)	0.5666

Source: C. Vowles (1951)

Table 2
Atomic and structural contribution to calculate Δ_T in equations 4.3 and 4.4

Group	Δ_T Riedel	Δ_T Lydersen
$-CH_3$, $-CH_2-$	0.016	0.020
$-CH_2-$ (in ring)	....	0.013
-CH (with vertical bond above)	0.013	0.012
(in ring)	....	0.012
-C- (with vertical bonds above and below)	0.003	0.00
(in ring)	....	(-0.007)*
H	0.00	0.00
=C-H (with vertical bond above), $=CH_2$	0.015	0.018
=C-H (with vertical bond above) (in ring)	....	0.011
=C- (with vertical bond above), =C=	0.003	0.00
(in ring)	....	0.011
≡C-, ≡CH	....	0.005
Ring atom	-0.005	
-F	0.015	0.018
-Cl	0.013	0.017
-Br	0.010	0.010
-I	....	(0.012)
-O-	0.020	0.021
(in ring)	....	(0.014)
-OH (alcohols)	0.070	0.082
-OH (phenols)	0.029	(0.035)
CO	0.046	0.040
(in ring)	....	(0.033)
-CHO	0.046	0.048
-COO-	0.039	0.047
-COOH	0.070	0.085
$-NH_2$	0.027	0.031
NH	0.027	0.031
(in ring)	....	(0.024)
N	0.012	0.014
(in ring)	....	(0.007)
-CN	0.053	(0.007)
-CN	0.053	(0.060)
-SH, -S-	0.012	0.015
-S- (in ring)	....	(0.008)
=S	....	(0.003)
=O (except for combinations above)	....	(0.02)
$-NO_2$	....	(0.055)

* Numbers in parantheses are based upon too few experimental data to be reliable

Sources: L. Riedel (1952), A.L. Lydersen (1955).

Table 3
Methyl-group contributions to a and b by hydrogen substitutions in saturated aliphatic hydrocarbons

Group	$a^{0.626}$	$b^{0.76}$
Methane (base value)	9,509	17.371
Primary methyl-group substitution	7,065	6.497
Methyl substitutions:		
1(1)	6,338	6.583
1(2)	6,417	6.686
1(3)	5,742	5.450
1(4)	5,968	4.870
2(1,1)	5,629	6.475
2(1,2)	5,261	5.879
2(1,3)	5,250	5.065
2(1,4)	5,189	4.351
2(2,2)	4,865	4.663
2(2,3)	4,758	3.882
2(2,4)	4,750	3.354
2(3,3)	5,137	3.481
3(1,1,1)	4,895	5.817
3(1,1,2)	4,575	4.455
3(1,1,3)	4,307	3.781
3(1,1,4)	4,087	3.471
3(1,2,2)	4,082	3.658
3(1,2,3)	3,814	2.984
3(1,3,3)	3,719	2.200
3(2,2,2)	3,770	2.401

Contributions to a and b involved in removal of hydrogen atoms from saturated aliphatic hydrocarbons to produce unsaturates

Group	$a^{0.626}$	$b^{0.76}$
First double bond between:		
1-1	-1,941	-2.126
2-1	-1,518	-1.940
2-2	-1,204	-2.123
3-1	-801	-1.692
3-2	-561 *	-1.858*
Second double bond between:		
3-1	-601 *	-1.158*
3-1	-713 *	-1.814*
3-2	-428 *	-1.811*
3-2	-670 *	-1.978*
3-3	-637 *	-2.142*
3 - 2-1 (**)	-792	-1.549
3 - 2-2	-610	-1.68 *
4 - 2-1	-750	-1.615
Triple bond between:		
1-1	-2.143	-3.876
1-2	-942	-3.260
2-2	-628 *	-3.443*

* Estimated values
** Arrow points away from the carbon atoms involved in formation of the double bond and toward the type of carbon adjacent.

Source: G. Thodos (1955)

Table 4
Atomic and structural contributions in the calculation of ¥ in Equation 4.7

Carbon	1.02
Hydrogen	1.95
Oxygen	3.00
Hydroxyl gropu (includes H and O)	23.10
Nitrogen	2.45
Sulphur	6.70
Fluorine	3.63
Chlorine	7.45
Bromine	9.38
Iodine	10.95
C=C	2.50
C=O	2.50
C=S	2.50
S=O	2.50
C≡C	5.59
C≡N	29.10
N≡N	2.68
5- or 6-membered ring	-0.14

Source: C. Vowles (1951)

Table 5
Atomic and structural contributions/constants for calculating ß in Eq. 4.7

Hydrogen	0.00
Carbon	0.23
Fluorine	0.23
Chlorine	0.32
Bromine	0.54
Iodine	0.87
Oxygen	0.18
Nitrogen	0.14
Sulphur	0.27
C=C	-0.06
C=O	-0.12
C≡N	0.00
O-H (alcohols)	-0.11
OH (aromatic alcohols)	-0.21
OH (acids)	0.00
NH	-0.02
C≡C	-0.14
C=S	-0.05
Each ring atom	-0.043
Condensed rings	-0.10

Source: L. Riedel (1949)

Table 5 (continued)

$-CH_3$, $-CH_2-$	0.227
$-CH_2-$ (in ring)	0.184
$-\overset{\vert}{\underset{\vert}{C}}H$	0.210
When in ring	0.192
$=\overset{\vert}{C}H$, $=CH_2$	0.198
$=\overset{\vert}{C}H$ in ring	0.154
$-\overset{\vert}{\underset{\vert}{C}}-$	0.210
When in ring	(0.154) *
$=\overset{\vert}{C}-$, $=C=$	0.198
When in ring	0.154
≡C-, ≡CH	0.153
-F	0.224
-Cl	0.320
-Br	(0.50)
-I	(0.83)
-O-	0.16
When in ring	(0.12)
-OH (alcohols)	0.06
-OH (phenols)	(-0.02)
CO	0.29
CO in ring	(0.2)
-CHO	0.33
-COO-	0.47
-COOH	(0.4)
$-NH_2$	0.095
NH	0.135
When in ring	(0.09)
N-	0.17
When in ring	(0.13)
-CN	(0.36)
-SH, -S-	0.27
-S- in ring	(0.24)
=O (except for combinations above)	(0.12)
$-NO_2$	(0.42)
=S	(0.24)

* Numbers in parantheses are based upon too few experimental data to be reliable

Source: A.L. Lydersen (1955).

Table 6
Estimation of A in Eq. 4.7

Family	Value of A
Alcohols	0.65
Phenols	0.665
Nitriles	0.68
Carbozylic acids	0.685
All other compounds	0.80

Source: Vowles (1951)

Table 7
Recommended atomic and structural contributions to calculate the parachor

CH_2 in $-(CH_2)_n$:			
n < 12		40.0	
n > 12		40.3	(1)
C		9.0	
H		15.5	
In OH		10.0	
In HN		12.5	
O		19.8	
O_2 in esters		54.8	
N		17.5	
S		49.1	
P		40.5	
F		26.1	
Cl		55.2	
Br		68.0	
I		90.3	
Se		63	
Si		31	
Al		55	
Sn		64.5	
As		54	
Ethylenic bond:			
Terminal		19.1	
2,3 position		17.1	
3,4 position		40.6	
Carbonyl bond in ketones:			
Total number of carbon atoms =	3	22.3	
	4	20.0	
	5	18.5	
	6	17.3	
	7	17.3	
	8	15.1	
	9	14.1	
	10	13.0	
	11	12.6	
Single bond		0.0	
Semipolar bond		0.0	

Table 7 (continued)

Singlet linkage	-9.5	
Hydrogen bridge	-14.4	(2)
Chain branching, per branch	-3.7	(3)
Secondary-secondary adjacency	-1.6	
Secondary-tertiary adjacency	-2.0	
Tertiary-tertiary adjacency	-4.5	
Alkyl gropus: (4)		
1-Methylethyl	133.3	
1-Methylpropyl	171.9	
1-Methylbutyl	211.7	
2-Methylpropyl	173.3	
1-Ethylpropyl	209.5	
1,1-Dimethylethyl	170.4	
1,1-Dimethylpropyl	207.5	
1,2-Dimethylpropyl	207.9	
1,1,2-Trimethylpropyl	243.5	
Position differences in benzene:		
Ortho-meta	1.8-3.4	
Meta-para	0.2-0.5	
Ortho-para	2.0-3.8	
Ring closure:		
3-membered ring	12.5	
4-membered ring	6.0	
5-membered ring	3.0	
6-membered ring	0.8	
7-membered ring	4.0	

Source: R. Quale (1953)

(1) Somewhat greater beyond 20 $-CH_2-$
(2) Includes any ortho effect
(3) Varies from -2.1 to -6.5
(4) Complete group, including branching.

Table 8
Values for C and D recommended by Herzog

Groups of compounds covered	C	D
Saturated and unsaturated hydrocarbons (acrylic, unsubstituted, not including methane)	3.0477	-0.6528
Aromatics and cyclics, excluding biphenyl and condensed ring systems	3.8584	-0.9215
Aliphatic amines, esters, halogen and sulphur compounds (excluding CH_3F and CS_2), inorganic halogen and sulphur compounds (excluding F_2), and NH_3	3.4271	-0.7829
Aliphatic acids, alcohols, and anhydrides	2.9929	-0.5718
Aliphatic ethers and ketones	3.3777	-0.7810
Aliphatic nitriles and HCN	2.6387	-0.4627

Source: R. Herzog (1944)

Table 9
Atomic and structural contributions to the molar refraction RD

	Increment
C	2.418
H	1.100
$-CH_2-$	4.618
O (hydroxyl)	1.525
O (ether)	1.643
O (carboxyl)	2.211*
Cl	5.967
Br	8.865
I	13.90
N (primary amine)	2.322
N (secondary amine)	2.502
N (tertiary amine)	2.840
N (nitrile)	5.516
3-membered ring	0
4-membered ring	0
5-membered ring	0
6-membered ring	0
7-membered ring	0
Double bond	1.733
Semipolar double bond	1.733
Triple bond	2.398

* Includes allowance for double bond

Source: F. Eisenlohr (1910)

Table 10
Atomic and atomic-group contributions to V_c

	Increment, cm^{3}7g mole
Base value	40
$-CH_3$, $-CH_2-$	55
$-CH_2-$ in ring	44.5
$-\overset{\vert}{\underset{\vert}{C}}H$	51
When in ring	46
$=\overset{\vert}{C}H$, $=CH_2$	45
$=\overset{\vert}{C}H$ in ring	37
$-\overset{\vert}{\underset{\vert}{C}}-$	41
When in ring	(31) *
$=\overset{\vert}{C}-$, $=C=$	36
When in ring	36
$\equiv C-$, $\equiv CH$	(36)
-F	18
-Cl	49
-Br	(70)
-I	(95)
-O-	20
When in ring	(8)
-OH (alcohols)	(18)
-OH (phenols)	(3)
CO	60
CO in ring	(50)
-CHO	73
-COO-	80
-COOH	80
$-NH_2$	28
NH	(37)
When in ring	(27)
N-	(42)
When in ring	(32)
-CN	(80)
-SH, -S-	55
-S- in ring	(45)
=O (except for combinations above)	(11)
$-NO_2$	(78)
=S	(47)

* Numbers in parantheses are based upon too few experimental data to be reliable

Source: A.L. Lydersen (1955).

Table 11A
Equations for calculating the oil/water partition coefficient (log K_{ow}) from the molecular weight (M), molar volume (V_m), and molar refraction (RD)

Class of compounds	n	r	Equation: log K =
Hydrocarbons	37	0.81	1.0 + 0.030 M
	35	0.90	-0.04 + 0.032 V_m
	37	0.77	1.00 + 0.088 RD
Saturated alcohols	7	0.98	-2.68 – 0.043 M
	7	0.87	-3.60 + 0.040 V_m
	7	0.98	-2.80 + 0.12 RD
Aldehydes	7	0.99	-3.95 + 0.063 M
	7	0.98	-4.15 – 0.054 V_m
	7	0.97	-3.79 – 0.22 RD
Amines	13	0.80	-3.30 + 0.042 M
	12	0.95	-4.81 + 0.052 V_m
	13	0.83	-4.25 + 0.16 RD
Chlorinated hydro-carbons	20	0.76	0.33 + 0.014 M
	19	0.83	0.67 + 0.029 V_m
	20	0.83	-0.15 + 0.088 RD
Miscellaneous organic	91	0.80	-3.50 + 0.053 V_m

Table 11B
Equations for calculating the water solubility (log S) from the molecular weight (M), molar volume (V_m), and molar refraction (RD)

Class of compounds	n	r	Equation: log S (mM/litre) =
Hydrocarbons	38	-0.46	0.61 - 0.011 M
	36	-0.58	1.52 - 0.016 V_m
	38	-0.45	0.60 - 0.03 RD
Alcohols	17	-0.81	4.78 - 0.024 M
	13	-0.92	6.30 - 0.035 V_m
	17	-0.82	6.00 - 0.12 RD
Nitro compounds	11	-0.74	3.97 - 0.021 M
	11	-0.89	6.15 - 0.046 V_m
	11	-0.89	4.44 - 0.103 RD
Amines	25	-0.58	4.88 - 0.026 M
	23	-0.43	4.68 -0.020 V_m
	25	-0.61	4.90 - 0.080 RD

n = number of compounds
r = regression coefficient

Source: Filou et al. (1973).

Table 12
Equations "calibrated" on sets of hydrophobic compounds: log K_{oc}/log K_{ow}

log K_{oc} = 0.72 log K_{ow} + 0.49 (predominantly methylated and halogenated benzenes, 13 compounds, r^2 = 0.95)
log K_{oc} = 1.029 log K_{ow} - 0.18 (13 pesticides, r^2 = 0.91)
log K_{oc} = 0.989 log K_{ow} - 0.346 (5 PNAs, r^2 = 0.997)
log K_{oc} = log K_{ow} - 0.317 (PNAs and substituted PNAs, 22 compounds, r^2 = 0.98)

Sources: Brown and Combs (1982), Burchill et al. (1981), Hance (1969) and Harris and Sheets (1965).

Table 13
Values of the collision integral Ω based on the Lennard-Jones potential

1.30 * T_p	Ω	1.30 * T_p	Ω	1.30 * T_p	Ω
0,30	2,662	1,65	1,153	4,0	0,8836
0,35	2,476	1,70	1,140	4,1	0,8788
0,40	2,318	1,75	1,128	4,2	0,8740
0,45	2,184	1,80	1,116	4,3	0,8694
0,50	2,066	1,85	1,105	4,4	0,8652
0,55	1,966	1,90	1,094	4,5	0,8610
0,60	1,877	1,95	1,084	4,6	0,8568
0,65	1,798	2,00	1,075	4,7	0,8530
0,70	1,729	2,1	1,057	4,8	0,8492
0,75	1,667	2,2	1,041	4,9	0,8456
0,80	1,612	2,3	1,026	5,0	0,8422
0,85	1,562	2,4	1,012	6	0,8124
0,90	1,517	2,5	0,9996	7	0,7896
0,95	1,476	2,6	0,9878	8	0,7712
1,00	1,439	2,7	0,9770	9	0,7556
1,05	1,406	2,8	0,9672	10	0,7424
1,10	1,375	2,9	0,9576	20	0,6640
1,15	1,346	3,0	0,9490	30	0,6232
1,20	1,320	3,1	0,9406	40	0,5960
1,25	1,296	3,2	0,9328	50	0,5756
1,30	1,273	3,3	0,9256	60	0,5596
1,35	1,253	3,4	0,9186	70	0,5464
1,40	1,233	3,5	0,9120	80	0,5352
1,45	1,215	3,6	0,9058	90	0,5256
1,50	1,198	3,7	0,8990	100	0,5130
1,55	1,182	3,8	0,8942	200	0,4644
1,60	1,167	3,9	0,888	400	0,4170

Source: Hirschfelder et al. (1954)

Table 14
Lennard-Jones force constants calculated from viscosity data

Compound	0.79 * T_C	q
Acetylene	185	4.221
Air	97	3.617
Argon	124	3.418
Arsine	281	4.06
Benzene	440	5.270
Bromine	520	4.268
i-Butane	313	5.341
n-Butane	410	4.997
Carbon dioxide	190	3.996
Carbon disulfide	488	4.438
Carbon monoxide	110	3.590
Carbon tetrachloride	327	5.881
Carbonyl sulfide	335	4.13
Chlorine	357	4.115
Chloroform	327	5.430
Cyanogen	339	4.38
Cyclohexane	324	6.093
Ethane	230	4.418
Ethanol	391	4.455
Ethylene	205	4.232
Fluorine	112	3.653
Helium	10.22	2.576
n-Heptane	282	8,88
n-Hexane	413	5.909
Hydrogen	33.3	2.968
Hydrogen chloride	360	3.305
Hydrogen iodide	324	4.123
Iodine	550	4.982
Krypton	190	3.61
Methane	136.5	3.822
Methanol	507	3.585
Methylene chloride	406	4.759
Methyl chloride	855	3.375
Mercuric iodide	691	5.625
Mercury	851	2.898
Neon	35.7	2.789
Nitric oxide	119	3.470
Nitrogen	91.5	3.681
Nitrous oxide	220	3.879
n-Nonane	240	8.448
n-Octane	320	7.451
Oxygene	113	3.433
n-Pentane	345	5.769
Propane	254	5.061
Sulphur dioxide	252	4.290
Water	356	2.649
Xenon	229	4.055

Source: Hirschfelder et al. (1949)

Table 15
Additive-volume increments to calculate V_b

	Increment, cm^3/g mole	
	Schroeder	Le Bas
Carbon	7	14.8
Hydrogen	7	3.7
Oxygen	7	
Doubly bound	....	7.4
Methyl esters and ethers	....	9.1
Ethyl esters and ethers	....	9.9
Higher esters and ethers	....	11.0
Acids	....	12.0
Joined to S, P, N	....	8.3
Nitrogen	7	
Primary amines	....	10.5
Secondary amines	....	12.0
Fluorine	10.5	8.7
Chlorine	24.5	21.6
Bromine	31.5	27.0
Iodine	38.5	
Sulphur	21	25.6
Ring	-7	
3-membered	....	-6.0
4-membered	....	-8.5
5-membered	....	-11.5
6-membered	....	-15.0
Naphthalene	....	-30.0
Anthracene	....	-47.5
Double bond between carbon atoms	7	
Triple bond between carbon atoms	14	

Source: Partington (1949)

Table 16
Log S - log K_{ow} regression equations (log S = -a log K_{ow} + b + c)

Coefficients of equation for S (mol/l)			Regression ec. statistics		Temp.	Range of	Chemical classes
a	b	c(T_m in °C)	No.		(°C)	log K_{ow}	represented
1.339	-0.978	-	156	0.935	Var.	0.3-4.7	Alcohols, ketones, esters, alkyl halides, alkynes, alkenes, benzene derivates, alkanes
2.38	6.90	-	11	-0.81	-	4.0-5.0	Phosphate esters
0.9874	0.7178	-0.0095 T_m	35	0.995	25	1.0-6.5	Halobenzenes
0.88	-0.012	-0.01 T_m	32	0.9895	25	3.3-7.1	Polynuclear aromatic hydrocarbons
0.962	0.50	-	9	0.937	20	1.4-3.4	Halogenated 1- and 2-carbon hydrocarbons (8 with Cl, 1 with Br)
1.12	1.30	-0.017 T_m	27	0.96	25	0.9-4.9	Wide variety
1.16	0.824	-	36	0.994	Var.	0.9-6.4	Wide variety
1.0	0.26	-0.01 T_m	38	0.91	Var.	-0.5-7.4	Wide variety
0.944	-	-	62	0.990	25	0.4-5.5	Hydrocarbons, halogenated hydrocarbons, aldehydes, ketones, esters, alcohols
1.000	0.501	-0.010 T_m	45	-	Var.	2.0-7.1	Hydrocarbons and halocarbons
0.980	-0.08	-	8	-	20	1.5-4.3	Phthalates
1.016	0.515	-	111	0.931	Var.	-0.9-5.1	Wide variety
0.944	0.323	-0.01 T_m	164	0.978	Var.	2.1-11.2	Polynuclear aromatic hydrocarbons, halogenated aromatics

Source: Jørgensen (1988)

Table 17
Regression equations for the relation S/log K_{ow}

N is number of compounds in data set used to obtain estimation, r^2 is the square of correlation coefficient.

Equataion	Units of S	N	r^2	Chemical classes represented
$\log S = -1.37 \log K_{ow} + 7.26$	µmol/l	41	0.903	Mixed classes; aromatics and chlorinated hydrocarbons well represented
$\log S = -0.922 \log K_{ow} + 4.184$	mg/l	90	0.740	Mixed classes; pesticides well represented
$\log S = -1.49 \log K_{ow} + 7.46$	µmol/l	34	0.970	Mixed classes; severalpesticides
$\log 1/S = 1.113 \log K_{ow} - 0.926$	mol/l	41	0.935	Alcohols
$\log 1/S = 1.229 \log K_{ow} - 0.720$	mol/l	13	0.960	Ketones
$\log 1/S = 1.013 \log K_{ow} - 0.520$	mol/l	18	0.980	Esters
$\log 1/S = 1.182 \log K_{ow} - 0.935$	mol/l	12	0.880	Ethers
$\log 1/S = 1.221 \log K_{ow} - 0.832$	mol/l	20	0.861	Alkyl halides
$\log 1/S = 1.294 \log K_{ow} - 1.043$	mol/l	7	0.908	Alkynes
$\log 1/S = 1.294 \log K_{ow} - 0.248$	mol/l	12	0.970	Alkenes
$\log 1/S = 0.996 \log K_{ow} - 0.339$	mol/l	16	0.951	Aromatics (benzene and benzene derivatives)
$\log 1/S = 1.237 \log K_{ow} + 0.248$	mol/l	16	0.908	Alkanes
$\log 1/S = 1.214 \log K_{ow} - 0.850$	mol/l	140	0.912	All chemicals represented by ??
$\log 1/S = 1.339 \log K_{ow} - 0.978$	mol/l	156	0.874	All chemicals represented by ??
$\log S = -2.38 \log K_{ow} + 12.90$	µmol/l	11	0.656	Phosphate esters
$\log S = -0.962 \log K_{ow} + 6.50$	µmol/l	9	0.878	Halogenated 1-& 2-carbon hydrocarbons (8 with Cl,1 with Br)

Source: Hansch et al. (1968), Kenaga and Goring (1980) and Saeger et al. (1979)

Table 18
Estimation of B, C, D, E and F in equation 4.22

	B	C	D	E	F
-H	0.066	0.018	-0.027	-0.019	-0.019
$-CH_3$	0.142	-0.020	-0.016	-0.023	-0.015
$-CH_2-$	0.076	-0.038	0.011	-0.004	0.003
>CH-	0.003	-0.058	0.053	0.018	0.015
>C<	-0.075	-0.076	0.091	0.043	0.034
-CH=CH-	0.147	-0.043	0.028	0.010	0.003
$-CH=CH_2$	0.212	-0.025	0.000	-0.009	-0.015
$>C=CH_2$	0.147	-0.043	0.028	0.010	0.003
-C≡CH	0.171	0.074	0.027	0.002	-0.012
$-C_6H_5$	0.467	-0.007	0.012	0.007	-0.017
=CH- (aromatic)	0.088	0.002	-0.007	0.001	-0.003
-naphthyl	0.766	0.018	-0.026	0.024	-0.028
-cyclohexyl	0.489	-0.148	0.004	-0.029	-0.009
-F	0.078	0.088	0.009	-0.019	-0.020
-Cl	0.165	0.087	-0.024	-0.012	-0.021
-Br	0.213	0.095	-0.033	-0.008	-0.020
-I	0.302	0.103	-0.056	-0.010	-0.031
$-CF_3$	0.150	0.017	0.035	-0.037	-0.013
$-CCl_3$	0.410	0.015	-0.009	-0.017	-0.017
-OH	0.202	0.324	-0.012	-0.015	0.003
-O-	0.044	0.155	0.06[illegible]	0.019	-0.022
-C=O-	0.135	0.246	0.06[illegible]	0.023	-0.021
-CH=O	0.219	0.244	0.010	-0.014	-0.027
-COO-	0.167	0.170	0.062	0.015	-0.027
-COOH	0.323	0.342	-0.011	-0.017	0.008
$-NH_2$	0.167	0.269	0.037	0.027	-0.014
-NH-	0.082	0.251	0.095	0.056	-0.010
-N-	-0.006	0.189	0.125	0.069	0.014
-CN	0.241	0.269	-0.007	-0.023	-0.041
-N= (pyridine)	0.102	0.183	0.031	-0.011	-0.020
$-NO_2$	0.238	0.241	-0.012	-0.027	-0.037
$-CONH_2$	0.444	0.499	-0.019	-0.039	-0.012
-S-	0.136	0.130	0.028	0.032	-0.020
-SH	0.231	0.155	-0.026	-0.011	-0.013
molecule	-0.5065	-0.056	0.007	0.031	0.028

Source: Cramer (1980) and (1980a)

Table 19
Substituent constants

Substituent	∂ meta	∂ para
a. Alkyl substituents		
CH_3	-0.069	-0.170
CH_2H_5	-0.043	-0.151
C_3H_7		-0.126
$CH(CH_3)_2$		-0.151
C_4H_9		-0.161
$CH_2CH(CH_3)_2$		-0.115
$CH(CH_3)C_2H_5$		-0.123
$C(CH_3)_2$	-0.120	-0.197
$(CH_2)_2CH(CH_3)_2$		-0.225
$C(CH_3)_2C_2H_5$		-0.190
CF_3	0.415	0.551
CH_2Cl		0.184
CH_2CN		0.007
CH_2CH_2COOH	-0.027	-0.066
b. Hydroxy and alkoxy groups		
OH	-0.002	-0.357
OCH_3	0.115	-0.268
OC_2H_5	0.150	-0.250
OC_3H_7		-0.268
$OCH(CH_3)_2$		-0.286
OC_4H_9		-0.320
OC_5H_{11}		-0.340
$O(CH_2)_5CH(CH_3)_2$		-0.265
$OCH_2C_6H_5$		-0.415
OC_6H_6		-0.028
O^-	-0.708	-0.519
c. Amino and amido groups		
NH_2	-0.161	-0.660
$NHCH_3$	-0.302	-0.592
NHC_2H_5	-0.240	
NHC_4H_9	-0.344	
$N(CH_3)_2$	-0.211	-0.600
$NHCOCH_3$		-0.015
$NHCOC_6H_5$	0.217	0.078
$NHNH_2$	-0.020	-0.550
$NHOH$	-0.044	-0.339
NH_3^+	0.634	
$NH_2CH_3^+$	0.958	
$NH_2C_2H_5^+$	0.958	
$N(CH_3)_3^+$	0.904	0.859
d. Carboxy, carbalkoxy, acyl groups and their derivatives		
$COOH$	0.355	0.265
$COOCH_3$	0.315	
$COOC_2H_5$	0.398	0.522
$CONH_2$	0.280	
CHO	0.355	0.216
$COCH_3$	0.306	0.516

Table 19 Continued

Substituent	∂ meta	∂ para
d. Carboxy, carbalkoxy, acyl groups and their derivatives		
COC_6H_5		0.459
CN	0.678	0.628
COO-	0.104	0.132
e. Miscellaneous groups		
NO_2	0.710	0.778
NO		0.123
F	0.337	0.062
Cl	0.373	0.227
Br	0.391	0.232
I	0.352	0.276
IO_2	(0.70)	(0.76)
SCH_3	0.144	-0.047
$SOCH_3$	0.551	0.567
SO_2CH_3	0.647	0.728
SCN		0.699
SeCN		0.664
$B(OH)_2$	0.006	0.454
$Si(CH_2)_3$	-0.121	-0.072
C_6H_5	0.218	0.009
$N{=}NC_6H_5$		0.640
$CH{=}CHC_6H_5$	0.141	
AsO_3H^-		-0.019
PO_3H^-	0.228	0.238
SO_3^-		0.381
SO_2NH_2		0.621
f. Fused-ring systems		
3,4-$(CH_2)_3$		-0.259
3,4-$(CH_2)_4$		-0.477
3,4-$(CH)_4$		0.170
3,4-CH_2O_2		-0.159

Table 20. Reaction constants for equilibrium reactions

Acid/reaction	Solvent	T	R	log kho
ArCOOH	H_2O	25	1.000	4.203
$O_2NArCOOH$	H_2O	25	0.905	2.206
$H_3CArCOOH$	H_2O	25	1.430	3.875
HOArCOOH	H_2O	25	1.103	3.997
$ArCH_2COOH$	H_2O	25	0.489	4.297
$ArCH_2CH_2COOH$	H_2O	25	0.212	4.551
ArCH=CHCOOH	H_2O	25	0.466	4.447
ArCH=CHCOCOOH	H_2O	25	-0.054	1.971
$ArAsO(OH)_2$	H_2O	18-25	1.050	3.540
$ArAsO_2OH^-$	H_2O	22	0.874	8.491
ArSeOOH	H_2O	25	0.905	4.740
alfa-ArCH=NOH	H_2O	25	0.857	10.695
ArOH	H_2O	25	2.113	9.847
$ArNH_3^+$	H_2O	25	2.767	4.557
$ArCOCH_3 + H_2$ = $ArCHOHCH_3$	H_2O	25	1.630	3.913
$ArSO_2Cl + H_2O$ - $ArSO_2OH$	H_2O	15	-0.300	2.945
$(ArCONOCOC_6H_5)^- + NH_3$ - ArNCO	H_2O	30	-2.593	4.587
$(ArCOONCOC_6H_5)^- + NH_3$ - $ArCOO^-$	H_2O	30	0.867	4.634
$ArO^- + CH_3OSO_3^-$ - $ArOCH_3$	H_2O	100	-0.813	4.121
$ArOSO_2OH + H^+$ - ArOH	H_2O	48.7	0.596	3.937
$ArOPO(OC_2H_5)_2$ + erythrocyte cholinesterase	H_2O, pH 7.6	37	4.058	1.727
$ArCH_2Cl + OH^-$ (0.0506N) - $ArCH_2OH$	H_2O	30	-0.333	5.484
$ArNH_2$ + 2-Cl-5-$NO_2C_5H_3N$ - 2-ArNH-5-$NO_2C_5H_3N$	H_2O	65	-3.687	4.740
$ArN=CHC_6H_5 + H^+$ - $ArNH_2$	H_2O	-0.1	0.424	5.825
$ArN=CHC_6H_5 + CH_3COOH$ - $ArNH_2$	H_2O	-0.1	0.384	0.992
$ArCOCH_3 + Br_2$ - $ArCOCH_2Br$	H_2O	25	0.417	3.522
$Ar(C_6H_5)C=NH + H_2O$ - $ArCOC_6H_5$	H_2O	0	1.704	2.125
$ArCH=NC_6H_5 + H^+$ - ArCHO	H_2O	-0.1	-0.216	5.840

APPENDIX 3

PARAMETERS FOR VARIOUS ELEMENTS AND CHEMICAL COMPOUNDS AND $L_W = f(\text{latitude})$

Table 1
Concentration factors (CF), (ww) means that CF is based upon wet weight, (dw) means that CF is based on dry weight

Component	Species	CF	Concentration in water/soil	Conditions
Ag	Daphnia magna	26 (ww)	0.5 mg l^{-1}	-
Ag	Phytoplankton	620-15,000 (ww)	wide range	-
Al	Brown algae	1550 (ww)		-
Aldrin	Buffalo fish	30,000 (ww)	0.007 μg l^{-1}	-
Aldrin	Catfish	1590 (ww)	0.044 μg l^{-1}	-
Aldrin	Earthworms	0.57 (ww)	0.37 mg/kg dry weight	
Aldrin	Oyster	10 (ww)	0.05 μg l^{-1}	-
As	Salmo gardneri egg	18.5 (ww)	0.05 mg l^{-1}	279.5°K 33 days
Atrazine	Gambusia affinis	17 (ww)	$0.5*10^{-7}$ mol l^{-1}	18 days
Atrazine	Scenedesmus acutus	31.8 (dw)	$0.5*10^{-7}$ mol l^{-1}	303 K, 8 h pH 7-8,
Au	Brown algae	270 (ww)	wide range	-
Ba	Brown algae	260 (ww)		-
Cd	Brown algae	890 (ww)	wide range	-
Cd	Zooplankton	6000 (ww)	10^{-4} mg m^{-3}	-
Cd	32 Freshwater plant species	1620	wide range	-
Chlordane	Algae	302 (ww)	6.6 ng l^{-1}	-
Chlorinated naphthalene	Chloroccum sp.	120 (ww)	100 μg l^{-1}	24 hours
Co	32 Freshwater plant species	4425 (dw)	wide range	-
Cr	Fish species	10 (ww)	wide range	Freshwater
Cr	Molluscs	21,800 (ww)	wide range	Marine sp.
Cs	Salmo trutta	1020 (ww)	wide range	Soft water 6.6 g fish
Cu	Chorda filum	560 (dw)	$2.5*10^{-7}$ g l^{-1}	Seawater
Cu	Ulva sp.	47000-56000 (dw)	low	Seawater
DDE	Mytilus spp.	310000 (ww)	5.1 ng l^{-1}	Seawater
DDT	Algae	500 (ww)	0.016 ng l^{-1}	-
DDT	Crab	144 (ww)	50 μg l^{-1}	Seawater
DDT	Crayfish	97 (ww)	0.1 μg l^{-1}	-

Table 1 (continued)

Component	Species	CF	Concentration in water/soil	Conditions
DDT	Oyster	70,000	0.1 μg l^{-1}	-
DDT	Sea squirt	160,000 (ww)	0.1 μg l^{-1}	Seawater
DDT	Snail	480	50 μg l^{-1}	-
DDT	Trout	200 (ww)	20 μg l^{-1}	-
Dieldrin	Algae	4091	0.011 ng l^{-1}	-
Dieldrin	Catfish	4444 (ww)	0.009 μg l^{-1}	-
Dieldrin	Trout	3300 (ww)	2.3 μg l^{-1}	-
Fe	Brown algae	17,000 (ww)	-	-
Fe	Zooplankton	144,000 (ww)	0.01 mg m^{-3}	Seawater
Heptachlor	Bluegill	1130 (ww)	50 μg l^{-1}	-
Hexabro-mobiphenyl	Salmo salar	1.73 (ww)	all	5.3 g fish 48 h
Hexachlo-robenzene	Salmo salar	690 (ww)	all	288°K 6 g fish
Hg	Daphnia magna	50 (ww)	2 mg m^{-3}	10 weeks
Hg	Zooplankton	650 (ww)	0.02 mg m^{-3}	Freshwater
Methoxychlor	Earthworms excl. lumbricus terrestris	4.17 (ww)	2.66 mg kg^{-1} (dw)	-
Methoxychlor	Oyster	10 (ww)	<1.0 μg l^{-1}	Conc. in sedi-ment: <0.01 mg kg^{-1} (ww)
Mirex	Chlorococcum sp.	10000 (ww)	10 pptr	lab. exp.
Ni	Macrophyte	1050 (ww)		Marine org. typical value
Ni	Molluscs	4000 (ww)		Marine org. typical value
Ni	Molluscs	650 (ww)		Freshwater typical value
Ni	Ulva fenestrata	200 (dw)	$7.6*10^{-7}$ g l^{-1}	Sea of Japan
Ni	Zooplankton	560 (ww)		Marine org. typical value
Sb	Phytoplankton	280-3200 (ww)	-	-
Sb	Plankton	50 (ww)	-	-
2,5,3'-4'-Tetra-chloro-biphenyl 288K	Salmo salar	389 (ww)	-	Uptake from water, steady state CF, 48 h, 6.24 g fish

Table 1 (continued)

Component	Species	CF	Concentration in water/soil	Conditions
Toxaphene	Aspergillus sp.	$1.7*10^4$ (ww)	-	Equilibrium CF 301 K, 24 h
Toxaphene	Bacillus subtilis	$3.4*10^3$ (ww)	-	Equilibrium CF 301 K, 24 h
Toxaphene	Chlorella pyrenoidosa	$1.7*10^4$ (ww)	-	Equilibrium CF 288 K, 200 h, 170 FTC
Toxaphene CF,	Flavobacterium harrisonii	$5.2*10^3$ (ww)	-	Equilibrium 301 K, 24 h
Toxaphene CF,	Seston	$6.6*10^3$ (ww)	-	Equilibrium 301 K, 24 h
Trimethyl-naphthalene	Rangia cuneata	26.7 (ww)	0.03 mg l^{-1}	24 h
Pb	Brown algae	70,000 (ww)	all	-
Pb	Fucus vesiculosus	870 (dw)	Very low	Seawater
Pb	Zooplankton	1500 (ww)	2 mg m^{-3}	-
PCB	Salmo salar	282 (ww)	wide range	5.29 g fish, 24 h
PCB	Yellow perch	17,000 (ww)	1.0 µg l^{-1}	Freshwater
Ra	Brown algae	370 (ww)	wide range	-
Se	Phytoplankton	900-5500 (ww)	wide range	-
V	Brown algae	250 (ww)	-	-
W	Brown algae	87 (ww)	wide range	-
Zn	Phytoplankton	8900-75,000 (ww)	all	-
Zn	Pike	1250 (ww)	low	Freshwater

Table 2
L_W values for latitude 20°N[a,b]

∂ center (nm)	Spring	Summer	Fall	Winter
299	5.16 (-4)	6.78 (-4)	3.74 (-4)	1.99 (-4)
304	3.47 (-3)	4.14 (-3)	1.00 (-2)	1.67 (-3)
309	1.09 (-2)	1.23 (-2)	8.37 (-3)	5.97 (-3)
314	2.18 (-2)	2.39 (-2)	1.70 (-2)	1.29 (-2)
323	3.23 (-2)	3.45 (-2)	2.55 (-2)	2.04 (-2)
340	3.99 (-1)	4.22 (-1)	3.19 (-1)	2.62 (-1)
370	5.34 (-1)	5.64 (-1)	4.35 (-1)	3.63 (-1)
400	7.74 (-1)	8.17 (-1)	6.31 (-1)	5.28 (-1)
430	1.08	1.14	8.83 (-1)	7.40 (-1)
460	1.36	1.44	1.12	9.40 (-1)
494	1.39	1.46	1.14	9.64 (-1)
537	2.47	2.61	2.05	1.73
588	2.56	2.71	2.13	1.80
638	2.59	2.75	2.17	1.84
691	2.59	2.74	2.17	1.85
747	2.48	2.62	2.08	1.78
800	2.38	2.51	2.00	1.71

Table 2
L_W values for latitude 40°N[a,b]

∂ center (nm)	Spring	Summer	Fall	Winter
299	1.25 (-4)	3.32 (-4)	4.84 (-4)	5.68 (-6)
304	1.49 (-3)	2.78 (-3)	7.29 (-4)	1.49 (-4)
309	6.51 (-3)	9.91 (-3)	3.38 (-3)	1.09 (-3)
314	1.57 (-2)	2.13 (-2)	8.44 (-3)	3.61 (-3)
319	2.68 (-2)	3.42 (-2)	1.48 (-2)	7.42 (-3)
323	2.71 (-2)	3.35 (-2)	1.51 (-2)	8.25 (-3)
340	3.56 (-1)	4.24 (-1)	2.06 (-1)	1.25 (-1)
370	4.89 (-1)	5.74 (-1)	2.93 (-1)	1.87 (-1)
400	7.10 (-1)	8.33 (-1)	4.28 (-1)	2.74 (-1)
430	9.92 (-1)	1.16	6.02 (-1)	3.87 (-1)
460	1.26	1.47	7.68 (-1)	4.94 (-1)
494	1.28	1.50	7.88 (-1)	5.07 (-1)
537	2.28	2.67	1.41	9.06 (-1)
588	2.35	2.75	1.46	9.30 (-1)
638	2.41	2.81	1.51	9.77 (-1)
691	2.43	2.81	1.53	1.01
747	2.33	2.69	1.48	9.84 (-1)
800	2.25	2.59	1.43	9.56 (-1)

[a] Based on GCSOLAR calculation
[b] Units of $L_∂$ are 10^{-3} einsteins cm^{-2} day^{-1}. Multiplication of $L_∂$ by $f_∂$ in units of M^{-1} cm^{-1} gives rate constant in units of day^{-1}.

Source: Jørgensen et al., 1988

Table 3
Photolysis of chemicals in water

Chemical/ structure	Conc. (M)	pH or other sol- vents	Photo- lysis wave- length (nm)	Quan- tum yield	Sunlight half life[a] ($t_{1/2}$, days^{-1})	Pro- cess type[b]	Products (% yields)/ comments
Aromatic chemicals:							
Napthalene	1.2(-4)	-	313	1.5(-2)	550	-	Solvent effect[c]
1-Methylnapthalene	1.2(-4)	-	313	1.8(-2)	180	-	
2-Methylnapthalene	8.8(-5)	-	313	5.3(-3)	410	-	
Phenanthrene	8.4(-7)	-	313	1(-2)	59	-	
Anthracene	2.3(-7)	-	366	3(-3)	4.5	-	Solvent effect[c]
9-Methylanthracene	3.1(-7)	-	366	7.5(-3)	0.79	-	No oxygen effect
9,10-Dimethyl- anthracene	1.3(-7)	-	366	4.0(-3)	-	-	
Pyrene	3.9(-7)	-	313, 366	3.1(-3)	4.2	-	Solvent effect[c]
Fluoranthene	6.6(-7)	-	313, 366	1.2(-4) (313)	160	-	ø drops sharply at 366 nm
Chrysene	8.2(-9)	-	313	2.8(-3)	13	-	
Napthacene	2.2(-9)	-	436	1.3(-2)	0.20	-	Solvent effect[c]
Benzanthracene	7.9(-8)	1% AN	313, 366,sun	3.2(-3)	0.20 E Mar	Ox	O_2 needed, 7,12 quinone (30)
Benzo[a]pyrene	1.1(-8)	1-20% AN	313, 366,sun	8.9(-4)	0.045 M Dec	Ox	O_2 needed; 3,6- & 1,6-quinones (>90)
Quinoline	6.3(-6)	-	313, sun	3.3(-4)	1.1 L Jun	-	
Benzoquinoline	4.1(-5)	1%MeOH	313,sun	1.4(-2)	0.58 M Jun	-	O_2 retards
Carbazole monomer	6.0(-6)	1% AN	313,sun	7.6(-3)	0.12 M Jan	Ox	Carbazole and dimer with HO; no O_2 effect
Dibenzocarbazole products	3.4(-8)	0.1% AN	313, 366,sun	2.8(-3)	2.2 min L Jan	Ox	No O_2 effect, two dihydroxy
Benzothiophene	7.5(-6)		Sun	-	1.2 L May	-	
Dibenzothiophene	2.7(-6)	1% AN	313,sun	5.0(-4)	3.9 E Ma-	-	No O_2 effect; 5 products
1-Napthaleneacetic acid droxy-methyl-nap-	5.38(-4)	pH 8	>290, sun	-	<1	Ox	MeNapthalene, hy- thalene, napthoic acid
4-Fluoroanisole	1.5(-4)	75% t-BuOH	280	0.49	-	S	4-Hydroxyanisole
N-Phenyl-O-i- propyl-carbamate	1(-4)	-	>300	1.0[d]	>254[d]		
N-Methyl-O- napthyl-carbamate	4(-4)	pH 5.5-6	313	0.0055	2	-	Non-volatile, colored

Table 3 Continued

Chemical/ structure	Conc. (M)	pH or other solvents	Photolysis wavelength (nm)	Quantum yield	Sunlight half life[a] ($t_{1/2}$, $days^{-1}$)	Process type[b]	Products (% yields)/ comments
Chloroaromatic chemicals:							
Chlorobenzene	1.57(-4)	1% AN	260-290	0.37		S	Phenol; no benzene
2-Chlorobiphenyl	8.00(-6)	10% AN	260-290	0.27		S	2 Hydroxybiphenyl, no biphenyl
4-Chlorobiphenyl	1.00(-3)	80% AN	254	0.0037	2.5(4)e		
2,4-Dichlorobiphenyl	1.00(-3)	80% AN	254	0.22	2700e		
2,4,6-Trichlorobiphenyl	1.00(-3)	80% AN	254	0.25	6200e		
2,2',5,5'-Tetrachlorobiphenyl	1.00(-3)	80% AN	254	0.017	2.5(4)e		
Decachlorobiphenyl	1.00(-3)	80% AN	254	0.21	156e		
2-,3- or 4-Chloro-acids	6.43(-4)	pH 8	254,S	-		S,R	Hydroxybenzoic acids or benzoic acid
4-Chlorophenoxyacetic acid	2.15(-3)	pH 8	S	-	7.0	S,OX	Hydroxyphenoxy acetate, 4-chlorophenol
2,4-Dichlorophenoxyacetic acid	2.0(-3)	pH 7	254,S	0.044	50	S,Ox	Dichlorophenol, 2- or 4-hydroxyphenoxyacetic acid
2-,3- or 4-Chlorophenylacetic acids	6(-4)	pH 8	354,S	-	30	S,Ox	Benzyl alcohol, hydroxyphenylacetic acid
Methyl-2,4-dichlorophenoxy acetate	5.0(-6)	pH 7.8	>290	0.031	14	S,Ox	Dichlorophenol, hydroxychlorophenylacetate esters
n-Butoxyethyl-2,4-dichloro-phenoxy acetate	3.0(-6)	pH 7.8	>290	0.056	8	S,Ox	Dichlorophenol, hydroxychlorophenylacetate esters
2,4-Dichlorophenol	>1(-3)	-	>290	-		S,Ox	Phenol dimers
3,5-Dichloroaniline	5(-6)	pH 4-10	313,sun	0.052	3 min		2-Chloro-5-aminophenol
Methoxychlor [p-$MeOC_6H_4]_2CHCCl_3$	1(-4) to 1(-3)	NW	>290,sun	-	10-100	E	1,1-Bis(p-methoxy phenyl)-2,2-dichloroethylene (60%)
1,1-Bis(p-methoxyphenyl)-2,2-dichloethylene	3.9(-8)	NW	>290,sun	0.3	0.094	Ox,I	4,4'-Dimethoxybenzeophenone (15%), photoisomer (35%)
DDT, $(p\text{-}ClC_6H_4)_2\text{-}CHCCl_3$	NW	>290	-	-	5		No photolysis in pure water
DDE, $(p\text{-}ClC_6H_4)_2C{=}CCl_2$	3(-9)	-	>290,sun	0.3	0.94	Ox,I	4,4'-Dichlorbenzophenone (15%), photoisomer (20%)
3,4-Dichloropropionanilide, 3,4-$Cl_2C_6H_4NHCOC_2H_5$	1(-3)	6	>290	-	5	R,S	Hydroxy- and dihydroxypropionalde chloropropionanilide and propionanilide

Table 3 Continued

Chemical/ structure	Conc. (M)	pH or other sol-vents	Photo-lysis wave-length (nm)	Quan-tum yield	Sunlight half life[a] ($t_{1/2}$, days^{-1})	Pro-cess type[b]	Products (% yields)/ comments
<u>**Chloroaromatic chemicals (cont.):**</u>							
Chloropropham, (3-Cl-$C_6H_4NCO_2$-i-Pr)	4(-4)	5.5-6.9, NW	313,sun	1.0[d]	>121[d]		
N'-(4-Chloro-2-methylphenyl)-N, N-dimethylform-amidine, 4-Cl-2-MeC_6H_3N=$CHNMe_2$	1.5(-3)	7.1, NW	>286	-		S	N-4-chloro-2-me-thylophenyl) form-amide (95%)
3-(4-Chlorophenyl)-1,1-dimethylurea, 4-ClC_6H_4-$NHCONMe_2$	7(-4)	8	Sun	-	>156	Ox	3-(4-Chlorophenyl -1-methylurea
3,3'-Dichlorobenz-idine		7.8	254,sun	0.43	2 min	R	3-Chlorobnenzidine (30%). No O_2 effect no reaction in hexane
3-Chlorobenzidine		7.8	300,sun	0.70	-	R	Benzidine
<u>**Nitroaromatic chemicals:**</u>							
Nitrobenzene	12.5	-	313,sun	1.8(-4)	32 Mar		
1,2-DNB	1.99	0.3% AN	313	8.7(-5)			
1,3-DNB	1.99	0.3% AN	313	<2(-5)			
1,4 DNB	1.99	0.3% AN	313	5.5(-5)			
2-NT	2.19	1% AN	313	2.23(-3)			
3-NT	2.18	1% AN	313	1.58(-2)			
4-NT	2.18	1% AN	313	4.60(-3)			
2,3-DNT	5.49	1% AN	313	6.84(-4)			
2,4-DNT	5.60	1% AN	313	6.56(-4)			
25-DNT	5.50	1% AN	313	6.72(-3)			
2,6-DNT	5.49	1% AN	313	1.88(-3)			
3,4-DNT	5.54	1% AN	313	7.0(-5)			
3,5-DNT	5.51	1% AN	313	<4.0(-5)			
2,3,4-TNT	1.32	0.3% AN	313	1.63(-3)			
2,3,5-TNT	1.33	0.3% AN	313	4.27(-3)			
2,3,6-TNT	1.34	0.3% AN	313	4.0(-4)			
2,4,5-TNT	1.32	0.3% AN	313	2.37(-3)			
2,4,6-TNT	1.0-3.0	1% AN, NW	313, 366,sun	2.14(-3)	80 July	Ox/R	Complex mixture of nitroso, hydroxyl amines, aldehyde, acid, azoxy; sensi-tized in natural wa-ters and with ace-tone
3,4,5-TNT	1.33	0.3% AN	313	4.0(-5)			
RDX $(C_3H_3N_3)(NO_2)_3$	3.85	1.3% AN	313	0.16	4 Mar		NO_2^-, NO_3^-, HCHO

Table 3 Continued

Chemical/ structure	Conc. (M)	pH or other sol- vents	Photo- lysis wave- length (nm)	Quan- tum yield	Sunlight half life[a] ($t_{1/2}$, days^{-1})	Pro- cess type[b]	Products (% yields)/ comments
Nitroaromatic chemicals (cont.):							
HMX	1.70	0.5% AN	313,sun	0.15	4.6 June		
Trifluralin	$1*10^{-4}$	-	sun	-	12 min	Ox/R	Benzimidizoles
o-Nitrobenz- aldehyde	$1*10^{-2}$ - 0.1	-	313, 366,sun	0.5	10 min	Ox/R	o-nitrosobenzoic acid
p-Nitrobenz- aldehyde	$1*10^{-3}$	-	313, 366	0.017	5	Ox/R	p-nitrosobenzoic acid
Carbonyls:							
n-PrCOMe	1.0	Deaerated	-	0.71	-	I, II	Acetone, propane (with H-donor)
MeCOCOMe	$1*10^{-3}$	-	313, 366	1.8, 0.19	0.35	I, Ox	Acetic acid
PhCOMe	$2.5*10^{-5}$ -		-	>300	<0.01	>30	-
PhCOPh Hydrobenzophenones	$2.8*10^{-5}$		-	>300	<0.001	>100	I,Ox ø increased 5 x by Mn and 500 x by 0.001 M EtOH
HNC(S)NH molinate	$5*10^{-4}$	NW	Sun	-	24-60	Ox	Glycine and sulphate sensitized only
	$5*10^{-4}$	NW	Sun	-	7	Ox	See text, sensitized only

AN = acetonitrile
NW = Natural water
e = estimated values for mid June, other values are measured in June to August unless noted otherwise.
E, I, Ox, R and S are elimination, isomerization, oxidation, reduction and substitution.
[d] Assumed value used to estimate maximum rate and minimum half life.
DNB, NT, DNT and TNT: dinitrobenzene, nitrotoluene, dinitrotoluene, and trinitrotoluene
[b] Half life in summer unless noted otherwise

Sources: Mill et al. (1981), Zepp and Schlotzhaven, Cornelisse and Havinga (1975), Crosby and Tang (1969), Havinga and Cornelisse (1976), Crosby and Wong (1973), Crosby and Tutass (1966), Zepp et al. (1975), Zepp et al. (1976), Su and Zabik (1972), Banerjee et al. (1978), Bunce et al. (1980), Tse et al., (1982), Le Maire (1967), Encinas et al. (1980), Soderquist et al. (1977) and Jørgensen and Gromiec (1988).

APPENDIX 4

TOXICITY DATA
AND
ESTIMATION METHODS

Table 1 Definition of toxicity indices

Critical range - the range of concentrations in mg/l below which all fish lived for 24 hr and above which all died. Mortality is genen as a fraction indicating the death rate (e.g. 3/4).

No effect level - implies that the animals remained in good condition. In most experiments, blood and urine tests were made. The urine tests included specific gravity, pH, reducing sugars, bilirubin, and protein. The blood tests included hemoglobin concentration, packed cell volume, mean corpuscular Hb content, a white and differential cell count, clotting function, and the concentration of urea, sodium, and potassium. Control tests for hematological examination were made on the group of animals before exposure. No effect level in this context means: no toxic signs, autopsy - organs normal, blood and urine tests - (if made) normal.

The merian tolerance (TLm) - this term has been accepted by most biologists to designate the concentration of toxicant or substance at which 50% of the test organisms survive. In some cases and for certain special reasons, the TL_{10} or TL_{90} might be used. The TL_{90} might be requested by a conservation agency negotiating with an industry in an area where an important fishery exists, and where the agency wants to establish waste concentrations that will definitely not harm the fish. The TL_{10} might be requested by a conservation agency which is buying toxicants designed to remove undesirable species of fish from fishing lakes.

LD_{50} (lethal dose fifty) - a calculated dose of a chemical substance which is expected to kill 50% of a population of experimental animals exposed through a route other than respiration. Dose concentration is expressed in milligrams per kilogram of body weight.

LC_{50} (lethal concentration fifty) - a calculated concentration which, when administered by the respiratory route, is expected to kill 50% of the population of experimental animals during an exposure of 4 hours. Ambient concentration is expressed in milligrams per litre.

MAC (maximum allowable concentration) - a value in accordance with environmental legislation. Often dependent on time. This relation may be expressed as follows: $\log C = 1.8 - 0.7 \log t + 0.068 \log t^2$, where C is the MAC (in mg/m^3) and t is the time (in hours).

NC (narcotic concentration) - NC_{50}: median narcotic concentration.

Table 2
Lethal doses 50% Mortality (LD_{50})

Component	Concentration (mg per kg body)	Species	
Acetaldehyde	193	Rat	(s.o.d.)
Acetonitrile	170-185	Rat	(a.o.)
Acetonitrile	180	Guinia pig	(a.o.)
Acrylamide	250	Rat	(s.o.d.)
Acrylonitrile	78	Rat	(a.o.)
Acrylonitrile	35	Mice	(a.o.)
Ag	100	Mouse	(oral)
Ag as oxide	2820	Rat	-
Al	770	Mouse	-
Al	3700	Rat	-
Alkylamine	106	Rat	(a.o.)
Allylchloride	700	Rat	(oral)
2-Aminopyridine	200	Rat	(a.o.)
2-Aminopyridine	50	Mouse	(a.o.)
As	9	Mouse	-
As as As_2O_5	8	Rat	-
As as As_2O_3	45	Rat	-
B as borax	4500	Rat	-
B as boric acid	2660	Rat	-
Ba as chloride	500	Mouse	-
Ba as chloride	150	Rat	-
Ba as carbonate	800	Rat	-
Be as chloride	86	Rat	-
Benzaldehyde	5000	Rabbit	(s.c.)
Benzene	5600-5700	Rat	-
Benzene sulphonic acid	400-3200	Mouse	(a.o.)
Benzoic acid	1700	Rat	(a.o.)
p-Benzoquinone	130-300	Rat	(oral)
Bi	13	Rat	(intraven.)
n-Butanol	4200-4400	Rat, rabbit	(oral)
Butylglycolether	320	Rabbit	(s.o.d.)
n-Butylether	7400	Rat	(a.o.)
Butyraldehyde	5900	Rat	(a.o.)
Butyronitrile	50-100	Rat	(a.o.)
Ca as acetate	4280	Rat	(oral)
Caprioc acid	6440	Rat	(oral)
CCl_4	4620	Mouse	-
Cd as oxide	72	Rat	-
Ce as nitrate	4200	Rat	-
Chloral	50-400	Rat	(oral)
Chlorobenzene	2200	Rat	(s.o.d.)
Chlorobenzene	2830	Rabbit	(s.o.d.)
Chlorodimethylether	500	Rat	(s.o.d.)
2-Chloropropane	3000	Guinea pig	(s.o.d.)
CN^- as Na-salt	3	Mouse	-
Co as chloride	80	Mouse	-
Cr as chloride	1870	Rat	-
Cu as chloride	4000	Rat	-

Table 2 (continued)

Component	Concentration (mg per kg body)	Species	
Cu as chloride	140	Rat	-
Cyanuric acid	10000	Rat, rabbit	(s.o.d.)
Cyclohexanol	2200-2600	Rat	(s.o.d.)
Diacetone alcohol	4000	Rat	(s.o.d.)
Dichloroacetic acid	4500	Rat	(s.o.d.)
Dichloroisopropylether	240	Rat	-
Diethanolamine	1820	Rat	(a.o.)
Diethylamine	540	Rat	(a.o.)
Dimethoxymethane	5000	Guinea pig	(s.c.)
Dimethylsulphate	50	Rabbit	(a.o.)
Dimethylsulphate	440	Rat	(a.o.)
Diphenylether	4000	Rat	-
Epichlorohydrine	40	Rat	(s.o.d.)
Ethanolamine	2700	Rat	(a.o.)
Ethylamine	400	Rat	(oral)
Ethylenebromide	420	Female mice	(s.o.d.)
Ethylenebromide	150	Male rat	(s.o.d.)
Ethyleneimine	4	Rabbit	(s.o.d.)
Ethyleneimine	4-12	Rat	(s.o.d.)
Ethyleneimine	8	Cat	(s.o.d.)
Fe(III) as nitrate	3250	Rat	-
Fe(II) as sulphate	1480	Rat	-
Formaldehyde	260	Guinea pig	(s.o.d.)
Formaldehyde	100-800	Rat	(s.o.d.)
Furfural	50-100	Rat	(a.o.)
Ge as oxide	750	Rat	-
Hexachloroethane	325	Dogs	(i.d.)
Hf as chloride	112	Mouse	(intraven.)
Hg(II) as chloride	37	Rat	(oral)
Isobutyric acid	400-800	Inhalation	(s.o.d.)
La	35	Rat	(intraven.)
Lactonitrile	21	Rat	(a.o.)
Li as carbonate	710	Rat	(oral)
Methylacrylate	200	Rabbit	(a.o.)
Methylamine	100-200	Rat	(a.o.)
Methylbenzoate	3400	Rat	(a.o.)
Methylchloroformate	<50	Rat	(a.o.)
Methylhydrazine	330	Rat	(a.o.)
Methylisobutylketone	2080	Rat	(s.o.d.)
Methylispropenylketone	180	Rat	(s.o.d.)
Methylsalicylate	700	Guinea pig	(a.o.)
Mg as chloride	2800	Rat	-
Monochloroacetic acid	80	Guinea pig	(oral)

Table 2 (continued)

Component	Concentration (mg per kg body)	Species	
Ni as fluoride	130	Mouse	(intraven.)
2-Octanol	>3200	Rat	(s.o.d.)
Pb as acetate	120	Rat	(oral)
Pentachlorophenol	70-100	Rabbit	(oral)
Pentachlorophenol	27-78	Rat	(oral)
Propeneoxide	690	Guinea pig	(a.i.)
ß-Propiolactone	50-100	Rat	(s.o.d.)
Propyleneimine	19	Rat	(a.o.)
Se as sulphide	38	Rat	-
Sn(II) as chloride	41	Mouse	(intraven.)
Strychnine	0.98	Mouse	(oral)
Te as Na-salt	20	Mouse	(oral)
Tetramethyllead	80-106	Rat	(s.o.d.)
Tetramethyllead	88	Rat	(intraven.)
Tetramethyllead	90	Rat	(i.p.)
Th as chloride	114	Mouse	(intraven.)
Tl as oxide	44	Rat	(oral)
Triethanolamine	9110	Rat	(a.o.)
U as oxide	6	Mouse	(intraven.)
V(II) as chloride	540	Rat	(oral)
V(IV) as chloride	160	Rat	-
Vinylacetate	2900	Rat	(a.o.)
Vinyl-2-pyridine	100-200	Rat	(a.o.)
Zn as acetate	2460	Rat	-

i.d. = intravenous dose
i.p. = intraperitoneal
s.o.d. = single oral dose
s.c. = subcutane
a.o. = acute oral

Table 2B
Lethal Concentration 50% Mortality (LC_{50})

Component	Concentration µg l^{-1}	Duration	Species
Acrylonitrile	$3.6*10^6$	1 h	Crangon crangon
Acrylonitrile	25000	24 h	Crangon crangon
Acrylonitrile	6000	96 h	Crangon crangon
Ag as nitrate	30	4 d	Daphnia magna
Al as chloride	3900	2 d	- -
Alkyl benzene sulphonate	25000	38 h	Tilapia (fish)
Ammonia	280	24 h	Salmo salar
Ba as chloride	14500	2 d	Daphnia magna
Benzene	20000	24-48 h	Bluegill sunfish
sec-Butanol	$4.3*10^6$	24 h	Goldfish
p-tert-Butyltoluene	3000	24 h	Goldfish
Cd as chloride	65	2 d	- -
Cd (hard water)	17	5 d	Salmo gairdneri
Chloramine (NH_2Cl)	100	24 h	Phytoplankton
Chlorine	100	24 h	- -
Co as chloride	1100	2 d	Daphnia magna
Cr(IV)	50	2 d	- -
Cr(IV)	32-6000	2 d	Phytoplankton
p-Cresol	21000	24 h	Crucian carp
Cu as chloride	9.8	2 d	Daphnia magna
1,5,9-Cyclodecatriene	4000	24 h	Goldfish
1,3,5-Cycloheptatriene	15000	24 h	Goldfish
Dibutyl-phthalate	50000	2 d	Goldfish
Diethyleneglycol	$>5*10^6$	no ind.	Goldfish
Diethyleneglycolmono-butylether	$2.7*10^6$	24 h	Goldfish
Diethyleneglycolmono-ethylether	$>5.0*10^6$	24 h	Goldfish
Epichlorhydrine	23000	24 h	Goldfish
Ethanolamine	$1.9*10^5$	24 h	Goldfish
Ethanolamine	$1.7*10^5$	96 h	Goldfish
Glycerol	$>5.0*10^6$	24 h	Goldfish
n-Heptane	$5.6*10^6$	no ind.	Gambusia affinis
Hg as chloride	5	2 d	Daphnia magna
Isopropanol	$1.4*10^6$	48 h	Brown shrimps
Isopropanol	$1.15*10^6$	96 h	Brown shrimps
2-Isopropoxyethanol	$>5.0*10^6$	24 h	Goldfish
Mesityloxide	540000	24 h	Goldfish
Methylethylketone	$5.0*10^6$	24 h	Goldfish
Methylisobutylketone	460000	24 h	Goldfish
Mn(II) as chloride	9800	2 d	-

Table 2B (continued)

Component	Concentration $\mu g\ l^{-1}$	Duration	Species
Naphthalene	33000	20°C, 24h	Chlorella vulgaris
Pentachlorophenol	210	24 h	Fathead minnow
Pentachlorophenol	200	96 h	Fathead minnow
Phenol	$5.6*10^6$	15°C, 3 min	Crangon crangon
Phenol	400000	15°C, 27 min	Crangon crangon
Phenol	80000	15°C, 3 h	Crangon crangon
Phenol	30000	15°C, 72 h	Crangon crangon
Phenol	25000	15°C, 96 h	Crangon crangon
Phenol	$1.8*10^7$	15°C, 9 min	Gobius minutus
Phenol	320000	15°C, 27 min	Gobius minutus
Phenol	20000	15°C, 3 h	Gobius minutus
Phenol	10000	15°C, 48 h	Gobius minutus
Phenol	9000	15°C, 96 h	Gobius minutus
Polyelectrolytes	345000	48 h	- -
Polyelectrolytes	>8000	48 h	Salmonoid fish
Polyoxyethylene	14500	48 h	- -
Sn(II) as chloride	55000	2 d	Daphnia magna
Sr as chloride	125000	2 d	- -
Tetrapropylene benzene-sulphonate	$5.6*10^7$	15°C, 9 min	Crangon crangon
Tetrapropylene benzene-sulphonate	$7.0*10^6$	15°C, 9 min	Gobius minutus
Tetrapropylene benzene-sulphonate	25000	15°C, 3-96 h	Gobius minutus
Toluene	58000	24 h	Fish
2,4,6-Trinitroresorcinol	2580	96 h	Fish
2,4,6-Trinitrotoluene	4000-5000	6 h	Fish
m-Xylene	16000	24 h	Goldfish
o-Xylene	55000	20°C, 1 d	Chlorella vulgaris
o-Xylene	13000	24 h	Goldfish
p-Xylene	18000	24 h	Goldfish
Zn as chloride	100	2 d	- -
Zn as chloride	6000	15 h	Salmonoid fish

Sources: Jørgensen and Johnsen (1989), and Jørgensen et al. (1988).

Table 3
Equations for calculation of MAC

$$\log MAC = -1.2 - 0.012\, t_{melt} + \log M$$

$$\log MAC = 0.40 - 0.01\, M + \log M$$

$$\log MAC = -0.4 - 0.006\, t_{boil} + \log M$$

Source: Liublina and Golubev (1967)

Table 4
Corrections to log MAC for the chemical structure of compounds

Compounds	Correction
Saturated aliphatic hydrocarbons	+0.5
Saturated aliphatic ketones, alcohols, ethers, and esters	+0.5
Saturated cyclic hydrocarbons containing a benzene ring (excluding benzene and the first member of a homologous series	+0.5
Straight-chain compounds containing a triple bond	-0.5
Aliphatic amines	-1.0
Aniline and its derivatives	-1.0
Acid anhydrides	-1.0
Cyclic compounds containing an NO_2 group in a side chain	-1.0
Compounds containing the ONO_2 group in the straight chain	-1.0
Compounds having a double or triple bond together with an active element or group (Cl, Br, F, NO_2, OH) in the straight chain	-1.0
Substances containing an epoxy group	-1.5
Organophosphorus compounds	-1.5
Aldehydes	-1.5
Substances splitting off the CN group	-2.0

Source: Liublina and Filov (1965)

Table 5
Equations relating toxicity indices (in mM/litre) of volatile organic compounds to some of their constants

$$\log LC_{50} = -0.02 - 0.009\, t_{boil}$$

$$\log LC_{50} = 0.08 - 0.011\, M$$

$$\log LC_{50} = -1.6 - 0.010\, t_{melt}$$

$$\log NC_{50} = 0.75 - 0.013\, M$$

$$\log NC_{50} = 0.49 - 0.039\, RD$$

$$\log NC_{50} = 0\text{-}345 - 0\text{-}0094\, t_{boil}$$

Source: Filov et al. (1973)

Table 6
Equations for the calculation of LC_{50} from molecular weight (M) and boiling point (t_{boil}) for individual classes of organic compounds

Class of compounds	n	r	Equation: $\log LC_{50}$ = (mM/litre)
Hydrocarbons	42	-0.90	2.17 - 0.026 M
	42	-0.94	0.79 - 0.011 t_{boil}
Alcohols	14	-0.90	1.25 - 0.016 t_{boil}
Ethers	7	-0.96	1.74 - 0.020 M
	7	-0.98	0.74 - 0.011 t_{boil}
Ketones	14	-0.82	1.16 - 0.015 t_{boil}
Amines (various)	23	-0.80	-0.60 - 0.010 M
	22	-0.79	-1.00 - 0.005 t_{boil}
Nitriles and cyanide compounds	10	-0.91	-1.30 - 0.014 M
Nitro compounds	13	-0.71	0.71 - 0.020 M
Heterocyclic compounds	17	-0.79	2.80 - 0.041 M
Chlorohydrocarbons	40	-0.60	0.20 - 0.012 M
	38	-0.71	-0.10 - 0.011 t_{boil}
Bromohydrocarbons	16	-0.74	-2.06 - 0.020 M
Aldehydes			1.3 - 0.027 M
			- 0.008 t_{boil}
Phenols (LD_{50})(mM/kg)	25	0.55	0.18 + 0.0036 M
Miscellaneous organic compounds	277	-0.52	0.08 - 0.011 M
including those listed above	249	-0.62	-0.02 - 0.009 t_{boil}

n = number of observations; r = regression coefficient

Source: Filov et al. (1973)

Table 7
Equations for calculating the threshold concentration (C_{min}) from molecular weight (M) and boiling point (t_{boil}) for particular classes of organic compounds

Class of compounds	n	r	Equation: log C_{min} = (mM/litre)
Hydrocarbons	13	-0.89	-0.95 - 0.010 t_{boil}
	13	-0.82	-0.51 - 0.015 M
Ketones	4	-0.97	-0.72 - 0.011 t_{boil}
Fatty amines and anilines	8	-0.71	-2.33 - 0.005 t_{boil}
Various amines	13	-0.63	-2.29 - 0.008 M
Miscellaneous organic compounds	92	-0.41	-1.88 - 0.006 t_{boil}
including those listed above	100	-0.55	-1.31 - 0.011 M

n = number of observations; r = regression coefficient

Source: Filov et al. (1973)

Table 8
Equations for calculating the MAC from the molecular weight (M) separately for volatile (V) and nonvolatile (NV) compounds
(For comparison, data are also given for volatile and nonvolatile compounds together)

Class of compounds		n	Equation: log MAC = (mM/m^3)
Saturated alcohols	V	10	3.0 - 0.040 M
	NV	8	-0.5 - 0.005 M
	All	18	1.0 - 0.015 M
Nitro compounds without unsaturated bonds in open chains	V	10	1.40 - 0.023 M
	NV	20	-1.35 - 0.0039 M
	All	30	0.15 - 0.013 M
Amines and their nonchlorinated derivatives	V	14	-0.30 - 0.013 M
	NV	26	-1.2 - 0.0046 M
	All	40	-1.04 - 0.006 M

n = number of observations

Source: Filov et al. (1973)

Table 9 A
Relations of toxicity indices

LC_{50}	~	0.5 LC_{100}
L_0	~	0.08 LC_{100}
MAC	~	0.0005 LC_{100}
MAC (mg/m^3)	~	0.5 LC_{100} (mg/litre)
LC_{100}	~	2 LC_{50}
LC_0	~	0.15 LC_{50}
MAC	~	0.0013 LC_{50}
MAC (mg/m^3)	~	1.3 LC_{50} (mg/litre)
MAC	~	0.066 C_{min}

Source: Filov et al. (1973)

Table 9 B
Correlations between toxicity indices of volatile organic compounds and equations for the preliminary calculation of unknown indices from those which are known (MAC in mM/m³), LC_{50} in mM/litre, LD_{50} in mM/kg)

n	r	Equation		
178	0.72	log MAC	=	0.72 log LC_{50} - 0.31
84	0.82	log C_{min}	=	0.76 log LC_{50} - 1.62
112	0.66	log MAC	=	0.88 log LD_{50} - 2.29
92	0.59	log LC_{50}	=	0.84 log LD_{50} - 0.33
36	0.68	log C_{min}	=	0.68 log LD_{50} - 3.4

n = number of observations; r = regression coefficient

Source: Filov et al. (1973)

Table 9 C
Correlations between log LC_{50} and log MAC and between log LD_{50} and log MAC and equations for calculating MACs from LC_{50} and LD_{50} for some classes of organic compounds

Class of compounds	n	r	Equation: log MAC =
Hydrocarbons	26	0.72	0.80 log LC_{50} - 0.04
Alcohols	11	0.86	0.94 log LC_{50} + 0.11
Aldehydes	5	0.98	0.43 log LC_{50} - 0.86
Nitro compounds	8	0.85	0.64 log LC_{50} - 0.14
Amines	15	0.53	0.86 log LC_{50} - 0.27
Heterocyclic compounds	11	0.85	0.42 log LC_{50} - 0.94
Chlorohydrocarbons	30	0.86	0.87 log LC_{50} - 0.36
Alcohols	5	0.91	0.98 log LD_{50} - 1.96
Nitro compounds	6	0.98	0.89 log LD_{50} - 2.70
Amines	24	0.40	0.37 log LD_{50} - 2.12

n = number of observations; r = regression coefficient

Source: Filov et al. (1973)

Table 10
Elements: Abundance and Biological Activity

Symbols used:

a = elements formed by radioactive decay of uranium and thorium. Have short physical half-lives and their crustal abundance are too low to be measured accurately.
b = very low, unmeasureable
ra = radioactive
cs = carcinogenic, suspected only.

s = stimulatory
cp = carcinogenic, proven
en = essential nutrient, established
ep = essential nutrient, probably or required under special conditions
t1 = toxic
t2 = very toxic

Element	Symbol	Atomic number	Crustal abundance weight (%)	Abundance in hydrosphere (mg/l)	Abundance in atmosphere (vol ppm)	Biological activity	Threshold limit (mg/ m^3 in air in 8 hours)
Actinium	Ac	89	manmade	manmade		ra	
Aluminium	Al	13	8	0.01		cs	
Americum	Am	93	manmade	manmade		ra	
Antimony	Sb	51	0.00002	0.0005		s t2	
Argon	Ar	18	0	0.6	9300		
Arsenic	As	33	0.00020	0.003		cs s t2	
Astatine	At	85	manmade	manmade		ra	
Barium	Ba	56	0.0380	0.03		s t1	0.5
Berkelium	Bk	97	manmade	manmade		ra	
Beryllium	Be	4	0.0002	b		cp (s) t2	
Bismuth	Bi	83	4E-7	2E-5		t1	
Boron	B	5	0.0007	4.6			
Bromine	Br	35	0.00040	65		t1=Br2	
Cadmium	Cd	48	0.000018	0.001			0.2
Calcium	Ca	20	5.06	400		en	
Californium	Cf	98	manmade	manmade		ra	
Carbon	C	6	0.02	28	CO_2=330	en	
Cerium	Ce	58	0.0083	0.0004		s	
Cesium	Cs	55	0.00016	0.0005			
Chlorine	Cl	17	0.019	18,980		Cl(-)=en Cl_2=t1	
Chromium	Cr	24	0.0096	5E-5		en cp s t1	0.1(CrO_3)
Cobalt	Co	27	0.0028	0.0005		en cp	
Copper	Cu	29	0.0056	0.003		en s t1	
Curium	Cm	96	manmade	manmade		ra	
Dysprosium	Dy	66	0.00085	b		s	
Einsteinium	Es	99	manmade	manmade		ra	
Erbium	Er	68	0.00036	b		s	
Europium	Eu	63	0.00022	b			
Fermium	Fm	100	manmade	manmade		ra	
Fluorine	F	9	0.0460	1.3		ep s	

Table 10
Elements: Abundance and Biological Activity (continued)

Element	Symbol	Atomic number	Crustal abundance weight (%)	Abundance in hydros-phere (mg/l)	Abundance in atmos-phere (vol ppm)	Biological activity	Threshold limit (mg/ m^3 in air in 8 hours)
Francium	Fr	87	manmade	manmade		ra	
Gadolinium	Gd	64	0.00063	b			
Gallium	Ga	31	0.00063	b			
Germanium	Ge	32	0.00013	7E-5		s	
Gold	Au	79	2E-7	4E-6		s t1	
Hafnium	Hf	72	0.0004	b			
Helium	He	2	0	5E-6	5.2		
Holmium	Ho	67	0.00016	b		s	
Hydrogen	H	1	0.14	H2O	CH_4=1.5 H_2=0.5	en	
Indium	In	49	0.00002	0.02		s t2	
Iodine	I	53	0.00005	0.06		I(-)=en I_2=t1	
Iridium	Ir	77	2E-8	b		t1	
Iron	Fe	26	5.80	0.01		en	
Krypton	Kr	36	0	0.0003	1		
Lanthanium	La	57	0.0050	0.0003			
Lead	Pb	82	0.0010	3E-5		t2 cp s	0.2
Lithium	Li	3	0.0020	0.17		s	
Lutetium	Lu	71	8E-5	b			
Magnesium	Mg	12	2.77	1350		en	
Manganese	Mn	25	0.100	0.002		en cs	5
Mendelevium	Md	101	manmade	manmade		ra	
Mercury	Hg	50	2E-6	3E-5		s t2	
Molybdenium	Mo	42	0.00012	0.01		en	5-15
Neodymium	Nd	60	0.0044	b			
Neon	Ne	10	0	0.0001	18		
Neptunium	Np	93	manmade	manmade		ra	
Nickel	Ni	28	0.0072	0.002		ep cp s	
Niobium	Nb	28	0.0072	0.002		ep cp s	
Nitrogen	N	7	0.0020	0.5	780,900	en	
Nobelium	No	102	manmade	manmade		ra	
Osmium	Os	76	2E-8	b		ra	
Oxygen	O	8	45.2	H_2O	209,500	en	
Palladium	Pd	46	3E-7	b		cs	
Phosphorus	P	15	0.1010	0.07		en	
Platinum	Pt	78	5E-7	b		t1	0.002
Plutonium	Pu	94	manmade	manmade		ra	
Polonium	Po	84	a	b		ra	
Potassium	K	19	1.68	380		en	
Praseody-mium	Pr	59	0.0015				
Promethium	Pm	61	manmade				Protach-tinium
Radium	Ra	88	a	1E-10		ra	
Radon	Rn	86	a	0.6E-15		ra	

Table 10
Elements: Abundance and Biological Activity (continued)

Element	Symbol	Atomic number	Crustal abundance weight (%)	Abundance in hydrosphere (mg/l)	Abundance in atmosphere (vol ppm)	Biological activity	Threshold limit (mg/m^3 in air in 8 hours)
Rhenium	Re	75	4E-8	b			
Rhodium	Rh	45	1E-8	b		cs	0.001
Rubidium	Rb	37	0.0070	0.12		s	
Rutherium	Ru	44	1E-8	b			
Samarium	Sm	62	0.00077	b			
Scandium	Sc	21	0.0022	4E-5		cs	
Selenium	Se	34	5E-6	0.004		cp s en t2	
Silicon	Si	14	27.2	3		ep cs	
Silver	Ag	47	8E-7	0.0003		cs t1	0.01
Sodium	Na	11	2.32	10,556		en	
Strontium	Sr	38	0.045	8			
Sulphur	S	16	0.030	885		en	
Tantalum	Ta	73	0.00024	b			5
Technetium	Tc	43	manmade	manmade		ra	

APPENDIX 5

FUGACITY MODEL

Table 1 A
Volumes and densities of the unit world

Compartment	Volume (m^3)	Density (kg/m^3)	Organic carbon content (fraction)
Air	$6.0 * 10^9$ (6000-m high)	1.21	-
Water	$7.0 * 10^6$ (10-m deep)	1000	-
Soil	$4.5 * 10^4$ (15-cm deep)	15000	0.02
Sediment	$2.1 * 10^4$ (3-cm deep)	1500	0.04
Suspended sediment	35.0 (5-ppm vol)	1500	0.04
Biota	7.0 (1-ppm vol)	1000	-

Table 1 B
Definition of fugacity capacities

Compartment		Definition of Z (mol/m^3 Pa)
Air (subscript A)	1/RT	R = 8.314 Pa m^3/mol K T = Temperature (K)
Water (subscript W)	1/H or C^s/P^s	C^s = aqueous solubility (mol/m^3) P^s = vapor pressure (Pa) H = Henry's law constant (Pa m^3/mol)
Solid sorbent (e.g., soil, sediment, particles) (subscript S)	$K_{sw}p_s/H$	K_{sw} = partition coefficient (l/kg) p_s = density (kg/l)
Biota (subscript F)	$K_{fw}p_f/H$	K_{fw} = bioconcentration factor (l/kg) p_f = density (kg/l)

Table 2
Volumes of other environments

Compartment	Urban	Rural	Lake	Marsh	River
Air	$1.0*10^9$	$6.0*10^9$	$6.0*10^9$	$6.0*10^9$	$1.0*10^9$
Water	0	10^5	10^7	$1.8*10^6$	10^7
Soil	10^5	$1.35*10^5$	0	10^4	0
Sediment	0	5,000	30,000	45,000	50,000
Suspended sediment	0	1	50	90	100
Biota	0	0.1	10	18	10
Water area fraction	0	0.1	1.0	0.9	1.0

APPENDIX 6

BIODEGRADABILITY

BIODEGRADABILITY CORRELATIONS

Phenols:
$bd = 822 + 3.68*c - 1.94\ mw + 28.5*d - 22.1*e - 69.4*s - 4.45*10^4*\ (mw)^{-1}$ (cc = 0.93)

Aromatic acids:
$bd = 800 + 183*c - 14.6*mw - 117*d - 480*e + 207*s - 6.85*10^5\ (mw)^{-1}$ (cc = 0.87)

Other aromatic compounds:
$bd = 622 + 7.61*c - 2.09*mw + 11.3*d + 17.8*e - 35.4*s - 3.33*10^4\ (mw^{-1})$ (cc = 0.84)

All aromatic compounds:
$bd = 121 + 3.13*c - 0.55*mw + 13.5*d + 5.36*e - 11.0*s$ (cc = 0.53)

Alcohols:
$bd = 119 + 53.5*c - 4.42*mw + 58.1*e + 9.99*s - 1.64*10^3\ (mw^{-1})$ (cc = 0.54)

Aromatic acids:
$BOD5 = 3.82 - 2.62*c - 0.032*mw - 0.66*d + 1.26*q$ (cc = 0.87)

Alcohols:
$BOD5 = 1.34 - 0.121*c - 0.493*d - 0.88*e - 0.012*mw - 0.533*r$ (cc = 0.67)

Amines and amides:
$BOD5 = 1.58 - 0.091*c - 0.28*d + 0.015*e - 0.038*s + 0.31*r - 1.15*a - 0.0081*mw$
(cc = 0.61)

Phenols:
$BOD5 = 1.93 - 0.82*c - 0.166*d - 1.39*e + 0.63*s + 0\ 0039*mw$ (cc = 0.95)

Aliphatic acids:
$BOD5 = 0.97 + 0.219*c - 0.040*d + 0.054*e + 0.183*s - 0.014*mw - 0.014*q$ (cc = 0.79)

Symbols used:

bd = mg COD removed / (g dry inoculum*h)
c = number of aliphatic carbon atoms
d = number of double bonds
e = number of -O- and -N=
s = number of substitutes
mw = molecular weigth
cc = coefficient of correlation
r = number of rings and (chains-1)
a = number of amides
q = s + e + r
BOD5 = BOD_5 as g pr. g of compound

REFERENCES

References of the general Chapters 1-6, Chapter 15 and the appendices 1-6

Aoyama, I., Yos. Inoue and Yor. Inoue, 1978. Simulation analysis of the concentration process of trace heavy metals by aquatic organisms from the viewpoint of nutrition ecology. Water Research, 12: 837-842.

Arp, P.A., 1983. Modelling the Effects of Acid Precipitation on Soil Leachates: A Simple Approach. Ecological Modelling, 19: 105-117.

Banerjee, S., Yalkowsky, S.H. and Valvani, S.C.,1980. Water solubility and octanol/water partitiion coefficients of organics. Limitations of the solubility-partition coefficient correlations. Environ. Sci. Technol., 14; 1227.

Banerjee, S., Sikka, H.C., Gray, R., and Kelly, C.M., 1978. Photodegradation of 3,3-dichlorobenzidine. Environ. Sci. Technol., 12: 1425.

Bartell, S.M., Gardner, R.H., and O'Neill, R.V., 1984. The Fates of Aromatics Model. Ecol. Modelling, 22: 109-123.

Bonner, J.T., 1965. Size and Cycle. An Essay on the Structure of Biology. Princeton University Press, Princeton, NJ, 219 pp.

Bosserman, R.W., 1980. Complexity measures for assessment of environmental impact in ecosystem networks. In: Proc. Pittsburgh Conf. Modelling and Simulation. Pittsburgh, PA, April 20-23.

Bro-Rasmussen, F., and Christiansen, K., 1984. Hazard assessment - a summary of analysis and integrated evaluation of exposure and potential effects from toxic environmental chemicals. Ecol. Modelling, 22: 67-85.

Brown, D.S., and Combs, G., 1982. A model for predicting sorption of organic cations in soils and sediments. J. Environ. Qual., submitted.

Bunce, N.H., Bergsma, J.P., Bersma, M.D., Graaf, W.D., Kumar, Y., and Ravanal, L., 1980. Structure and mechanism in the photoreduction of aryl chlorides in alkane solvents. J. Org. Chem., 45: 3708.

Burchill, S., Greenland, D.J., and Hayes, M.H.B., 1981. Adsorption of organic molecules, in: The Chemistry of Soil Processes (D.J. Greenland & M.H.B. Hayes eds.). John Wiley & Sons, Chichester. 221.

Callahan, M.A., and Slimak, M.W., 1979. Water-related environemtal fate of 129 pollutants. Report No. EPA-440/4-79-029a and b U.S. Environmental Protection Agency, Washington, D.C.

Chiou, C.T., Schmedding, D.W., and Manes, M., 1982. Partitioning of organic compounds in octanol/water systems. Environ. Sci. & Technol., 16,4.

Chiou, C.T., Freed, V.H., Schmedding, D.W., and Kohnert, R.L., 1977. Partition coefficient and bioaccumulation of selected organic chemicals. Environmental Science and Technology 11: 475-478.

Cornelisse, J. and Havinga, E., 1975. Photosubstitution Reactions of Aromatic Compounds. Chem. Rev., 75, 353.

Costanza, R., and Sklar, F.H. (Baston Rouge, LA. USA), 1985. Articulation, accuracy and effectiveness of mathematical models: a review of freshwater wetland applications. Ecol. Modelling, 27: 45-69.

Cramer, R.D., 1980a. BC(DEF) Parameters. II. An empirical Structure-based Scheme for the Prediction of some Physical Porperties. J. Am. Chem. Soc., 102, 1849.

Cramer, R.D., 1980. CD(DEF) Parameters. I. The Intrinsic Dimensionality of Intermolecular Interaction in the Liquid State. J. Am. Chem. Soc., 102, 1837.
Crosby, D.G., and Wong, A.S., 1973. Photodecomposition of p-chlorophenoxyacetic acid. J. Agric. Food Chem. 21: 1049.
Crosby, D.G. and Tang, C.S., 1969. Photodecomposition of napthalene acetic acid. J. Agric. Food Chem., 17, 1291, 1969.
Crosby, D.G., and Tutass, H.O., 1966. Photodecomposition of 2,4-dichlorophenoxyacetic acid. J. Agric. Food Chem. 14: 596.

Dilling, W.L., 1977. Interphase transfer processes. II. Evaporation rates of chloromethanes, ethanes, ethylene, propanes and propylenes from dilute aqueous solutions. Environ. Sci. Technol., 11, 405.

Eaton, J.G., Parrish, P.R., and Hendricks, A.C., (eds.). Special Technical Publications, No. 707: 78-113. American Society for Testing and Materials, Philadelphia.
Eisenlohr, F., 1910. Z. physik. Chem.; 75: 585.
Encinas, M.V., Lissi, E.A., and Sciano, J.C., 1980. Photochemistry of aliphatic ketones in polar solvents. J. Phys. -chem., 84: 948.
Exner, O., 1972. The Hammett Equation - The Present Position in Advances in Linear Free Energy Relationspips. V.B. Chapman & J. Shorter (eds.). Plenum Press, London. 2.

Fagerstroem, T., and Aasell, B., 1973. Methyl mercury accumulation in an aquatic food chain. A model and implications for research planning. Ambio 2: 164-171.
Fenchel, T., 1974. Intrinsic rate of natural increase: the relationship with body size. Oecologia, 14: 317-326.

Gillet, J.W., et al., 1974. A conceptual model for the movement of pesticides through the environment. National Environmental Research Center, US Environmental Protection Agency, Corvallis, OR Report EPA 660/3-74-024, p.79.
Gromiec, M.J., and Gloyna, E.F., 1973. Radioactivity transport in water. Final Report No. 22 to US Atomic Energy Commission, Contract AT (11-1)-490.

Halfon, E., 1984. Error analysis and simulation of Mirex behaviour in Lake Ontario. Ecol. Modelling, 22: 213-253.
Halfon, E., Unbehauen, H. and Schmid, C., 1979. Model order estimation and system identification theory to the modelling of 32P kinetics within the trophogenic zone of a small lake. Ecol. Modelling, 6: 1-22.
Hance, R.J., 1969. Influence of pH, exchangeable cation and the presence of organic matter on the adsorption of some herbicides by montmorillonite. Can. J. Soil Sci., 49, 357.
Hansch, C., and Leo, A.J., 1979. Substituent Constants for Correlational Analysis in Chemistry and Biology. John Wiley & Sons, New York, 1979.
Hansch, C., Quinlan, J.E., and Lawrence, G.L. (1968). The Linear Free-Energy Relationships between Partition Coefficients and the Aqueous Solubility of Organic Liquids. J. Org. Chem. 33: 347-50.
Harris, J.R.W., Bale, A.J., Bayne, B.L., Mantoura, R.C.F., Morris, A.W., Nelson, L.A., Radford, P.J., Uncles, R.J., Weston, S.A., and Widdows, J. A Preliminary model of the dispersal and biological effect of toxins in the Tamar estuary, England, Ecol. Modelling, 22: 253-285.
Harris, C.I. and Sheets, T.J., 1965. Influence of soil properties on adsorption and phytotoxicity of CIPC, diuron, and simazine. Weeds 13: 215.
Herzog, R., 1944. Ind. Eng. Chem., 36: 997.
Havinga, E., and Cornelisse, J., 1976. Aromatic photosubstitution reactions. Pure Appl. Chem., 47, 1.
Hirschfelder, J.O., Curtiss, C.F., and Bird, R.B., 1954. Molecular Theory of Gases and Liquids. John Wiley & Sons. 2: 354.

Irish, D.D., 1963. Aliphatic halogenated hydrocarbons. In: Industrial Hygiene and Toxicology, Vol. 2, 2nd ed., Patty, F.A., (ed.). Interscience, New York, 1288.

Jaffe, H.H., 1953. A reexamination of the hammett equation. Chem. Rev., 53: 191.

Jenne, E.A., 1976. Trace element sorption by sediments and soils - sites and processes. In: Symposium on Molybdenum. Chappell, W., Petersen, K. (eds.) Vol. 2. New York: Marcel Dekker. p.425-553.

Jørgensen, S.E., 1988. Fundamentals of Ecological Modelling. Elsevier, Amsterdam, New York, Oxford, Tokyo.

Jørgensen, S.E., 1984. Parameter estimation in toxic substance models. Ecological Modelling, 22: 1-13.

Jørgensen, S.E., 1979. Modelling the distribution and effect of heavy metals in aquatic ecosystems. J. Ecol. Model., 6: 199-223.

Jørgensen, S.E., 1979 (editor-in-chief; editorial board: M.B: Friis, J. Hendriksen, L.A. Jørgensen, S.E. Jørgensen and H.F. Mejer). Handbook of Environmental Data and Ecological Parameters. International Society of Ecological Modelling, Copenhagen.

Jørgensen, S.E., 1976. A model of fish growth. J. Ecol. Model., 2: 303-313.

Jørgensen, S.E. and Johnsen, F., 1989. Principles of Environmental Science and Technology, 2. edition. Elsevier, Amsterdam, New York, Oxford, Tokyo.

Jørgensen, S.E. and Gromiec, M., 1988. Submodels in Water Quality.

Jørgensen, S.E., Nors Nielsen, S. and Jørgensen, L.A., 1938. Handbook of Ecological Parameters and Ecotoxicology. CRC. Boca Rouge. Florida U.S.A.

Jørgensen, S.E., Jørgensen, L.A., Kamp Nielsen, L., and Mejer, H.F., 1981. Parameter Estimation in Eutrophication Modelling. Ecol. Modelling, 13: 11 -129.

Jørgensen, S.E., and Mejer, H., 1977. Ecological buffer capacity. J. Ecol. Model., 3: 39-61.

Kenaga, E.E., and Goring, C.A.I., 1980. Relationship between water solubility, soil sorption, octanol-water partitioning and concentration of chemicals in biota. In: Aquatic Toxicology. J.G. Eaton, P.R. Parrish, and A.C. Hendricks (eds.). American Society for Testing and Materials, Philadelphia, 78.

Kenaga, E.E., and Goring, C.A.I. Relationship Between Water Solubility, Soil Sorption, Octanol-Water Partitioning, and Bioconcentration of Chemicals in Biota. Pre-publication copy of paper dated October 13, 1978, given at American Society for Testing and Materials. Third Aquatic Toxicology Symposium, October 17-18, New Orleans, La. (Symposium papers were published by ASTM, Phipladelphia, Pa., as Special Technical Publication (STP) 707 in 1980.)

Kirchner, T.B., and Whicker, F.W., 1984. Validation of PATHWAY, a simulation model of the transport of radionuclides through agroecosystems. Ecol. Modelling, 22: 21-45.

Kohlmaier, G.H., Sire, E.O., Brohl, H., Kilian, W., Fishbach, U., Plochl, M., Muller, T., And Ynsheng, J., 1984. Dramatic development in the dying of German spruce-fir forests: In search of possible causeeffect relationships. Ecological Modelling, 22: 45-65.

Lam, D.C.L., and Simons, T.J., 1976. Computer model for toxicant spills in Lake Ontario. Environmental Biogeochemistry vol. 2. Metals transfer and ecological mass balances (ed. J.O. Nriago). Ann Arbor, MI: Ann Arbor Science, pp. 537-549.

Larson, R.J., and Perry, R.L., 1981. Use of the electrolytic respirometer to measure biodegradation in natural waters. Water Res., 15: 697.

Lassiter, R.R., 1978. Principles and constraints for predicting exposure to environmental pollutants. U.S. Environmental Protection Agency, Corvallis, OR Report EPA 118-127519.

Lassiter, R.R., 1975. Modelling dynamics of biological and chemical components of aquatic ecosystems. EPA-660/3-75-012, U.S. Environmental Protection Agency, Washington, D.C.

LeMaire, J., 1967. Photoenolization of biacetyl. J. Phys. Chem., 71: 2653.

Leung, D.K., 1978. Modelling the bioaccumulation of pesticides in fish. Center for Ecological Modelling Polytechnic Institute, Troy, NY Report 5.

Liublina, E.I., 1965. Relationships between toxicity and physicochemical properties of metals. In: I.D. Gadaskina, A.A. Golubev, and E.T. Lykhina, eds., Voprosy Obshchei i Chastnoi Promyshlennoi Toksikologii, pp.

Liublina, E.I., and Golubev, A.A., 1967. Instruktsiya po Ustanovleniyu Rastchetnym Sposobom Orientirovochnykh Predelno Dopustimykh Kontsentrasii Promyshlennykh Yadov v Vozdukhe

Rabochikh Pomeshchenii [Instructions for Establishing by Calculation of Approximate Maximum Allowable Concentrations in the Air of Work Areas], 2nd ed., Institut Gigiyeny Truda i Profzabolevaniy, Leningrad.

Liublina, E.I., and Filov, V.A., 1965. Physicochemical properties of organic substances and their toxicity indices. In: I.D. Gadaskina, A.A. Golubev, and E.T. Lykhina, eds., Voprosy Obshchei i Chastnoi Promyshlennoi Toksikologii, pp. 7-16. Institut Gigiyeny Truda i Profzabolevaniy, Leningrad.

Lu, P.-Y., and Metcalf, R.L., 1975. Environmental fate and biogradability of benzene derivatives as studied in a model aquatic ecosystem. Environmental and Health Perspectives 10: 269-284.

Lydersen, A.L., 1955. Estimation of Critical Properties of Organic Compounds. Coll. Eng., Univ. Wisconsin, Eng. Expt. Sta. Rept. 3, Madison, Wis., April.

Mackay, D., Bobra, A., Chan, D.W., and Shiu, W.Y., 1982. Vapor pressure correlations for low-volatility envirnmental chemicals. Environ. Sci. Technol., 16: 645.

Mackay, D., Shiu, W.Y., Bobra, A., Billington, J., Chau, E., Yeun, A., Ng, C., and Szeto, F., 1982. Volatilization of Organic Pollutants from Water, EPS Rep. No. 600/63-82-019, NTIS No. PB 82-830-934, Athens, Georgia.

Mackay, D., and Paterson, S., 1981. Calculating fugacity. Environ. Sci. Technol., 15: 1006.

Mackay, D., Mascarenkas, R., and Shiu, N.Y., 1980. Aqueous solubility of polychlorinated biphenyls. Chemosphere, 9, 257.

Matthews, A.P., Am. J. Physiol., 10, p. 280 (1904), cited in J.R.E. Jones, J. Exp. Biol., 16, 425-437 (1939).

Metcalf, R.L., Sangha, G.K., and Kopoor, I.P., 1975. Model ecosystem for the evaluation of pesticide biodegradability and ecological magnification. Environmental Science and Technology 5: 709-713.

Mill, T., Mabey, W.M., Lan, B.Y., and Baraze, A., 1981. Photolysis of polycyclic aromatic hydrocarbons in water. Chemosphere, 10: 1281.

Miller, P.C., Stoner, W.A., and Tieszen, L.L., 1976. A model of stand photosynthesis for the wet meadow tundra at Barrow Alaska. Ecology, 57: 411-430.

Neely, W.B., Branson, D.R., and Blau, G.E., 1974. Partition coefficient to measure bioconcentration potential of organic chemicals in fish. Environmental Science and Technology 8: 1113-1115.

Nihoul, J.C.J., 1984. A non-linear mathematical model for the transport and spreading of oil slicks. Ecol. Modelling, 22: 325-341.

Nyholm, N., Nielsen, T.K. and Pedersen, K., 1984. Modelling heavy metals transport in an arctic fjord system polluted from mine tailings. Ecol. Modelling, 22: 285-325.

Odum, E.P., 1959. Fundamentals of Ecology (2nd edition). Saunders, Philadelphis, PA.

Partington, J., 1949. An Advanced Treatise on Physical Chemistry. Vol. 1, Fundamental Principles - The Properties of Gases. Longmans, Green & Co., Ltd., London.

Polson, A., (1950). J. Phys. Colloid Chem., 54: 649.

Pritchard, P.H., Bourquin, A.W., Fredrickson, H.L., and Maziarz, T., 1979. System design factors affecting environmental fate studies in microcosms. In: Proceedings of the Workshop: Microbial Degradation of Pollutants in Marine Environments, EPA-600/9-79-012. A.W. Bourquin & R.H. Pritchard (eds.). US: Environmental Protection Agency, Gulf Breeze, Fla. 251.

Quale, O.R., (1953). Chem. Revs. 53: 439.

Reid, R.C., Prausnitz, J.M., and Sherwood, T.K., 1977. The Properties of Gases and Liquids, 3rd ed., McGraw-Hill, New York.

Riedel, L. (1955). Chem.-Ing.-Tech., 27: 475.

Riedel, L. (1949). Elektrochem., 53: 222.

Saeger, V.W., Hicks, O., Kaley, R.G., Michael, P.R., Mieure, J.P., and Tucker, E.S. (1979).

Environmental Fate of Selected Phosphate Esters. Environ. Sci. Technol., 13: 840-44.

Schwarzenbach, R.P. and Imboden, D.M., 1984. Modelling concepts for hydrophobic pollutants in lakes. Ecol. Modelling, 22: 171-213.

Seip, K.L., 1978. Mathematical model for uptake of heavy metals in benthic algae. Ecological Modelling 6: 183-198.

Shaw, W.H.R., 1961. Cation toxicity and the stability of transition-metal complexes. Nature, 192: 754-755.

Smith, J.H., Mabey, W.R., Bohonos, N., Holt, B.R., Lee, S.S., Chou, T W., Bomberger, D.C., and Mill, T., May 1978. Environmental pathways of selected chemicals in freshwater systems. III. Laboratory studies, Report No. EPA-600/7-78-074, U.S. Environmental Protection Agency, Athens, Ga.

Soderqist, C.J., Bowers, J.B., and Crosby, D G., 1977. Dissipation of molinate in a rice field. J. Agric. Food Chem., 25: 940.

SRI International,1978. A study of industrial data on candidate chemicals for testing, Report No. EPA-560/5-78-002, U.S. Environmental Protection Agency, Washington, D.C.

Su, G.C.C., and Zabik, M.J., 1972. Photochemistry of bioactive compounds. J. Agric. Food Chem., 20: 320.

Thodos, G. (1955). AIChEJ J., 1: 165.

Thomann, R.V., 1984. Physio-chemical and ecological modelling the fate of toxic substances in natural water systems. Ecol. Modelling, 22: 145-171.

Thomann, R.V., et al., 1974. A food chain model of cadmium in westwen Lake Erie. Water Research 8: 841-851.

Tse, D., Mabey, W.R., and Winterle, J.S., 1982. Unpublished results on EPA contract 68-02-2981.

Uchrin, C.G., 1984. Modelling transport processes and differential accumulation of persistent toxic organic substances in groundwater systems. Ecol. Modelling, 22: 135-145.

Veith, G.D., Defoe, D.L., and Bergstedt, B.V., 1979. Measuring and estimating the bioconcentration factor of chemicals in fish. J. Fish. Res. Board Can., 36: 1040-1048.

Verschueren, K., 1977. Handbook of Environmental Data on Organic Chemicals, Van Nostrand Reinhold, New York.

Vowles, C., 1951. S.M. thesis in chemical engineering. Massachusetts Institute of Technology.

Weiss, A., Amstutz, G.C. (1966). Ion-exchange reactions on clay minerals and cation selective membrane properties as possible mechanisms of economic metal concentration. Miner. Deposita 1: 60-66.

Welte, D., 1969. Organic geochemistry of carbon. In: Handbook of Geochemistry. Wedepohl, K.H. (ed.). Berlin, Heidelberg, New York: Springer.

Windholz, M. (ed.), 1976. The Merck Index, 9th ed., Merck & Co. Publishers, Rahway, N.J.

Wolff, C.J.M., and Van der Heidje, H.B., 1982. A model to assess the rate of evaporation of chemical compounds from surface waters. Chemsphere 11: 103.

Zeigler, B.P., 1976. Theory of Modelling and Simulation. Wiley, New York, NY, 435 pp.

Zepp, R.G., and Schlotzhauer, P.F., 1979. Photoactivity of selected aromatic hydrocarbons in water. In: Polynuclear Hydrocarnons. Jones, T.W. and Leber, T., eds. Ann Arbor Science Publishers, Ann Arbor, Michigan. p.141.

Zepp, R.G., Wolfe, N.L., Azarraga, L., Cox, R.H., and Pape, C.W., 1977. Photochemical transformation of DDT and methoxychlor degradation products by sunlight. Arch. Environ. Contam. Toxicol., 6: 305.

Zepp, r.G., Wolfe, M.L., Gordon, J.A., and Baughman, G.L., 1975. Dynamics of 2,4-d-esters in surface waters; hydrolysis, photolysis and vaporization. Environ. Sci. Technol., 9: 1144.

LIST OF AUTHORS

Badeck, F.W. Institut für Physikalische und Theoretische Chemie, Johann Wolfgang Goethe Universität Frankfurt, Niederurseler Hang, 6000 Frankfurt 50.

Chesters, Gordon. Professor of Soil Science, Chairman of the Water Chemistry Program, and Director of the Water Resources Center, 1975 Willow Drive, University of Wisconsin-Madison, Madison, WI, U.S.A. 53706.

Cirillo, Mario C. ENEA, Comitato Nazionale per la ricerca e per lo sviluppo dell'Engergia Nucleare e delle Energie Alternative, S.P. Anguillarese, 301, 00100 Roma A.D. Italy.

Clerici, Giancarlo. ENEA, Comitato Nazionale per la ricerca e per lo sviluppo dell'Engergia Nucleare e delle Energie Alternative, S.P. Anguillarese, 301, 00100 Roma A.D. Italy.

Dahab, Aboul, Dr. Oceanography Department Faculty of Science, Moh. Bay, Alexandria, Egypt.

Grydersen, U. The Copenhagen College of Engineering, Prinsesse Charlotte-gade 38, DK-2200 København N, Denmark.

Halfon, Efraim. Lakes Research Branch, National Water Research Institute, Canada Centre for Inland Waters, Burlington, Ontario, Canada L7R 4A6.

Halling-Sørensen, Bent. The Royal Danish School of Pharmacy, Dept. of Chemistry AD, 2 Universitetsparken, DK-2100 Copenhagen, Denmark.

Jørgensen, Sven Erik. The Royal Danish School of Pharmacy, Dept. of Chemistry AD, 2 Universitetsparken, DK-2100 Copenhagen, Denmark.

Kohlmaier, G.H. Institut für Physikalische und Theoretische Chemie, Johann Wolfgang Goethe Universität Frankfurt, Niederurseler Hang, 6000 Frankfurt 50.

Manzi, Donatella. ENEA, Comitato Nazionale per la ricerca e per lo sviluppo dell'Engergia Nucleare e delle Energie Alternative, S.P. Anguillarese, 301, 00100 Roma A.D. Italy.

Marani, Allessandro. Associate Professor, Dipartimento di Scienze Ambientali, Universitá di Venezia, Venice Italy.

Oliver, Barry G. Lakes Research Branch, National Water Research Institute, Canada Centre for Inland Waters, Burlington, Ontario, Canada L7R 4A6.

Plöchl, M. Institut für Physikalische und Theoretische Chemie, Johann Wolfgang Goethe Universität Frankfurt, Niederurseler Hang, 6000 Frankfurt 50.

Siebke, K. Institut für Physikalische und Theoretische Chemie, Johann Wolfgang Goethe Universität Frankfurt, Niederurseler Hang, 6000 Frankfurt 50.

Wientzek, C. Institut für Physikalische und Theoretische Chemie, Johann Wolfgang Goethe Universität Frankfurt, Niederurseler Hang, 6000 Frankfurt 50.

INDEX

SOFTWARE ON
ECOTOXICOLOGICAL PARAMETERS

ISEM has recently developed a new software which is available at ISEM Headquarters for US$ 298. It is a software which estimates a series of ecotoxicological parameters on the basis of the chemical formula for the for the considered compounds, as shown in this book, Chapters 4 and 5. If some physical data are also known, for instance, water solubility and the boiling point, the estimation will be more accurate. The following parameters useful for ecotoxicological models are estimated:

1. Vapor pressure at room temperature
2. Vapor pressure as function of the temperature
3. Critical data
4. Water solubility (if known)
5. Partition ratio octanol/water solubility
6. Henry's Constant
7. Biodegradability
8. Concentration factor
9. Biomagnification factor
10. Adsorption isotherms (adsorption to soil of given composition)
11. LD_{50} and LC_{50} values
12. MAC value.

The estimation is of course given with a certain standard deviation - in most cases less than 30% relative in accordance with a wide spectrum of compounds used to evaluate the standard deviation of the estimation methods. If more input data is available than just the chemical formula, the estimation methods become more accurate and the average of the various values found by use of different pathways is found. The software is very user friendly as its use is completely explained by a menu.
Contact ISEM Secretariat, Langkærværge 9, DK-3500 Værløse, Denmark.